P9-AFO-786

Ecological Studies

Analysis and Synthesis

Edited by

W.D. Billings, Durham (USA) F. Golley, Athens (USA)
O.L. Lange, Würzburg (FRG) J.S. Olson, Oak Ridge (USA)
H. Remmert, Marburg (FRG)

Volume 70

James A. Larsen

The Northern Forest Border in Canada and Alaska

Biotic Communities and Ecological Relationships

With 73 Illustrations

Springer-Verlag
New York Berlin Heidelberg
London Paris Tokyo

James A. Larsen
Box 1496
Rhinelander, WI 54501

Library of Congress Cataloging-in-Publication Data
Larsen, James Arthur
The northern forest border in Canada and Alaska : biotic communities and ecological relationships / James A. Larsen.
p. cm. – (Ecological studies ; v. 70)
Bibliography: p.
ISBN 0-387-96753-2
1. Forest ecology–Canada. 2. Forest ecology–Alaska. 3. Timberline–Canada. 4. Timberline–Alaska. 5. Biotic communities–Canada. 6. Biotic communities–Alaska. I. Title. II. Series.
QH106.L283 1988
574.5′2642′0971–dc19 88-24973

Typeset by Publishers Service, Bozeman, Montana.
Printed and bound by Quinn-Woodbine, Inc., Woodbine, New Jersey.
Printed in the United States of America.

9 8 7 6 5 4 3 2 1

ISBN 0-387-96753-2 Springer-Verlag New York Berlin Heidelberg
ISBN 3-540-96753-2 Springer-Verlag Berlin Heidelberg New York

Preface

It is enough to work on the assumption that all of the details matter in the end, in some unknown but vital way.

Edward O. Wilson, *Biophilia*

Advances in knowledge of northern ecology have been so rapid that to undertake a synthesis of all the literature now available would be a major enterprise, perhaps even a life's work, and so it must be considered permissible to fill in a few gaps, follow one's own inclinations, leaving comprehensive syntheses to those willing to undertake them. This is the rubric under which I have written, reporting some of the more interesting data I and others have obtained over the years, often diverging into discussions of plants, soils, climate, and faunal relationships which have perhaps not previously been dealt with extensively, or at least in quite the same way. This is purely intentional, since I find it difficult to summon up the needed enthusiasm, at this late hour, to write on topics which unfortunately for me have little attraction. I have thus written for the pleasure derived from depicting, perhaps at times as something of an impressionist, a fascinating biotic region, a captivating land, a collection of interesting ecological problems, environmental relationships to be discerned in part, perhaps understood to some small degree, perhaps one day to be modeled mathematically. As Leo Szilard once wrote: ". . . to be able to say even this much might be of some value" (Szilard, 1960).

There are, in fact, few conclusions given. Much of the material here presented is simply a record of observations made by others and myself, primarily during field excursions to sites in northern Canada. If nothing else, these data will serve as a baseline for future comparisons, as climate and perhaps other environmental conditions change, or succession occurs, or evolutionary development takes place. In support of this approach, too, I draw upon the view expressed by Robert H. MacArthur that works committed to principles are "doomed to early obsolescence," while pure observations are "never out of date."

With this in mind, I bring together, draw upon, and reiterate data gathered by myself and many others, to affect a certain degree of synthesis of the environmental and vegetational community influences at work in the northern regions, aware that, as M.P. Austin (1985) points out, vegetation science has no theoretical base at present, and that it is wise to avoid the almost irresistible temptation, summarized by G.W. Salt (1983), to subject every data pile to multivariate analysis in the hope that something of value may be sifted out. And since the data presented exist in a certain historical and geographical context, some attention is given to the historical and geographical background in which the vegetational communities of the northern lands exist.

It is my hope, also, that in stating these things I contribute in some measure to the region's continued preservation as one of the few remaining relatively untouched wilderness areas left on an overpopulated globe. My atavistic hope is that it will be long preserved as such, in some proportion if not as a whole, and that at least parts will remain a blank spot on the industrial, technological, and economic map; a place for a walk in the sun. There are too few such places remaining in the limited confines of the Earth.

Long listings of the names of persons to whom an author feels an obligation are somehow shadowed by the indecent hint that any shortcomings can, by this means, be somehow shifted or at least shared. The numerous shortcomings of this book, and of some I am conscious, are the product of my efforts alone. I do, however, feel some need to express gratitude to a few. Those who gave support, advice, and encouragement over the years are Reid A. Bryson, Robert A. Ragotzkie, Grant Cottam, John W. Thomson, and John T. Curtis. Those who helped with taxonomy include George Argus, A.E. Porsild, Howard Crum, John W. Thomson, James Zimmerman, Hugh Iltis, Frank Crosswhite, and N.H. Russell. Those who gave freely of advice and assistance on statistical matters include John E. Kutzbach, Charles Hutchins, Robert Ream, T.J. Blasing, and Umesh Agrawal. These individuals helped not only with matters related to this present volume but previous books and scientific papers. I am deeply grateful to them all; I owe to them a debt that I could never repay. As for this volume, I am grateful to Curtis J. Sorenson, who helped prepare the keys for interpretation of the aerial photographs, and whose doctoral thesis on northern soils provided much of the material for the chapter on one of the most important facets of the boreal and arctic environments.

Acknowledgments

Gratitude is expressed for permission to use various materials herein, and these materials and authorizing agencies are as follows:

Canadian National Air Photographic Library, for permission to use the aerial photographs presented in Chapter 4. A listing of the data for individual photographs is given in Appendix E.

Methuen, Inc., for permission to use materials (herein Figs. 5.1a,b and 8.1) from the reference cited under Larsen (1974), *Arctic and Alpine Environments*, Jack Ives and Roger Barry, editors.

Permission to use data excerpted and condensed from various publications was given by:

Serge Payette for data from *Tree-Line Ecology* (1983), Morisette, Payette, and DeShaye, Collection Nordicana No. 47, Laval University.

Lands Directorate, Environment Canada, and J.S. Rowe, for data excerpted from *An Ecological Land Survey of the Lockhart River Area*, Ecological Land Classifications Series No. 16, Environment Canada, reference cited under Bradley et al. (1982).

Ecological Society of America for material cited from Larsen (1965), which appeared in *Ecological Monographs*, and to the Society and E.A. Johnson for data excerpted from *Ecology*, referred under Johnson (1981).

Contents

Appendixes

Illustrations

Tables

1. Background and Setting

Roughly speaking, the causes of small-scale short-term phenomena in the biosphere are the subject matter of ecology; and the causes of large-scale long-term phenomena are the subject matter of biogeography.

E.C. Pielou, *Ecological Diversity* (Wiley Interscience, 1975, p. 112).

A fundamental tenet of biogeography is that rather close relationships can be seen to exist between climate and the composition and structure of vegetational communities —the species making up the communities (composition) and relative abundance with which each is represented (structure). This has been recognized since at least the days of Humboldt and has been held responsible in part for the global distribution of forest, savanna, prairie, desert, and so on. More recently, climatic adaptations are invoked to explain certain characteristic physiological responses of plant species to temperature and water regimes as maintained in laboratory, greenhouse, and test plot studies.

There are, it is true, descriptive accounts by naturalists that, to a limited degree, delineate the distinctive environmental tolerance limits of each species, but detailed explanations in terms of biochemistry and genetic structure have yet to emerge, although they are beginning to do so. From recent progress in these fields it is evident that a comprehensive knowledge of the biochemical aspects of adaptation will eventually be made available, but presently we can say only that, while we know the manner with which the species respond, we often do not yet know, in explicit biochemical terms, why they do so in such apparent and distinctive ways.

In the discussion that follows, neither relationships between climate and vegetation nor physiological and genetic characteristics of the species are principal topics. Rather, the topic under consideration is the composition of plant communities of the ecotonal region between forest and tundra in central northern Canada. It is the purpose here to delineate the habitat preferences of certain of the more common species. Physiological and biochemical explanations for habitat preference must await a better understanding of these aspects of plant life.

A basic assumption, therefore, is that plant species, and animal species as well, are climatically controlled in the sense that each species is found, under natural conditions, growing within a specific climatic range, to which it is said to be adapted. Much is known about the climatic tolerances of plant and animal species as a result of simple correlations with climatic records. The present-day range of a species is often taken as evidence that the climate of the area is favorable for the species and that neither the climate nor the tolerance limits of the species have changed appreciably in recent time.

For example, this is borne out for the region under consideration—central northern Canada—by the evidence at Ennadai Lake, a narrow body of water 45 miles long and trending from the southwest to the northeast, located in the southwest corner of Keewatin. The north end of the lake is in tundra, the south end covered by forest. Here at Ennadai Lake a series of forest fires occurred centuries ago, destroying forest trees and leaving a charcoal layer in certain favorable places, most frequently along sandy shorelines where eskers abut the shore near the north end of the lake. A covering of sand now lies over the charcoal layer, preserving it and indicating that the forest once covered the landscape but failed to regenerate after the fire, or series of fires, presumably due to changed climate. Then windblown sand from the nearby eskers covered the charred remnants of spruce trees and preserved also the podzol soil that had been formed under the influence of the forest—a distinctive soil type which has now become a paleosol or fossil soil under the circumstances (Bryson et al., 1965; Larsen, 1965).

Investigation has indicated that, in its former extent, the forest stretched more than 100 miles north of the present day forest border at the south end of Ennadai Lake. Subsequent dating of charred material from the paleosols at the north and south ends of Ennadai Lake, employing the C^{14} dating technique, gave an age of 860 $\pm$ 135 years at the south end of the lake and 1070 $\pm$ 180 years at the north end. These dates would place the southward retreat of the forest border at about the close of the so-called Little Climatic Optimum. The various dates of forest advance and retreat in southern Keewatin correlate with other major known periods of climatic change in North America and Europe. It seems apparent that the last forest burn in the Ennadai Lake area coincided with, or closely followed, a change in climate that resulted in conditions inimical to seed maturation; it was presumably colder, perhaps drier, with a shorter growing season that prevented effective widespread reproduction of the spruce (Clarke, 1940; Larsen, 1965; Elliott, 1979a,b, 1983; Black and Bliss, 1980; Payette and Gagnon, 1985). Published palynological literature delineating these major climatic shifts in some detail is now abundant (Nichols, 1967a,b,c, 1969, 1970, 1974, 1976; Ritchie, 1976, 1977, 1984, 1988; Ritchie and Hare, 1971; Jordan, 1975; Ritchie and Yarranton, 1978; Ritchie et al., 1987; Short and Nichols, 1977; Ritchie and MacDonald, 1986; Larsen and Barry, 1974; Barry et al., 1981; Kay and Andrews, 1983; Jacoby and Cook, 1981; Gagnon, 1982; Payette, 1983).

The period marked by the farthest northward postglacial advance of the forest border coincides with the so-called Hypsithermal (or Altithermal) episode, a time of warmth during which the Pleistocene continental ice disappeared. Eventually, after a period of retreat and readjustment to a subsequent somewhat cooler climatic regime, communities resembling in essential respects those existing today began to emerge. These events are recorded by pollen evidence in peat deposits and lake sediments subject to palynological investigation. There are, as well, relict groves of stunted spruce trees in a few favorable places (Clarke, 1940; Larsen, 1965) which remain to show that the forest occupied regions farther north in relatively recent time.

There is, of course, the question of how the northern flora existing during the Pliocene survived the Pleistocene epoch, in which glacial ice covered large portions of North America. Evidence seems strong that the boreal forest flora retreated before the advance of the ice, surviving to the south, and that a majority of the species of the arctic flora survived in far northern areas such as the unglaciated parts of Peary Land and the Arctic Archipelago as well as parts of Alaska, areas where these species found refuge during the times of maximum glaciation. At least a few of the arctic species may have survived south of the continental glacier in the central and eastern United States and in alpine regions of the western United States. A few species with widespread distribution in arctic regions have disjunct populations in, for example, the Mississippi Valley and the Driftless Area of Wisconsin, Iowa, Minnesota, and Illinois. These include *Rhododendron lapponicum*, *Sedum rosea*, and *Chrysoplenium iowense* (*C. alternifolium*), all of which are otherwise arctic or boreal and circumpolar in range. Evidence of *Dryas integrifolia* as well as *Rhododendron* and *Sedum* has been found in late-glacial sediments in northeastern Minnesota. Whether these species persisted south of the glacial ice throughout the Pleistocene or migrated into the region during the late-glacial or postglacial period is still a matter for conjecture. The dispersal rates for arctic species are sufficiently rapid to account for spread of the plants into all available habitats within a relatively short time following retreat of the glacier (Löve and Löve, 1974; Ives, 1974).

The continental glaciations and the location of the various refugia for the arctic and boreal flora have left their marks upon the morphology, physiology, and the patterns of distribution of the species in a number of discernible ways. There are patterns of distribution that lend themselves readily to the interpretation that glacial ice separated many species into eastern and western or northern and southern populations during the long glaciations. These populations developed distinct morphological and physiological characteristics by which they can still be differentiated (Bliss, 1962; Billings and Mooney, 1968; Savile, 1956, 1972).

Oxyria digyna is an example of an arctic-alpine species with northern arctic and southern montane populations, with morphological and physiological gradation between the extremes. Of tree species now found in the Great Slave Lake region, all with the possible exception of *Picea mariana* and *Populus balsamifera* have eastern or western varieties or quite similar eastern and western species within the genus. This same east-west differential appears to exist in some shrub populations. Many willow genera are represented by similar eastern and western species, often with intermediate forms found in areas between the two (Argus, 1965, 1966).

On the other hand, some wide-ranging herbaceous species show a remarkable uniformity across the continent, as do many of the shrubs such as *Salix bebbiana*, *Alnus crispa*, *Viburnum edule*, *Myrica gale*, *Potentilla fruticosa*, *Betula glandulosa*, *Salix planifolia*, and *Salix myrtillifolia*. This is perhaps the result of shorter life-cycles than is the case with trees and a more rapid fusion of eastern and western forms following glacial retreat. It is of some interest, however, that the exception, black spruce (*Picea mariana*), is a tree showing little if any segregation into eastern and western types, while white spruce (*Picea glauca*) has well-defined western varieties and a western alpine species, *Picea engelmannii*, that is very similar to *Picea glauca*. There is, thus, the reasonable explanation that a continuous band of *Picea mariana* forest with associated shrubs and herbs existed in the periglacial zone all across the continent during glacial times. In this case there would have been a continuous gene exchange and the species populations would have had little chance to segregate into distinctly eastern and western types. *Picea glauca*, on the other hand, with narrower tolerance limits, may, by contrast, have been split into distinct eastern and western, perhaps lowland and montane, populations during the glacial epoch (Ritchie and MacDonald, 1986), and the same most likely happened with certain other shrubs and herbaceous species.

It should be apparent that, in considering the human prospect on earth during future years and centuries, there may well be some considerable value in developing an ability to predict climatic trends and to anticipate what the effect upon global vegetation, oceans, and ultimately the commerce in, and between, nations will be when climatic change occurs. From geological and vegetational evidence it is apparent that climatic fluctuations of some magnitude have occurred time and again in the past million or so years, not only fluctuations on a major scale—resulting in glaciation of vast regions—but minor fluctuations less severe in total effect, although sufficient to disrupt agriculture and forestry, transportation, ocean fishing, and resource management whenever they occur again. It appears that some shifts in climate occur abruptly, and when they occur again some advance warning of the effects would be of great economic and social value.

With this evidence in mind, upon what do we base the assumption stated above, that no change of sufficient magnitude affecting the vegetation of the forest-tundra ecotonal region of central Canada has occurred within recent time? First, there are the meteorological records, which are not of exceedingly long duration but are of sufficient length to indicate that no rapid change is currently under way. There are, secondly, the trees at the northernmost limits of the forest-tundra ecotone, many of relatively great age, which are obviously capable of growing there even if reproduction by seed is infrequent at best. There is evidence, too, that these trees or their predecessors once grew as part of a forest cover that was forced to retreat but which has left relict clones that now have persisted for a great many years without further deterioration.

There is also another fascinating source of information concerning the nature of the vegetation during the past century or so—accounts by early-day explorers who kept a record of their travels and published journals that now provide a description of the vegetation in the region a century (at most a few centuries) ago, and from which we can infer whether or not marked vegetational changes have occurred in the interval. The next chapter is devoted to a brief reference to some of the descriptions of northern Canada by a few of these intrepid explorers. Then, in later chapters are presented, first, a general description of the landforms and vegetational communities found in a number of the areas, west to east, in which studies have been conducted by a number of individuals in recent years, and, second, a description and analysis of data obtained from communities sampled during the course of the work conducted preliminary to the preparation of this volume. (See Fig. 1.1 for study areas in which sampling was undertaken.)

The past few decades have resulted in a rapid extension of knowledge of the boreal forest, from ecological and biogeographical points of view as well as from the perspective of practicing foresters. Although the latter is not the focus of this volume, it should

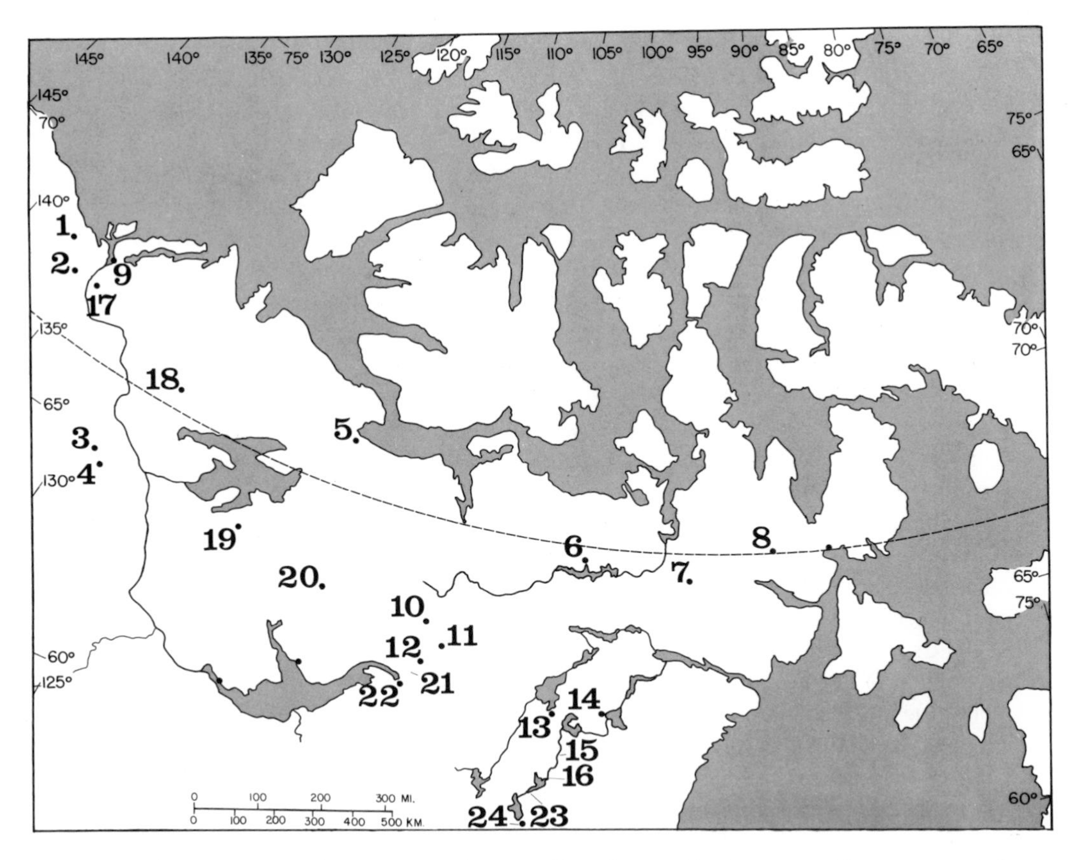

Figure 1.1. Areas mentioned in text where studies of vegetational communities were carried out; areas are listed according to predominant communities present: montane western tundra (1–4); continental tundra (5–8); northern forest-tundra with clumps of dwarfed trees (9–16); northern spruce forest near tree-line (17–24). Dots without numbers indicate other areas where studies were carried out, but the data were not employed in the analyses here. Two areas not shown are in northern Quebec: Shefferville and Rainy Lake (see Larsen, 1980). Names of the study areas are as follows: 1—Trout Lake; 2—Canoe Lake; 3—Florence Lake; 4—Carcajou Lake; 5—Coppermine; 6—Pelly Lake; 7—Snow Bunting Lake; 8—Curtis and Stewart Lakes; 9—Reindeer Station; 10—Aylmer Lake; 11—Clinton Colden Lake; 12—North Artillery Lake; 13—Dubawnt Lake; 14—Yathkyed Lake; 15—Kazan River; 16—North Ennadai Lake; 17—Inuvik; 18—Colville Lake; 19—"East Keller" Lake; 20—Winter Lake; 21—South Artillery Lake and Pike's Portage; 22—Fort Reliance; 23—South Ennadai Lake; 24—Kasba Lake. In all instances, study areas are near the named lakes or villages. In the areas, lakes or rivers are the only suitable landing sites for aircraft during summer, and, excepting for canoe travel, hence the only means of access at this time (with the exception now of Inuvik).

be mentioned that there are now virtually countless studies reported in various publications of the Canadian and Alaskan forest services. The basic outlines of major research programs under way have also now been admirably summarized in compendium volumes that are now appearing at an accelerated pace, notable among them *Resources and Dynamics of the Boreal Zone*, by Ross W. Wein, Roderick R. Riewe, and Ian R. Methven (editors and contributors), *Northern Ecology and Resource Management* by Rod Olson, Frank Geddes, and Ross Hastings (editors), and *Forest Ecosystems in the Alaskan Tundra* by K. Van Cleve, F. S. Chapin III, P. W. Flanagan, L. A. Viereck, and C. T. Dyrness. In these volumes a variety of topics are reviewed and original data presented by individuals conducting research in the field, describing the impressive programs sponsored by the various Canadian and Alaskan governmental, educational, and private industrial agencies, reporting new knowledge recently accumulated on resources of Alaskan and Canadian northern lands, including forest ecology, nutrient cycling, influence of fire, food and energy potentials, soils, effects of environmental factors such as snow, the ecology of such animals as moose, caribou, pine marten, Dall sheep, wood bison, bear species, as well as land-use problems, management, and research needs. These are extremely valuable books, and it is, thus, apparent that the importance of Canadian and Alaskan biological resources is now well and widely recognized. It is to the credit of everyone involved that such research is now being given the attention its importance merits. One wishes that the other major world ecosystems had been given at least marginally equal attention before they disappeared – the forests, for example, of those parts of the world in which native vegetation is fast disappearing or, worse, vanished long ago.

The spruce forests are the basis of much of the forest industries of Canada and Alaska as well as of northern European countries, and growth and regeneration of the species of spruce is a major focus of many of the research programs carried on. Important knowledge of the tolerance limits of black and white spruce can be obtained by studies of these species at the edge of their range, as Black and Bliss (1980) have pointed out: "Studies of the causal factors of species distribution are most effectively carried out near the species' limits. The study of *Picea mariana* (Mill.) BSP. at its northern limit is of particular interest because of its wide distribution within the boreal forest and its importance as a tree line species in numerous geographic regions." In their study, Black and Bliss found that seed production and germination were limited in black spruce at the forest border and the lower cardinal germination temperature of 15°C determined germination timing and success. Other studies have also shown that reproductive capability of black spruce is limited at the northern forest edge (Larsen, 1965, 1980; Elliott, 1979).

It is not the purpose of the chapters that follow to further elaborate upon the autecology of black and white spruce, from the point of view of either the ecologist or practicing forester, but to delineate, in an ecological and environmental perspective, certain bioclimatic and biogeographic aspects of the vegetational communities of the forest border region in central and western Canada primarily. This discussion will perhaps bear on certain practical problems, at least in an oblique manner, and it is hoped they do so, but it is primarily the purpose here to satisfy to some degree a natural curiosity that arises when one views the continental treeline and the adjacent plant communities and asks questions concerning the variations in community structure and composition, variations that are less visible, but perhaps from a scientific point of view just as interesting, ultimately as important, as the obvious zonation apparent in the tree species. While the practical applications may not be such as to stir the imagination of those more interested in immediately profitable enterprises than in aspects of theoretical biogeography, perhaps the knowledge will one day be not entirely without appreciable value, if solely for the amplification of certain ecological and biogeographical principles, the great significance of which may not be fully appreciated at the present moment but which will bear importantly upon revegetation and ecological restoration of areas destroyed by natural or man-made forces.

There are indications that a global climatic change may be imminent or pending, but the direction is, at least as yet, not entirely certain. A cooling may be in the early stages of development as a result of the Earth's axial tilt and orbital ellipticity, in turn countered by warming from natural or anthropogenic sources, increases in chemical emissions from industry, or as a result of certain land use practices. Which of these shall in fact dominate in years to come is, at present, uncertain; it remains to be seen whether trends correspond to any known planetary orbital or solar emission cycles or to changes in atmospheric composition. It is reasonable to conclude with Lamb (1966, 1972) that in no case has the significance of a suggested periodicity, not to mention changing atmospheric composition, been clearly established. Moreover, because many or most of the various forcing factors will act in a cumulative, perhaps exponential, manner, or, conversely, will compensate for one another to maintain an equilibrium, the end result may be more difficult to discern as to cause than might be expected at first glance. If a climatic change does occur, however, a collection of base-line data, such as might be obtained in the plant community structure and composition in an ecotonal region such as the northern forest border, might well be invaluable as an indicator of the direction and the magnitude of the change. It is with this in mind that the data in this volume are presented, and descriptive comments given a central focus.

2. Early Observations in the Northern Barrens

The northern coast of Canada and the Arctic Archipelago were occupied by Eskimos long before the arrival of the first European explorers, and it is evident from the archeological and historical record that there was habitation of the interior barrens of Keewatin by bands of Caribou Eskimos living along the Kazan River and elsewhere. Although they were nomadic and ranged over large areas, the aboriginal peoples were never present in such numbers that they would have had a measurable influence upon the vegetation of the region. Recent estimates place the early-day population of northern or Chipewyan Indians at little more than 4000, the northern coastal Eskimo at 6000, and the inland Caribou Eskimo at 1500 or fewer. The population today is about three times this total figure, but the people today are, for the most part, located in small population centers, no longer in nomadic bands that roam over the entire region. The upper Kazan River, for example, is now devoid of habitation except for the small meteorological station at Ennadai Lake and for a group of Eskimos who ascend the Kazan for perhaps 30 miles from Baker Lake during the summer. Thus, today, there are large areas once thinly occupied that now have no resident human inhabitants. The present-day disturbance of the vegetation is thus minimal, and can be reasonably compared with the conditions existing in the days when the first explorers crossed the interior of the continent. Records of these early explorers can often furnish valuable information on the nature of the vegetation a century or more ago, with which to make comparisons with the present day, even on the plants used for food and other purposes (Walker, 1984); it is often possible, too, to make inferences concerning the impact upon vegetation of climatic changes or the influences of introduced technology when and where it occurs.

The early-day explorations of the northern interior of Canada were without doubt the work of men who took great risks and traveled under arduous circumstances. Many are nameless who took leave of the last outpost along the river and were never seen again. It is, however, the survivors whose names come down to us, survivors who, for one reason or another, often because they were in government employ, kept a journal of their travels.

They give us in these journals an accurate, often a vivid, picture of the landscape as it appeared to them at that time. Some gave descriptions of the peoples encountered, most mention the animals, a few mention the vegetation, almost all took note of the glacially created landforms we now know as drumlins, moraines, eskers, roche-moutonierre, ice-rafted boulders, kettles, crag-and-tail hills, ripple till, all characteristic of areas where forest and tundra now occupy the landscape. They form the geological stage upon which the panorama of northern life has played—and continues to play—its fascinating role in the drama of life on earth.

But we shall deal with these things in greater detail later, and for the purpose at hand will begin with a small selection of the writings of a few of the Europeans who, in an age of intrepid voyagers and almost incredibly hardy explorers, first set foot on these wild and strangely beautiful yet unremittingly unforgiving lands.

The first explorer to cross the interior barren lands of central Canada was Samuel Hearne, whose account of his efforts to reach what is now Coppermine from the fort at the mouth of the Churchill River on Hudson Bay, is recognized as a classic among journals of exploration. Hearne made his journeys in the years 1769–72. The next to cross much of the land that Hearne traversed were the brothers J. B. and J. W. Tyrrell, who made the first canoe explorations of the major rivers, principally the Dubawnt and Kazan, in the 1890s. J. B. Tyrrell kept detailed geological notes for his sponsoring agency, the Canadian Geological Survey, and these were subsequently turned into most fascinating formal reports. J. W. Tyrrell made plant collections of great botanical interest and wrote somewhat more popular accounts of journeys undertaken in concert with his brother but also independently in other regions of Canada. J. B. Tyrrell compiled a valuable annotated edition of Hearne's journals, as well as the writings of another early-day explorer of northern Canada, Turnor.

In a most discerning analysis of Hearne's journey, Ball (1986) recently noted that observations by northern Indians, reported by Hearne, indicated a movement of the tree-line in northern Canada. In Hearne's words: "Indeed some of the older Northern Indians have assured me that they have heard their fathers and grandfathers say, they remembered the greater part of those places where the trees are now blasted and dead, in a flourishing state" (Hearne, 1772, p. 138). Hearne's map, moreover, indicates that the tree-line at the time of his journey was then farther west in the north/south section of western Canada and farther south in the east/west section of central Canada than it is today. In his observations, Ball compares the present-day position of the tree-line with the position given in Hearne's journal, indicating that, in some places, the present position is 100 to 200 kilometers (60–120 miles) north of its position in Hearne's day. "This location," Ball writes, "and subsequent movement appears to be logical in relation to the climatic conditions that occurred in the region as a result of the Little Ice Age and the warmer conditions thereafter."

Hearne's descriptions of the landscape and vegetation are brief, since most of his account is devoted to the events of his travels and to the details of the customs and habits of his Indian companions and those whom he encountered along the way. He also described the animal life, including perceptive and accurate descriptions of the life history and behavior of the caribou, moose, buffalo, beaver, and other animals. Hearne excused his lack of interest in botany in the following words: "The vegetable productions of this country by no means engaged my attention so much as the animal creation; which is the less to be wondered at, as so few of them are useful for the support of man." He does, however, provide brief descriptions of the more useful plants such as gooseberries, cranberries, crowberries (which he calls heathberries), cloudberries, currants, juniper berries, strawberries, blueberries, and partridgeberry (the alpine bearberry), as well as Labrador tea, the common bearberry, and various grasses and shrubs. Of the area around Prince of Wales's Fort (Churchill, Manitoba), where Hearne held command for a period of 10 years, he writes that: "The forest trees that grow on this inhospitable spot are very few indeed; pine, juniper, small scraggy poplar, creeping birch, and dwarf willows, compose the whole catalogue. Farther westward the birch tree is very plentiful; and in the Athapuscow country, the pines, larch, poplar, and birch, grow to great size; the alder is also found there." By pines, Hearne undoubtedly refers to both species of spruce. The Athapuscow country to which he makes reference is the region around Great Slave Lake which he covered extensively in his journey to the Coppermine country.

While the band of Indians with which Hearne traveled often retreated into forested country for better hunting and shelter, the route of much of his journey to and from Coppermine was accomplished to the north of the continental forest border in the barren lands. And of these, Hearne writes in his report: "With regard to that part of my instructions which directs me to observe the nature of the soil, the productions thereof, etc., it must be observed, that during the whole time of my absence from the Fort, I was invariably confined to stony hills and barren plains all the summer, and before we approached the woods in the fall of the year, the ground was always covered with snow to a considerable depth; so that I never had an opportunity of seeing any of the small plants and shrubs to the Westward. But from appearances, and the slow and dwarfy growth of the woods, etc. (except in the Athapuscow country), there is undoubtedly a greater scarcity of vegetable productions than at the Company's most northern settlement; and to the eastwards of the woods, on the barren grounds, whether hills or valleys, there is a total want of herbage except moss, on which the deer feed; a few dwarf willows creep among the moss; some wish-a-capucca and a little grass may be seen here and there, but the latter is scarcely sufficient to serve the geese and other birds of passage during their stay in those parts, although they are always in a state of migration, except when they are breeding and in a moulting state."

The wish-a-capucca to which Hearne refers are the two species of *Ledum* which probably he did not distinguish; the one which is known as Labrador tea (*Ledum groenlandicum*) grows abundantly in the forest regions and extends for a distance into the barrens. His description of the paucity of vegetation on the barrens is somewhat exaggerated, of course, there being a multitude of plant species in most areas, many more than the mosses, dwarf willows, Labrador tea, and grass to which Hearne limits himself. And during the summer, actually, food for geese and grazing animals is relatively abundant. This latter deficiency in observation seems to be one which Hearne makes habitually, for elsewhere he refers to the large number of animals present, obviously more than could have subsisted on the vegetation he describes: "Though the land was entirely barren, and destitute of every kind of herbage, except wish-a-capucca and moss, yet the deer were so numerous that the Indians not only killed as many as were sufficient for our large number, but often several merely for the skins, marrow, etc. and left the carcasses to rot or to be devoured by the wolves, foxes, and other beasts of prey." By contrast, Hearne writes of the land around Island Lake (to which he gives the latitude as 60°45′ north, 102°25′ west) as being quite heavily wooded: "Many of the islands, as well as the main land round this lake, abound with dwarf woods, chiefly pines; but in some parts intermixed with larch and small birch trees. The land, like all the rest which lies to the north of Seal River, is hilly, and full of rocks; and though none of the hills are high . . . they in general show their snowy heads far above the woods which grow in the valleys, or those which are scattered about their sides."

Following Hearne, it would be many years before the exploration of the northern interior of Canada would be continued. Southward, however, well within the broad expanse of the boreal forest proper, exploration had long been underway and trade and travel along many of the major waterways had become relatively commonplace. Henry Ellis, for example, had wintered in 1746–47 at the mouth of the Hayes River. Of this area he wrote: "Its banks are low, and covered with large woods, chiefly spruce, fir, poplar, birch, larch, willow, etc., and abounds with deer, hares, rabbits, geese, ducks, partridges, pheasants, plover, swans, and many other fowl in their proper season, as also fish in great plenty. . . . In the southern parts, and where we wintered, the soil is very fertile; the surface being a loose dark mould, under which are layers of different coloured clays, pale, yellow, etc. Nigh the shores the land is low and marshy, covered with trees of various sorts, as spruce, larch, poplar, birch, alder, and willow; within land there are large plains, with little herbage on them except moss, and interspersed with tufts of trees and some lakes, as also some hills or islands, as they are called, covered with shrubby trees, and deep moss, the soil of a turfy nature." Like Hearne, Ellis took special note of some of the few plants whose berries were eaten and those which were employed for "scorbutic disorders" and for "promoting digestion."

Farther to the westward, Alexander Mackenzie was the first to record in detail the features of the country along the water route between Lake Athabasca and the Arctic Ocean, having followed the route along the Slave and Mackenzie Rivers in 1789. He, too, took greater interest in the animals than in the botanical features to be seen along these rivers, but occasionally would indicate the nature of the vegetational cover. His descriptions of the Peace River country were the first to be published, and it is of interest that he notes the soil appeared fairly productive. His description of one point along the river is of interest: "The country in general is low from our entrance of the river to the falls, and with the exception of a few open parts covered with grass, it is clothed with wood. Where the banks are very low the soil is good, being composed of the sediment of the river and putrefied leaves and vegetables. Where they are more elevated, they display a face of yellowish clay, mixed with small stones. On a line with the falls, and on either side of the river, there are said to be very extensive plains, which affords pasture to numerous herds of buffaloes. . . . In addition to the wood which flourished below the fall, these banks produce the cypress tree, arrow-wood, and the thorn. On either side of the river, though invisible from it, are extensive plains which abound in buffaloes, elks, wolves, foxes, and bears. . . . The whole country displayed an exuberant verdure; the trees that bear a blossom were advancing fast to that delightful appearance, and the velvet rind of their branches reflecting the oblique rays of a rising or setting sun, added a splendid gaiety to the scene, which no expressions of mine are qualified to describe. The east side of the river consists of a range of high land covered with the white spruce and the soft birch, while the banks abound with the alder and the willow."

Philip Turnor was a surveyor, sent to Canada by the Hudson's Bay Company to prepare maps of the region between James Bay and Great Slave Lake, and his contribution to the early development of this region lies not so much in his verbal descriptions as in those maps which were the first to be made of large tracts of country. These actually are far more vivid in their portrayal of the land than his journal descriptions which, at best, are often largely concerned with the state of the weather and the daily events of his travels, a common fault of many of the early writings.

Turnor does, however, present a most interesting account of the description provided by an Indian of the land to the northward of the east arm of Great Slave Lake. He points out that, according to the Indian's description, Artillery Lake (which is termed the "northern Indian great lake") is "very long but not very wide and is very full of islands which have woods upon them but none upon the shore being all rocky barren land."

To the northward of this, at what apparently is Aylmer Lake, there is a short "carrying place" to the headwaters of the Back River. This is called the Esquimay River, and it is, in Turnor's words, "a bold deep river without any falls in it but rather strong current and no woods growing about it."

The position of the forest border can be seen in very much the same position on Artillery Lake at that time as it is today, and even Crystal Island on Artillery Lake is well wooded in more protected spots while the shores of Artillery Lake proper are barren excepting for the extreme southern end and for a few miles northward along the western shore.

David Thompson was another of the great early-day surveyors who opened large regions to the collective eye of Canadians. Born in England, he came to Canada at 14 years of age and shortly thereafter learned surveying under Turnor. His writings are of particular interest, for they constitute valuable adjuncts to his maps, and they present vivid and accurate descriptions of not only the country through which he passed but many observations on the growth habits and behavior of plants and animals. Writing of the Canadian shield, he described both vegetation and soils: "The northern parts are either destitute of woods, or they are low and small; especially about Hudson's Bay where the ground is always frozen; even in the month of August, in the woods, on taking away the moss, the ground is thawed at most, for two inches in depth . . . All the trees in this frozen soil have no tap roots; their roots spread on the ground, the fibres of the roots interlace with each other for mutual support; and although around Hudson's Bay there is a wide belt of earth of about one hundred miles in width, apparently of ancient alluvium from the rounded gravel in the banks of the rivers, yet it is mostly a cold wet soil, the surface covered with wet moss, ponds, marsh, and dwarf trees . . . The rock region close westward of this coarse alluvium already noticed, in very many places, especially around its lakes, and their intervals, have fine forests of pines, firs, aspens, poplar, white and grey birch, alder and willow; all these grow in abundance, which makes all this region of rock and lake appear a dense forest, but the surface of the lakes cover full two-fifths, or more of the whole extent. The most useful trees are the white birch, the larch, and the aspen."

It is of interest, that as early as 1798 Thompson had interpreted the role of fire in maintaining the prairie-forest ecotone at the southern edge of the aspen parkland forest: "The grass of these plains is so often on fire, by accident or design, and the bark on the trees so often scorched, that their growth is contracted, or they become dry; and the whole of the Great Plains are subject to these fires during the summer and autumn before snow lies on the ground . . . Along the Great Plains, there are very many places where large groves of aspens have been burnt, the charred stumps remaining; and no further production of trees have taken place, the grass of the plains covers them; and from this cause the Great Plains are constantly increasing in length and breadth, and the deer give place to the bison. But the mercy of Providence has given a productive power to the roots of the grass of the plains and of the meadows, on which the fire has no effect . . . If these grasses had not this wonderful productive power on which fire has no effects, these Great Plains would many centuries ago, have been without Man, Bird, or Beast."

While scientific and geological examination of the regions of southern Canada and of both East and West continued unabated after the initial journeys of the first explorers, the vast interior plains of the central northern regions remained beyond the limits of exploratory ambition until the end of the 1880s when the Tyrrells canoed northward into the wilderness of the Dubawnt and Kazan Rivers, proving that the former empties into the Thelon and ultimately Baker Lake and that the latter flows at least to Yathkyed Lake, from which, because of the lateness of the season, they portaged across country to one of the smaller rivers flowing directly eastward into Hudson Bay.

The Tyrrell explorations are recounted in some of the earliest reports of the Canadian Geological Survey, as well as in a volume describing the Kazan and Dubawnt River journeys, all of which constitute fascinating sources of accurate and detailed information concerning not only the country but also the native peoples, birds, animals, and plant life. While the popular travel accounts are generally better reading, the detailed geological and vegetational observations of J. B. Tyrrell are of great value today for they clearly delineate the condition of the land in the presettlement days (if, indeed, the country can today be considered as "settled"). The descriptions of the conditions of travel are vastly understated in the geological reports, although they often are briefly indicated; more than one subsequent canoe traveler, using a Tyrrell geological journal report as a guide, has boldly struck forth into waters described by Tyrrell as "fairly rapid water" only to find himself in the midst of the most terrifying rapids he has ever seen. Only then does Tyrrell's gift for understatement become clearly apparent —as well as consummate skill as a canoeist (or consummate faith in his professional canoemen).

The reports, on the other hand, are valuable for their detailed description of the vegetation, and especially so for delineating the northern ranges of many of the dominant species of trees, for Tyrrell carefully observed the groves and clumps along the shores of the rivers, indicating the northernmost point at which the various species were observed. That climatic conditions have not changed appreciably since Tyrrell's day is indicated by the fact that the limit of trees—and sometimes the individual grove—can still be found on the same site where Tyrrell indicated they were to be found nearly a century ago. His observations also include peat banks along the lake shores, of palynological interest today, and his journals contain a multitude of other records which will be of enduring botanical, anthropological, as well as geological value. It is truly fortunate that the earliest geological explorations of many areas of northern Canada were made by men of such abilities and broad interests as Tyrrell and others, notably A. P. Low whose reports on the eastern region of Canada—particularly Labrador and Ungava—are equally valuable. The quiet accomplishments of these men rank with the highest in the history of world exploration.

It would constitute a most useful enterprise to bring together and annotate the many references to flora and vegetation to be found in Tyrrell's journals, and in those of the other early day explorers of remote regions of Canada, but in this brief review it is obviously not possible to undertake such a work. It is of some value , however, to indicate a few of the more interesting observations made during Tyrrell's descent in 1893 and 1894 of the Dubawnt and Kazan Rivers. For additional references, the interested reader is referred to the reports cited at the end of this volume (some of which, unfortunately, are rather difficult to obtain). In the introduction to his geological survey report, Tyrrell points out:

"Since a large portion of this region lies north of the country where fur-bearing animals are abundant, it had not been travelled over by fur-traders, or even by voyageurs or Indians in search of furs, and the characters of the lakes and streams were, therefore, unknown to any but the few Indian and Eskimo deer-hunters who live on their banks, and who come south once or twice a year to trade wolf or fox skins for ammunition and tobacco.

Tyrrell, then, was making observations in country which was totally undespoiled by the hand of man, and although his records were not quantitative in the modern ecological sense, and hence not strictly or statistically comparable to modern data, the gross visual comparisons that become possible have great value today and will become more so in the future. Already, in the case of the present-day location of the northern tree-line, Tyrrell's observations have been of value in establishing that the tree-line has not moved significantly either northward or southward of the position it had when Tyrrell passed through along his routes down the Dubawnt and the Kazan rivers. One of his most detailed descriptions concerns the possible mode of formation of the deep banks of peat seen along the shores of a number of the lakes. There is no obvious manner in which these could have been formed and the first question one asks is what processes are involved in their origin. The banks, in the fullest development seen by the author, extend some 10 or 12 feet above the shoreline of a lake, the peat having crumbled away from the face of the bank, giving the impression that the whole deposit must be moving slowly down the usually gentle declivity between hills. The peat itself, when surface detritus has been cleared away, shows apparent strata, indicating periods of rapid growth alternating with periods of slow growth or of deterioration by desiccation and, one supposes, possibly oxidation or bacterial activity. On occasion a small bole, presumably of a black spruce, will be seen in one of the strata, indicating that at one time or another a single, or perhaps a few, spruce occupied the surface of the gentle peat slope. Today these surfaces are almost always uniformly tundra, even though some small clumps of dwarfed spruce may exist nearby.

Describing these features, Tyrrell writes: "The drainage from the higher land accumulated at the bases of the hills on soil which was either impervious in itself or was rendered impervious by being permanently frozen. Moss, small spruce and larch, began to grow on this wet ground, and each winter the moss froze to the bottom, thawing again with the return of summer. It increased in thickness year by year, until it had reached such a depth that the lower part remained permanently frozen, the heat of summer not being sufficient to thaw it. Many of the swamps in the more northern portions of Canada would seem to be thus permanently frozen, but every summer they thaw to a sufficient depth to permit the continuous growth of this upper layer of moss and the overshadowing forest of stunted conifers. But in this region the summer heat is not sufficient to thaw the moss to such a depth as to allow trees to grow over the frozen substratum. The trees therefore die, and the moss, having the ice close beneath it and deprived of its overshadowing screen of trees, also dies, and the surface of this dead and dry peat bog soon becomes covered with such small plants as the country will produce."

However interesting they may be, these are unusual formations, found only near the forest border and never far northward into tundra. Over the vast expanse of the tundra proper, three major vegetational communities can usually be discerned. These might be called rock fields, tussock muskeg, and low meadow, following in order their topographic preference from the summits of the rolling rocky hills to the lower areas where water accumulates to a depth of several inches, at least for the first few weeks of spring and summer. As the tree-line or, more properly, the forest border is approached from the south, the forest is first seen to be absent from the tops of the hills and then, continuing northward, from slopes ever lower and lower. Finally it is found only in narrow ravines between the highest hills, usually on sites where a small continuous trickle of water runs beneath boulders. At some point, continuing northward, even these sites are too severe for spruce survival. The land becomes continuous tundra, principally of vegetational associations that occupy rock fields, the tussock muskeg of intermediate slopes, and low meadows. Even the familiar open muskeg areas so common in the spruce forest disappear; such areas, in tundra regions, retain their original character and remain shallow lakes. As Tyrrell describes the forest-tundra ecotone region: "We had now reached the northern edge of the forest, and hence forward any timber seen on this river [the Dubawnt] was in the form of scattered and often widely separated groves. With the disappearance of this stunted forest, mossy plains and bogs also almost disappeared, and they were nowhere to be found to extend beyond the extreme northern limit of trees."

The occasional tree of large size, or even a number of large individuals making up a small grove, will be present at times on special sites beyond the forest border. These always seem strangely out of place, as though they remain there for the express purpose of making rational ecological explanations seem absurd. One such grove exists at Ennadai Lake, nestled between high hills, perhaps 20 miles beyond that could be described as the forest border. It seems reasonable, however, that the special characteristics of this site—protected from winds and provided with a deep snow blanket in winter—make it possible for trees here to attain unusually large size. The same phenomenon was observed along the Dubawnt by Tyrrell: "For four miles beyond this strait [below Boyd Lake] the northeast shore is marshy or grassy, and then, near the mouth of a small brook, it is broken by morainic boulder ridges a hundred feet in height, the edge of the grassy plain and the foot of the hills meeting in a fairly well-defined line . . . At the mouth of the brook is a small grassy glade, wooded with white spruce, one tree of which was fifty feet high and thirty inches in diameter, two feet above the butt. Under the trees were ferns, raspberry bushes, etc., the last that we were destined to see that summer."

Since most, if not nearly all, of these unusual groves located well beyond the forest border are closely associated with a brook or at least an audible trickle of water beneath a bed of boulders or rocks and gravel, it is easy to make the assumption that the latter is an indispensible feature of the environment for the survival of these groves at high latitudes. Where such a feature is not present, usually some other depression or proximity to water of a lake or river is a paramount environmental feature. The conclusion that one logically comes to is that running water, or proximity to the lake, is somehow involved in warmer temperatures for root systems during critical periods of spring when sunlight is intense, air temperatures may be warm, yet the surface of the soil is still often encased in ice. At such times, water transport through roots is a necessity. A supply of running water to thaw the root systems, or proximity to a lake shore which might accomplish the same end, is a necessity for survival under these rigorous conditions. Even a steep hillside will furnish the appropriate conditions on rare occasions, and a few small groves are seen on southeast facing hills where snow melts early in spring and water runs beneath the snow from the earliest of warm days. The same kind of site was observed by Tyrrell in the journey along the Dubawnt:

"From the north end of Nicholson Lake, the river flows northward for two miles and a half down a heavy rapid . . . On the steep hillsides were some small groves of white spruce . . . while the little patches of snow here and there in every direction would have kept us reminded that we had reached a sub-arctic climate, if the almost constant cold rain and wind had not made us thoroughly alive to the fact. On the hillsides, Arctic hares were seen for the first time . . . The Telzoa River flows from the east side of the oblong lake . . . On the north bank of the river, half way between the above lake and Doobaunt* Lake, is the last grove of black spruce on the river, where the trees are so stunted that they are not as high as one's head . . . Back from the river is a stony plain, parts of which were whitened by the flowers of the Labrador tea (*Ledum palustre*) or the white tassels of the anemone (*Anemone parviflora*), while many of the knolls were pink with the beautiful little flowers of *Rhododendron lapponicum* . . . Probably it was near this grove of spruce that Samuel Hearne, our only white predecessor in this portion of the Barren Lands, had crossed the Doobaunt River in company with a large band of Indians a hundred and twenty-three years before."

In the following year (1894) Tyrrell traveled down the Kazan River, this time without the company of his brother, canoeing northward toward Reindeer Lake and up the Cochrane River, from which a portage took him to the headwaters of the Kazan. From there he descended the Kazan through Kasba, Ennadai, Angikuni, and Yathkyed Lakes, leaving the latter on its eastern shore and portaging to one of the rivers that flows eastward into Hudson Bay. Along the shores of Reindeer Lake, well within the wooded

*Tyrrell's spelling of Dubawnt.

country of the central boreal forest, he writes that: "The dark lichen-covered hills bear a scattered growth of black spruce, with an occasional stunted canoe birch, on the lower slopes, while a few small Banksian pines and aspen poplars grow on the sandy terraces, almost to the north end of the lake." Of Lac Du Brochet north of Reindeer Lake he writes: "To the south are some rather high hills, and to the north the country is low, swampy and wooded with black spruce, the low hills here and there being composed of gneiss . . . At the west end of the lake is a portage 300 yards long over a low swampy island covered with yellow cloudberries (*Rubus chamaemorus*)."

Ancient beach ridges were discovered by Tyrrell in the vicinity of the south end of Kasba Lake, indicating that the land had once been under a vast lake and that water had persisted for long periods of time at what are levels 50, 120, and 200 feet above the present-day surface of the lake. Subsequent examination of the Kazan valley from this point northward to the vicinity of Dimma Lake has indicated that apparently the retreating glacial ice blocked passage of the river, creating an ice-dammed lake that extended over a vast region and that has been given the name Glacial Lake Kazan. From the south end of Kasba Lake, Tyrrell describes the view of the country visible from a high hill: "Cranberries, blueberries, crowberries and willow berries (*Vaccinium vitis idaea* and *uliginosum*, *Empetrum nigrum*, and *Arctostaphylos arctica*) were found in abundance. A magnificent view is had from the summit of this hill. To the south-east are many shining lakelets, and gentle green slopes thinly wooded with dark spruce. To the north and north-east long lanes of water run between the wooded ridges, while a high sandy terrace marks an old shore-line of the lake. To the west Kasba Lake extends as a beautiful sheet of open water to the blue hills on its further shore." He continues his description of Kasba Lake at a point some 30 miles northward along the eastern shore: "Behind the beach the country is thinly wooded in its southern part, while farther north it rises gently in green grassy slopes to hills, some of which are several hundred feet in height . . . Their summits are almost bare, while on their sides and in the pit-like depressions between them, are some spruce and larch trees of moderate size."

Along the Kazan River, Tyrrell next encountered Ennadai Lake, the southern part of which is within the forested region but northward some 15 or 20 miles from the south end the forest ends rather abruptly and the barren lands are at hand. It is of interest that on the eskers which are found infrequently along the Ennadai shoreline, white spruce grows with some abundance and density, attaining surprisingly large diameters for an area at the edge of the trees. They also grow in particularly favorable sites between the hills where water trickles between boulders over which grow their roots. The greatest proportion of the forested area, however, is occupied by stunted black spruce. Larch also is present, but in much lesser density by comparison, although individual trees are often larger than the average large spruce. Of terrain and vegetation around Ennadai Lake, Tyrrell writes: "Ennadai Lake is a long narrow sheet of clear water, lying in a north-easterly and south-westerly direction, and at an elevation of about 1100 feet above sea-level. Its greatest length in a straight line from end to end is fifty miles, and its greatest breadth seven or eight miles. At its southern arm, near where the river enters it, a low ridge of boulders runs along its eastern shore, while an even sandy esker, a hundred feet in height, forms its western shore . . . The lower and more sandy hills are thinly wooded with larger white spruce, and some canoe-birch. The view from the top of these hills shows the open lake, almost without islands, stretching away to the north-east, while all around are gently sloping, thinly wooded hills . . . At the south end the hills are usually wooded, but within a few miles the forest disappears, or becomes confined to the ravines, and the hillsides are grassy or bare."

Near the northern end of the lake, Tyrrell was forced to remain camped for a period of three days by a storm (from August 11 to the 14th) with rain and snow. "The camp was pitched on a sloping hillside, where a little rill, trickling through the stones, nourishes on the wet ground a small grove of dwarf black spruce and larch. The surrounding country is quite barren."

In the 1960s the forest border lay in essentially the position established by Tyrrell more than 60 years before (Larsen, 1965). The position of the tree-line near the southern end of the lake is most convenient for purposes of study, since transport through the tree-line can be accomplished by means of relatively rapid and easy canoe travel and vegetational sampling can be carried on at all points from barren tundra of the northern end of the lake to the nearly complete forest cover at the south end. Detailed studies of the behavior of black spruce at this northern limit of forest indicated that the critical factors must be climatic since topography, soils, geology, and geological history are essentially identical at both ends of the lake.

Robert Bell was another of the early geologists who explored Manitoba. He was particularly observant of the role of fire in the northern forests: "In going northward, there is of course a gradual diminution in the size of the trees and the height of the forest, as well as in the number of species. Owing, however, to the fires which sweep over large tracts at different periods, it is seldom that one sees the full size to which the trees are capable of growing." Bell was aware of the fact that the number of tree species declines northward, and he writes that, northward from the Lake of the Woods: ". . . the different species of trees which are found growing at the boundary line disappear in the following order: – Basswood, sugar maple, yellow birch, white ash, soft maple, grey elm, white and red pine, red oak, black ash, white cedar, serrated-leaf poplar, mountain ash, balsam fir, white birch, Banksian pine, balm of Gilead, aspen, tamarack, white and black spruce, willows . . . ."

Westward of Manitoba the land has been explored by many individuals, among them R. G. McConnell, who provided the first broad description of the region between the Peace and Athabasca Rivers. He writes: "The greater part of this district may be described as a gently undulating wooded plain, diversified with numerous shallow lakes, muskegs, and marshes. Small prairie patches, manifestly due to forest fires, occur north of the west end of Lesser Slave Lake, at several points along the Loon and Wabiscaw rivers, also on the Peace River around Fort Vermillion and at other places, but their total area is relatively insignificant. The principal forest trees are the white and black spruces, *Picea alba* and *nigra*, the balsam-fir, *Abies balsamea*, the Banksian pine, *Pinus Banksiana*, the larch, *Larix americana*, the aspen, *Populus tremuloides*, the balsam-poplar, *Populus balsamifera*, and the canoe birch *Betula papyracea*.* The species of spruce occur along many of the river flats, and on the uplands they are found nearly everywhere except on the drier hills. The white spruce attains, in favorable localities, a diameter of two feet or more, but it is usually much smaller. It is the most valuable tree in the district. The Banksian pine grows thickly on the sandy and gravelly

*Several of these names have been changed during subsequent years.

ridges, while the aspen prefers a loamy soil and characterizes the best agricultural parts of the country. The larch, balsam, balsam-poplar, and birch, although found in every part of the district, are more scattered and do not form continuous forests like spruce, Banksian pine, and aspen."

In reading many modern works on plant geography and ecology, one is given the impression that the dominant species of the boreal forest is white spruce, intermixed with a fairly wide variety of other species, including balsam fir and white birch. While this is true of certain portions of the southern fringes of the boreal forest, it is not true of the main extent of boreal lands and, indeed, it seems doubtful that white spruce was ever as plentiful, even in the southern portions of the forest, as much published literature leads one to believe.

Extensive stands of white spruce are not numerous in central Canada or westward as far as the Mackenzie Valley. There are, it is true, a few regions where white spruce appears to be the dominant tree on upland habitats, and one of these is the area around Colville Lake to the northwest of Great Bear Lake. Another is the area of the Labrador Trough in central Quebec and Ungava. But in these areas there are special conditions—principally the existence of soils derived from sedimentary parent materials, richer in nutrients than soils from the granitic rocks, common elsewhere over the Canadian Shield. Another reason has been the fact that most early travel was by means of canoe. For canoe travelers, it would be natural to affirm that the northern forest is of white spruce. Without traveling at least a few hundred yards or perhaps a mile from the borders of the lakes and streams, one would easily receive the impression that white spruce was the dominant forest tree. Even so careful and widely traveled a botanist as Raup, when writing of the forest vegetation of the region around Great Slave Lake and northward, was led into error. In his report on the Athabasca and Great Slave Lake region (Raup, 1946) he presents a map indicating that the predominant forest tree from Great Slave Lake northward to tree-line is white spruce. Although white spruce is more common in this region than perhaps it is elsewhere, it is by no means dominant except along shores of the lakes. Even in many of these places black spruce constitutes the only tree visibly dominant over the landscape for many miles. One questions whether, even much farther south, the preferences of field botanists and foresters for larger and more botanically attractive open white spruce forests have not led to a false impression that white spruce is dominant over large regions that actually are more properly designated as mixedwood, a forest in which white spruce is present but where balsam fir and a host of other tree and shrub species attain greater importance in the total vegetational cover than white spruce. Perhaps selective forestry in many of these areas has been successful in removing white spruce to the advantage of these other species. Whatever the reasons, even in the southern boreal forest, white spruce is, at least today, much less prevalent than one might be led to believe from many reports now available.

From the early journals of exploration, too, one gains support for the idea that white spruce was never abundant. Tyrrell, again, is our authority, and in describing the general region northward from the Churchill River to Lake Athabasca he writes as follows: "The country is generally forested, though much of the timber is small black spruce (*Picea nigra*)* and tamarack (*Larix Americana*). Banksian pine (*Pinus*

---

*Tyrrell uses the nomenclature of his day.

Banksiana) forms thin park-like woods on the sandy plains. White spruce (*Picea alba*) forms some groves of fair size in the bottom lands near the Churchill River, but farther north it is rarely seen except in some particularly favorable localities. One small isolated grove of white spruce was found on a high sandy island in Hatchet Lake, standing out conspicuously in the midst of the surrounding forest of small black spruce. Poplar (*Populus tremuloides*) and birch (*Betula papyrifera*) are the only remaining forest trees of any importance. They are found chiefly in the vicinity of Churchill River, though small scattered trees were seen on the banks of Stone River."

Tyrrell continues with a description of the understory plants common in the forests of this region: "In places some of the more northern berries grow in great profusion, chief among which are the common huckleberry (*Vaccinium Canadense*) and the small cranberry (*Vaccinium Vitis-Idaea*). The former grows in the deciduous woods along the Churchill River, while the latter covers the dry slopes from the Saskatchewan northward. The blue huckleberry (*Vaccinium uliginosum*) grows on the banks of Cree and Stone rivers, but the bushes did not seem anywhere to bear much fruit. The raspberry (*Rubus strigosus*) grows on the richer ground by some of the streams. The yellow swamp-berry (*Rubus chamaemorus*) is found abundantly in the moss of the wet spruce and tamarack swamps. The crowberry (*Empetrum nigrum*) occurs on the drier land towards the north, and the Pembina berry (*Viburnum pauciflorum*) grows in the deciduous woods besides the streams, especially in the southern portion of the district."

No less arduous than Tyrrell's (and today no less amazing) were the explorations of the Labrador Peninsula by A. P. Low in the years 1892–95 (Low, 1896), and he provides very detailed accounts not only of the geology of the region but of the plants, animals, and much of great interest regarding the indigenous peoples. He summarizes his observations on the vegetation of Labrador in a few pages of his report, and many of the descriptive passages therein constitute, even today, the best available on the vegetation of northern regions: "The southern half of the Labrador Peninsula is included in the sub-arctic forest belt, as described by Prof. Macoun.* Nine species of trees may be said to constitute the whole arborescent flora of the region. These species are: —*Betula papyrifera*, Michx., *Populus tremuloides*, Michx., *Populus balsamifera*, Linn., *Thuya occidentalis*, Linn., *Pinus Banksiana*, Lam., *Picea alba*, Link., *Picea nigra*, Link., *Abies balsamea*, Marsh, and *Larix Americana*, Michx."

Writing of black spruce, Low points out: "*Picea nigra*, Link., is the most abundant tree of Labrador and probably constitutes over ninety percent of the forest. It grows freely on the the sandy soil which covers the great Archaean areas, and thrives as well on the dry hills as in the wet swampy country between the ridges. On the southern watershed the growth is very thick everywhere, so much so that the trees rarely reach a large size. To the northward, about the edge of the semi-barrens, the growth on the uplands is less rank, the trees there being in open glades, where they spread out with large branches resembling the white spruce. The northern limit of the black spruce is that of the forest belt; it and larch being the last trees met with before entering the barrens."

Low accurately assesses the importance of soil in the distribution of white spruce. He points out that although white spruce is found throughout the wooded areas of the

*The Forest Trees of Canada, John Macoun. *Trans Royal Soc. Canada* Sec. iv, 1894, pp. 5–7.

peninsula, it is not common and "its distribution is but little affected by climate or by height above sea level; it appears to depend altogether on the soil." It is probable that he underestimates the importance of climate in white spruce distribution, but he correctly infers that soil is much greater significance in its ecology than for many other species common in that region.

While botanical exploration of Canada has been relatively comprehensive, with few regions which have not been visited by at least a single botanical collector, Raup (1941) points out that studies of the "arrangement of the flora into plant formations and lesser communities" had not (at that time) been undertaken and the next major task was to begin studies of plant communities found over the Canadian landscape. "Although a great deal of work on the structure of arctic and subarctic plant communities has been done in the Old World, in America it is hardly more than begun. Botanical work in boreal America during this period of change has remained to a large extent floristic; and geographical problems which have grown out of floristic investigations are still uppermost." Those who initiated work include Raup (1946, 1947), who provided descriptions of the vegetation of the Athabasca and Great Slave Lake regions and southwestern Mackenzie. Porsild furnished many descriptions of the vegetation of western Canada and the Canadian Arctic Archipelago (Porsild, 1951, 1955). Polunin described the botany of the Canadian Arctic (1948). Following these early works, increasing numbers of studies have been conducted in this region (see references) and with growing economic importance and easier transportation, we can expect ever greater numbers to be forthcoming.

Accounts of the early explorers, however, will continue to be a valuable source of knowledge concerning the general condition of the vegetation at the time of exploration; not only the journals of those individuals mentioned above, but many others as well (among them: Anderson, 1956; Back, 1836; Blanchet, 1927, 1930; Douglas, 1914; Dowling, 1893; King, 1936; Hanbury, 1900, 1903; McConnell, 1891; Rasmussen, 1927; Pike, 1892; Pullen, 1852; Richardson, 1851; Mackenzie, 1801; Franklin, 1823, 1828; Low, 1889). A valuable compendium of references to plants useful for food taken from the journals of the early explorers, plus other noteworthy descriptions of conditions found during the initial explorations of much of northern Canada, is found in the most interesting volume by Walker (1984).

3. The Forest/Tundra Transition or Ecotone

It is not difficult to identify regions clearly "Arctic" and those that are unequivocally "Boreal," the former in high latitudes, treeless, snow-covered for much of the year, with short and very cool summers, the Boreal regions located adjacent to, and south of, the Arctic, dominated (at least usually) with a vegetation in which trees are a prominent feature, with longer summers and brief periods when summer days can be very warm. There are correspondingly quite obvious differences in plant and animal species occupying the two regions.

There is, however, great disparity among definitions as to where the Boreal regions end and the Arctic regions begin. It is apparent that physical and biotic features are sufficiently distinct to clearly warrant separation. The zone of transition between the two can be identified with a certain degree of accuracy, but the drawing of lines on maps to delineate Arctic, Transitional, and Boreal zones has always been a source of difficulty, not to say dispute, among geographers and others interested in the northern lands. For conceptual reasons there would be advantages—political and economic—if the regions could be clearly circumscribed, but physical, biotic, and ethnic parameters do not clearly coincide throughout the circumpolar range of the arctic/boreal transition zones, and thus any realistic definition will be either too general, long, or complex, with too many variations and exceptions to be taken into account for general acceptance.

As Hustich (1979) pointed out, the Arctic as a geographical concept has been most usefully defined either as the whole area north of the polar tree-line or as the region north of the July mean daily isotherm of 10°C. Use of climatic parameters to define the

Arctic is not, however, entirely satisfactory because the climatic lines do not invariably coincide with the biotic or ethnic regions and, moreover, their position varies from year to year as weather events and perhaps climatic change dictate. Climatic records, also, are often not of sufficient length to firmly establish a satisfactory average of climatic values.

Thus, in ecological terms the pragmatic course of action has been to delimit the southern edge of the Arctic as a border that can be observed quite easily in the vegetation as photographed from aircraft or satellite. Such a border is the northern tree-line, or, more accurately, the northern forest border, whether the forest be deciduous (birch, alder, poplar), as in parts mainly of northern Europe, or the coniferous (spruce, larch, pine) forest, as found in North America and Eurasia.

The tree-line is more difficult to define properly than the forest border, since the tree-line depends on how the word "tree" is defined; whether it be an individual tree of a certain minimum height or whether it is simply an individual of the tree species under consideration, no matter what its size. The northernmost individuals of tree species are more likely than not invariably small, dwarfed, decumbent, or deformed, and often are found growing as low clumps or mats trimmed by wind and snow abrasion. The forest border, however, can often be quite readily delineated on aerial or satellite photographs inasmuch as the forest canopy can be discerned clearly. Even if the trees are widely spaced, as in lichen woodland (coniferous trees with heavy ground lichen growth between), the community is usually identifiable. Even here, however, there may be some uncertainty as to whether the particular community under study should be considered forest, tundra, or forest/tundra ecotone.

It is apparent, thus, that there will probably never be universal agreement as to where the Arctic begins, and there will be a continual need for redefinition to suit individual requirements, however unsatisfactory this may be (Remmert, 1980). For the purposes here, a detailed definition is not necessary, and the literature can be consulted by individuals interested in a somewhat more complete discussion (Hustich, 1979; Remmert, 1980; Hañet-Ahti, 1981; Bradley et al., 1982; Payette, 1983) and the extensive references cited therein.

Tree-Line: General Character

The zone of transition from forest to tundra, or the northern forest/tundra ecotone, the region of North America under consideration in this volume, has been rather broadly defined as that land where, undisturbed by fire or other influence, including human activity, and on the scale of most of the selected aerial photo coverage presented herein (on conventional 9 × 9 inch photographs, one inch equals 5000–5600 feet), unbroken forest occupies less than 75% of the land surface above the water table (upland), *or* less than 75% of the area is unbroken tundra, i.e., 25% of the land area or more is occupied by an admixture of forest and tundra, and presumably with tundra occupying the windswept upland summits and with forest in more protected sites where snow cover is sufficient to protect trees in winter. Admittedly, there are arbitrary aspects to the definition, as a brief consideration of the percentage relationships will reveal. The advantage is, however, readily apparent when aerial photographs are observed, and there is usually little difficulty in determining whether or not a given photographic image is forest, tundra, or ecotone.

There are also, apparently, some areas in which this definition cannot be applied unambiguously because the trees extend over broad expanses of the landscape but are widely spaced, as is often the case in the lichen woodlands. For the purposes here, again, this is defined as forest. Experience dictates that such areas should be considered forest because the trees are recognizable as such and are not exceedingly dwarfed, decumbent, or deformed by climatic conditions and reproduction is occurring.

At the northern edge of the forest-tundra zone, and in particular within the Canadian Shield regions of Canada, there is a notable decline in the capacity of the trees to produce viable seeds (Clarke, 1940; Larsen, 1965; Hustich, 1979; Viereck, 1979; Black and Bliss, 1980; Elliott, 1979a,b; Payette, 1983) and if viable seed years do occur in the northern parts of these regions they do so infrequently; reproduction in *Picea mariana*, for example, is accomplished, when it occurs, by layering primarily (see Appendix C).

This conception of forest border is both a demonstrably useful description of what is seen in the field and on aerial and satellite photographs, and it provides a reasonably accurate approximation of the location of the forest/tundra ecotone, to which one might conceivably put a "±" number indicating the distance south to the edge of "forest tundra" as defined by some observers (Payette, 1983) and north to the end of the trees or tree species, or whatever one might consider to be the northern edge of the forest/tundra ecotone. It is, in short, a useful basis for theory and for biogeographic mapping and, one would suppose, for forest management practice.

If one prefers to exclude wetlands from the analysis, it is quite reasonable that this should be done. It is obvious that if trees do not grow in an area because the water table is near, at, or above the surface of the soil, then the reason for their absence on surfaces above the water table, if they are absent there also, or for their presence if this is the case, is not the same as that for their presence or absence on the wetland areas. The environmental conditions supporting or prohibiting the existence of forest on uplands are not those influencing growth on the wetland areas. Thus, on the relatively small scale of most aerial photographs and maps of the forest tundra, it is the presence or absence of forest on uplands that is considered diagnostic of environmental conditions, and in particular those conditions related to the climate.

This would, of course, be the case particularly on gently rolling country such as the Precambrian Shield, with young postglacial drift deposits, low exposed bedrock hills, and immature drainage systems, extending in a quite homogeneous manner over large areas. In such areas, wet lowlands are occupied either by *Picea mariana* forest on accumulations of peat overlying mineral substrata or, northward, by open muskeg in which shrubs and sedges are dominant. Northward, the hill summits are occupied increasingly by lichen woodland and, ultimately as one travels north, by tundra. In northern Quebec and Ungava this ecotonal region is quite broad, presumably the result of a rise in elevation southward, as well as the result of a broadening, in comparison to regions westward, of the climatic frontal zone between air masses of arctic and more southern origin, this in turn the result of the influence on broad-scale wind patterns by Hudson Bay and the Arctic Ocean to the west and north and the Atlantic on the north and east. In Keewatin and westward, the forest/tundra ecotone appears more compressed, and the transition from forest to tundra occupies a more restricted zone in terms of north-south extent.

These considerations presuppose a varied topography in which inordinately large areas are not given over to extreme lowlands and wet marshy flats, but rather to a fairly homogeneous expanse of rolling interfluves. For an excellent illustration of landscape in which approximately half the area is forested, with hilltops given over to treeless, lichen-heath-dwarf birch tundra vegetation, reference is given to an aerial photograph presented by Payette (1983, Fig. 6, p. 10) and to a photograph taken showing black spruce growing near the far northern edge of the forest/tundra ecotone at Ennadai Lake in which tundra occupies at least 75% of the land surface and probably more (Fig. 3.1, by the author).

In regard to the ecological significance of the ecotone between forest and tundra, the temptation to quote a sentence from Margalef (1968) cannot be resisted: "In fact, we do not really need marked frontiers. It is enough to observe that the composition of an

Figure 3.1. Forest-tundra with clumps of dwarfed trees about 15 miles north of continuous forest in the Ennadai Lake area. Both spruce (*Picea*) and tamarack (*Larix*) are represented in the photograph. A typical upland ridge with a rock field tundra vegetational community is in the foreground. Maximum tree height is about 20 feet or less.

ecosystem is different in two spots to assume that some change must occur between them—perhaps gradually, perhaps more steeply." The change that occurs between the forest and tundra in northern Canada is visually a striking one, and one in which the environmental differences are beginning to be elucidated. The environmental factors of primary importance in this zone of transition from forest to tundra have been summarized by Bradley et al. (1982):

> Treeline represents a striking break in the patterning of regional vegetation; the change from woodland to tundra on the upland interfluves takes place within a short distance and is expressed over a circumpolar range. It parallels no known discontinuities in lithology, surficial geology, or relief. Therefore, it appears to be climatically induced, and most ecologists accept it as such.

Ecological Limits

The existence of the forest border is thus generally accepted to be the consequence of an inability on the part of the tree species to exist under climatic conditions found northward; it then becomes logical to ask why this should be, and efforts to answer the question have been, to date, not exceedingly numerous but nevertheless quite revealing of the nature of the climatic differences between the two zones. In one such study, Black and Bliss (1980) present abundant data to show that, in their words: "The climatic gradient and fire intervals near tree line interact in controlling the success of *Picea mariana* in the lower Mackenzie Valley." Averring that studies of the causal factors of species distribution are most effectively carried out near the species' limits, they add that studies of black spruce at its northern limit are of particular interest because of its wide distribution and its importance at tree-line in many geographic regions.

Coincidence of tree-line with climatic parameters was noted by Brown (1943), who saw a correlation of 10°C mean July temperature with tree-line, Hopkins (1959) who found a degree day relationship, Bryson (1966) who noted a frontal zone coincidence between climate and tree-line, and Larsen (1971b, 1980) who used vegetational community data in correlations with air mass data, and noted, also, that others had, at one time or another, discerned the correspondence between vegetational zones and climatic factors (Larsen, 1965, 1980).

Outliers of spruce found today beyond the main forest-tundra zone are generally considered relict from past advances of forest (Larsen, 1965, 1980) and it is also apparent that these clumps or clones of spruce survive over the years by means of layering rather than by viable seed (Larsen, 1965; Elliott, 1979; Black and Bliss, 1980). The environmental factors controlling the limit of *Picea mariana* at tree line in northwestern Canada were studied by Black and Bliss, who ascertained the role of water relations, photosynthesis, germination, and fire in the range limit of spruce.

Along a 135 km transect from self-reproducing forest to the south, through forest-tundra, and into tundra to the north, there were a number of measurable changes in climate. At the north end of the gradient, mean daily summer temperature was 4°C lower than at the south end, the growing season was shorter, and wind abrasion and water stress were greater. Soil hummocks, with a microenvironment that greatly restricts *Picea mariana* seed germination, increase in size and occurrence to the north, and this, coupled with low temperatures, the effects of forest fires, and other factors, greatly reduces the capacity of the species to reproduce at the north end of the study area. The effect of fire is an interesting one. Forest fires occur with some frequency, and in the far northern area along the transect, spruce stands originating prior to 1850 failed to regenerate when burned. These stands, Black and Bliss indicate, probably represent a northward shift of the forest line in response to the warming trend of the eighteenth and nineteenth centuries but they are today within a zone beyond the limit of the ability of spruce to reproduce readily. Southward, along the transect, stands regenerate without apparent difficulty, provided the interval of time between burns allows seed production in the new forest. This inability of spruce to produce viable seeds was also noted by Larsen (1965, 1980) and Elliott (1979a,b) at the northern edge of the forest in the Ennadai Lake area.

Given the inability of *Picea mariana* to reproduce at the northern edge of the forest-tundra, what can account for the obvious fact that it is the dominant species at this northern limit of its range? The answer, Black and Bliss state, probably lies in its ability to survive the relatively frequent (100–200 year intervals) forest fires and, also, to tolerate the cold, wet, nutrient-poor soils (see also Strang and Johnson, 1981). The black spruce, also, possesses serotinous (or, more properly, semiserotinous) cones which are accumulated over a period of years, not to release seeds until the cones are opened by the heat of a fire. Other species, such a *Picea glauca* and *Larix laricina*, are limited by their inability to re-seed large burned areas since seed dispersal must occur over long distances; the region is relatively flat, and with few areas protected from fire there are few survivors when a burn occurs. In contrast to this, *Picea mariana* "floods the burned site immediately after fire with seed stored in cones of the previous years." These factors account for the survival of the *Picea mariana* forest to the edge of the forest-tundra, and layering accounts for reproduction in the forest-tundra zone itself, north to the edge of unbroken tundra, and possibly within the forest-tundra there are also the occasional favorable years when viable seeds are produced, some of which are released at times during favorable summers when at least a few may be able to germinate and grow, maintaining the scattered clumps and clones of trees seen at the northern edge of the ecotone between forest and tundra.

Regional Variation

In a discussion of the boreal forest flora and the vegetational communities, Rowe (1956) summarizes some broad relationships that exist: two centers of species richness are found, southern British Columbia, where the trees predominantly are coniferous, and southern Ontario, with a preponderance of broadleafed trees. From these centers, the numbers of species decrease toward the center of the country from south to north. There are, for example, approximately 30 tree species in British Columbia, 20 in Alberta (including the Rocky Mountains), and 10 in Saskatchewan. In southern Ontario, there are about 75 native tree species, half this number in the Clay Belt of central Ontario, and 12 near James Bay. In the ecotone between forest and tundra the number declines to three or four.

The zone of forest-tundra has been aptly described by Rowe (1972) as a relatively narrow region of transition between subarctic forest and tundra, an ecotonal region stretching across northern Canada from the Mackenzie Delta to James Bay, Ungava Bay, and the Atlantic Coast. In this broad geographical range there is some variety present in terms of geological substrata and landforms, but there is a relatively uniform vegetational pattern in which tundra is dotted with patches of stunted forest around lakes and along rivers, and northward the area given over to forest shrinks and tundra expands.

Dominants in the groves of trees are black spruce (*Picea mariana*), white spruce (*Picea glauca*), and larch or tamarack (*Larix laricina*), with white spruce found to be a dominant in many stands, especially on areas of flat alluvium along rivers and around lakes, in places around the shores of Hudson Bay, and along the Atlantic coast of Labrador. Black spruce and tamarack were more important inland although white spruce occurs on favorable sites throughout the ecotonal region.

As indicated above, in general, evidence points to climatic influences as the limiting factors preventing the trees from expanding their range into the tundra, and these can be summarized as follows:

- Low air temperatures during the growing season
- Short growing season
- Desiccating winds and low humidity
- Nutrient-poor, often unstable soils, frequently underlain by permafrost and with a shallow active layer
- Snow cover that is often blown away from exposed uplands, with drifting to considerable depth in protected sites (covering and protecting trees from snow abrasion and desiccation), thus creating a variety of conditions affecting fall, winter, and spring survival of seedlings
- All of the above as they relate to the incidence of forest fires and tundra fires

These are conditions found universally throughout the forest-tundra transitional ecotone, but there are regional and local variations in the appearance and structure of the vegetation in the various regions from Alaska and western Canada across the continent to Quebec, Ungava, and Labrador. These are the result, as would be expected, of differences from place to place in the prevailing climatic conditions, and to a lesser degree to local geological and topographic characteristics. Great similarities exist between East and West, however, because the features that are considered diagnostic of past glaciation (drumlins, eskers, moraines, ridges, and kames, however variously defined) are common in most of at least the eastern and central regions.

It is interesting, moreover, that there also are areas disjunct from the northern transition. One such area, possessing characteristics of the northern forest-tundra ecotone although geographically distant from it, exists in the southwestern Yukon, south of the Ogilvie Mountains and surrounding the Dawson Range and the Yukon River. It is a plateau region encircled by high mountains and within a rain shadow that gives the area climatic features typical of a steppe region. On valley slopes are found stands of white spruce, and in the valley bottoms dwarfed black and white spruce are commonly present. Because of the severity of the climatic conditions locally, the stunted forest trees resemble those at the northern forest-tundra transition. At 3000–4000 feet altitude, the stunted forest grades into dwarf birch and alpine tundra (Rowe, 1972, see photo, p. 50).

Community Descriptions

Published accounts of the plant cover of many of the more accessible areas of northern Canada are now quite numerous, and even areas of difficult access have increasingly been the subject of field studies. The significance of these becomes apparent when it is obvious that if studies are to be made of the vegetation of these areas before they are disturbed by human activities of one kind or another—oil, mineral, recreational development—these studies must be made soon before the opportunity is lost forever.

Recent accounts of some of this work and reasonably complete bibliographies are contained in such generally available works as those on northwestern Canada by Ritchie (1984, 1988) on eastern Canada by Payette (1983), and of earlier studies of the North American boreal zone in general by Larsen (1980). These studies, in large part, are the result of attempts to describe and classify the vegetation. Often the vegetational communities studied are grouped into "types" or "units" that have pragmatic utility in that they facilitate recognition of the kind of communities referred to, but they also compound semantic difficulties in that many different terms are employed for what appear to be similar entities, leading to an endless and unproductive definition of terms that tends to render the literature somewhat sterile and opaque. To compound the difficulty, different methodologies are employed to sample stands of the vegetation and to conduct statistical analysis of the data, and while these are the unavoidable consequence of an immature science, in which techniques are being rapidly developed, modified, improved, and then discarded in favor of newer and better ones, these events occur at the same time the methodologies are employed to obtain descriptive data on the communities in the various regions. The consequence in terms of ready comprehension of the results has been somewhat less than ideal.

A commendable effort to clarify the nomenclature of the studies on the vegetation of northwestern Canada and Alaska is made by Ritchie (1984, see p. 64), and it might be hoped that similar efforts to codify equivalent terms will be made by others. In summary, Ritchie gives the following as major vegetation-landform units of northwestern Canada:

FORESTS
- White spruce on recent alluvium
- White spruce on noncalcareous uplands
- White spruce on calcareous uplands
- White spruce-birch-poplar on uplands
- Balsam poplar on alluvium
- Black spruce-lichen on upland
- Black spruce-moss-shrub on lowlands
- Black spruce mires on lowlands
- Larch on montane limestone slopes

TUNDRA
- *Salix phlebophylla* on summits and ridges
- *Dryas-Carex scirpoides* on calcarous slopes and summits
- *Salix* on snow-patch surfaces
- *Betula-Ledum* on uplands and middle montane

- *Betula-Eriophorum* on lower slopes
- *Eriophorum vaginatum* on lowlands
- *Carex* meadows on lowlands

SHRUB AND THICKET
- *Alnus crispa-Salix* on flooded alluvial sites
- *Salix* along channels and streams

These units can, in a somewhat general way, be employed also for the vegetation of the forest-tundra zone eastward across Canada, with modification in terms of significance of the various groupings. For example, the white spruce forests become of lesser importance, as Ritchie points out, eastward across the continent, the black spruce-lichen woodland, relatively rare in the West, is widespread and an important feature in central and eastern Canada, as evidenced in the work of Payette (1983), Rowe (1984), and others.

All of the units are artificial, however, in the sense that they represent conceptual entities rather than actual field reality. In practice, ordination of a suitable number of sampled vegetational community stands will reveal a vegetational continuum, in which one community grades into another, gradually or abruptly as topography and other local environmental factors dictate. Deceptively, there may be a clustering of similar entities in one or more portions of the ordination, depending upon the communities of each for which data is included. The prevalence over the landscape of each community will be seen in the sampling if random selection has been made of sampling sites, not, on the other hand, if selection has been made by the investigator with a conscious or unconscious preference for one kind of community over others. Thus, unless selection of sites has been random, the vegetational communities sampled will have been selected on the basis of visual characteristics or some other quality that has been perceived by the investigator as being of significance. Since personal preference has been at work, vegetational units so selected will receive greater emphasis in the result, or a greater uniformity will appear in the data than actually exists in the field.

These are caveats with which field workers are familiar, and it is generally agreed that it is inordinately difficult to proceed with field work in a completely objective manner. On the other hand, some studies are performed with deliberate selection of vegetational stands on the basis of certain criteria so that the influence of variation in others can be discerned in the effect on the structure and composition of the vegetational communities.

One of the more interesting studies representing this latter approach is that of Johnson (1981), who obtained data during the course of studies of open boreal woodland near the northern forest border in the region south and west of Great Slave Lake. Selecting stands on the basis that the vegetation be rooted in mineral soil, that no standing water be in evidence, that the composition of the sampled community be visually homogeneous, and that burned areas of all ages be sampled but that areas disturbed by man be avoided (cutting, camps, cabins, and so on), Johnson found upon statistical analysis of the vegetational and environmental data (ordination of principal components analysis was employed), that topographic position, canopy coverage, and nutrient content of the soil were dominant factors in the environmental complex determining the character of the community. The history of fire, which occurs on average every

70–100 years, is the causal influence creating variation in community composition and structure in each of the six major vegetational units delineated by the study:

- Open black spruce-lichen
- Open jack pine-lichen
- Open black spruce-glandular birch-lichen
- Open black spruce-white birch-lichen
- Closed black and white spruce—feather moss
- Closed white spruce-white birch

And, as Johnson concludes, " . . . there is no evidence in stand ages or structure that the canopy closure gradient of the habitat ordination is a succession sequence." There is, he adds, " . . . great similarity of species between the old pre-fire and early postfire communities . . . . A comparison of vegetation from before and after fire reveals a change primarily in abundance and not in species composition." Data delineating environmental characteristics of the six habitat types are given in Table 3.1.

In other studies elsewhere, the vegetational communities are delineated into types according to the vegetational data, and definitive initial studies giving a good descriptive survey of vegetation in the areas have been conducted in the West along the Dempster Highway (Stanek et al., 1981); along a transect from south to north in Keewatin (Zoltai and Johnson, 1988); in interior Keewatin (Gubbe, 1976) and the Yukon (Wiken

Table 3.1. Environment of Communities

	Open Black Spruce-Lichen 1	Open Jack Pine-Lichen 2	Open Black Spruce-Gland. Birch-Lichen 3	Open Black Spruce-White Birch-Lichen 4	Closed Black and White Spruce-Feather Moss 5	Closed White Spruce-White Birch 6
SOIL						
Percent organic	1	1	3	4	18	28
Conductivity mg/cm	0.1	0.1	0.1	0.1	0.2	0.3
NO_3-N mg/g	1	1	1	1	2	3
pH	5.3	5.2	5.2	5.2	5.3	6.1
Phosphorus mg/g	2–3	2–3	2–3	8	16	8
Calcium mg/g	1.2	1.2	1.2	2.1	5.5	n.d.
RADIATION						
Potential shortwave, June equinox W/cm^2	375	400	375	360	350	330
Slope position: base = 0.1, top of slope = 0.9	.7–.8	.7	.7	.5	.4	.4

Figures are approximate and rounded: data from Johnson (1981).

et al., 1981), and, as discussed previously, the Lockhart River area to the east and north of Great Slave Lake (Bradley et al., 1982).

The vascular plants, as well as the mosses and lichens, appear to have preferences, as expressed in terms of their frequency, at different and individualistic areas along the environmental gradient, and also, as demonstrated in data presented by Johnson, at different periods in the time-span since the last fire. For example, *Cladonia gracilis* generally declines or remains at a stable frequency over a 250-year time span, *Cladonia mitis** increases steadily over the first 50–100 years and then remains stable, *Ledum groenlandicum* increases abruptly to 50 years and then declines and stabilizes in black spruce-lichen woodland and open black spruce-white birch-lichen woodland, but in closed white spruce-white birch woodland it increases steadily to about 100 years and then declines severely. In closed black and white spruce-feather moss communities, *Ledum* increases abruptly the first 10–15 years after a fire, then continues to increase, but more slowly, to 200 years. Other species each have their own individual response to the conditions existing after a fire in each of the six communities.

In descriptive terms, the open canopy black spruce-lichen and jack pine-lichen communities are found on horizontal surfaces or south-facing slopes; black spruce-white spruce-lichen and black spruce-white spruce-feather moss communities are found on northerly slopes. Soil substrata range from sandy outwash and till in the former to loamy sand, sandy loam, and silt loam till and lacustrine in closed black and white spruce-feather moss forest stands. Stand slope positions range from flat to convex upper slopes in open black spruce-lichen and jack pine-lichen, through concave upper and middle slopes, to concave lower slopes in closed black and white-spruce feather moss communities. Thus, in terms of habitat, there is a topography-canopy coverage-soil nutrient gradient. Fire changes the canopy coverage, but most species found in older stands are present in the first years after a fire. These change in abundance during the years after the fire, from shorter-lived, faster-growing, poorly-competitive species to longer-lived, slower-growing species capable of effectively competing in the older closed-canopy forest, Johnson found.

The species increasing in abundance quickly following fire were *Ledum groenlandicum*, *Vaccinium vitis-idaea*, and *V. uliginosum*, which were also present in more of the stands than any other vascular plants. Other species increasing in abundance quickly were the lichens *Cladonia cornuta* and *C. gracilis* and the mosses *Pleurozium schreberi*, *Polytrichum juniperinum*, and *Ceratodon purpureus*. Seedlings of black spruce and jack pine were also abundant.

In stands of moderate age following fire, *Epilobium angustifolium* appeared as did *Cladonia coccifera*, *Cetraria islandica*, and *Peltigera malacea*, as well as the moss *Polytrichum piliferum*. White spruce was also present. Species increasing in abundance later in the forests were black spruce (trees), white birch, *Empetrum nigrum*, *Geocaulon lividum*, *Arctostaphylos uva-ursi*, the lichens *Cetraria nivalis*, *Cladonia amaurocraea*, *C. arbuscula*, *C. (Cladina) mitis*, *C. rangiferina*, and mosses *Hylocomium splendens*, *Dicranum polysetum*, *D. fuscescens*, and *Ptilidium ciliare*.

**Cladonia mitis* = *Cladina mitis*.

The Lockhart River Area

It should be noted that the study of Johnson described above was conducted well south of the extreme northern edge of the forest border, but it is of value for describing the forests of this region and for purposes of comparison with northern border forests. Encompassing a larger area, east and west in the same region as described above, Bradley, Rowe, and Tarnocai (1982) conducted an ecological land classification of the region around and extending north- and south-eastward from the East Arm and McLeod Bay of Great Slave Lake, including Artillery and Clinton-Colden Lakes, as far as the Thelon River on the northeast, the Taltson and Tazin Rivers to the south, and MacKay Lake to the northwest. This area, the Lockhart River Map Area of the Canadian Topographic Map Series, is described in detail in the impressive report, based on ground surveys of vegetation, soils, surficial geology, fire history, and other characteristics of sites selected from satellite and high-altitude photographs as well as low-level reconnaissance and photographic flights. The report describes the features in five main units, on the basis of physiography and vegetation, with accompanying photographs of each: High Subarctic Forest Tundra, High Subarctic Shrub Tundra, Mid Boreal, High Boreal, Low Subarctic. Described are the vegetational communities in each unit along a topographic catena, from wet, through moist and dry, to dry and exposed; thus from sedge meadow marsh to rock lichen communities on hilltops. It was not the primary purpose of the study to record in detail the structure and composition of the various plant communities but rather to describe them in physiographic terms and present a listing of dominant species, thus to delineate land management possibilities and conservation needs in light of the likelihood of future disturbance as mines and transportation corridors are opened and as renewable resource use and recreational activities expand. In addition to the report cited above, the rationale of ecological land classification is presented in a companion report (Rowe and Sheard, 1981). In Table 3.2 are presented the listings of vascular plants found in more than one of the units from which data were obtained in the Lockhart River Area study.

Dominant lichens found in the Heath-Lichen-Grass units were *Cornicularia divergens*, *Cetraria cucullata*, *Alectoria nitidula*, *Alectoria ochroleuca*, and *Cetraria nivalis*; *Cladina mitis*, *Cladina rangiferina*, and *Cladina stellaris* were found abundant in Moss Lichen Woodland and Moss Forest. Found in Heath Lichen and Heath Lichen Woodland as well as Shrub Heath and Shrub Heath Woodland were many of the above species as well as *Stereocaulon paschale*, *Cladonia uncialis*, *Cladonia amaurocraea*, and *Cladonia deformis*.

Of the vascular plant species present in the units, only *Hierochloe alpina*, *Calamagrostis purpurascens*, *Luzula confusa*, and *Silene acaulis* were distinctly indicative of a single vegetational community or unit, in this instance the Heath-Lichen-Grass, which was found only in the far northern portion of the study area. These species all range far northward into the Canadian Arctic and the Canadian Arctic Archipelago and possess a range throughout the circumpolar arctic regions.

It should be seen that this and other earlier studies (Kershaw, 1977, for example) note the abundance of *Stereocaulon paschale* in woodlands of the northern boreal region. As Kershaw points out, all accounts of northern boreal lichen woodlands indicate that

Table 3.2. Species in Communities

	Heath-Lichen-Grass	Moss-Lichen Woodland	Shrub Heath (and Woodland)	Heath-Lichen (and Woodland)	Heath-Lichen Bog Bog Woodland	Sedge Fen	Sedge Meadow Marsh
Unit	2	3	6	7	8	9	10
Calamagrostis purpurascens	x						
Hierochloe alpina	x						
Luzula confusa	x						
Silene acaulis	x						
Arctostaphylos alpina	x			x			
Ledum decumbens	x		x	x	x		
Empetrum nigrum	x		x	x	x		
Vaccinium vitis-idaea	x	x	x	x	x		x
Picea mariana		x	x	x	x		
Picea glauca		x	x	x			
Betula papyrifera		x		x			
Betula glandulosa			x		x		x
Juniperus communis			x				
Loiseleuria procumbens			x	x			
Pinus banksiana				x			
Vaccinium uliginosum				x	x		
Rubus chamaemorus					x		
Andromeda polifolia					x	x	x
Carex spp.						x	x
Eriophorum spp.						x	x

Data obtained from Bradley et al. (1982) and adapted for use here.

Cladonia (Cladina) stellaris, *C. mitis*, and *C. rangiferina* occur in abundance on the forest floor; *Stereocaulon* lichen woodland, however, occurs extensively in the Northwest Territories, and in many areas there is an almost pure ground cover of *Stereocaulon*. Kershaw indicates that these woodlands appear to be quite different in environmental relationships than the *Cladonia* woodlands and recommended that the two distinct types of northern lichen woodlands be referred to specifically as *Stereocaulon paschale* or *Cladonia (Cladina)* woodland to differentiate clearly what appear to be differences not only in species composition but, concomitantly, differences in edaphic factors, microclimate, and recovery sequence after fire.

Other Pertinent Studies

In recent years, research on the boreal forest burgeoned into a full-fledged and coherent program, carried on by a diverse cast of Canadian government and university foresters, ecologists, botanists, and soils specialists, as well as palynologists, climatologists, and those in a great many other perhaps somewhat more peripheral disciplines in terms of forest ecology.

Some of the earlier studies are of great value today, and include those of Johnson and Vogel (1966) dealing with the Yukon Flats, Johnson et al. (1966) on northern Alaska, and Hultén (1968), on a definitive taxonomic work for Alaska and adjacent regions, also containing brief descriptions of various plant communities on the diverse landforms in the region. In the central zone are the studies of Ritchie (1956a,b, 1960, 1962, among others); Raup, in the Athabasca–Great Slave Lake region (1946), southwestern Mackenzie (1947), and the area adjacent to the southern part of the Alaskan highway (1950). Porsild made studies in the regions to the westward of the Mackenzie River, reporting on field excursions to the east slope of the Mackenzie Mountains (1945, 1950), the southwestern Yukon region (1966), and other areas of the Northwest Territories and the Arctic Archipelago (1943, 1955, 1957).

In the east, early studies include, for example, those by Wilton (1964) on the forests of Labrador and by Damman (1964) on some forest types in central Newfoundland and relationships with environmental factors. Hare (1950, 1954, 1959) described the general climatic relationships of the boreal forest zone and, in the latter publication, provided maps derived from aerial photographic coverage which show the extent of the various forest types and tundra in the Labrador Peninsula. Linteau (1955) gives a forest classification of the boreal zone in Quebec, providing species lists of value and interest, and Damman, in another publication (Damman, 1965), presents an interpretation for the unusual distributions of some of the plant species of Newfoundland. Maycock (1963) discussed the deciduous forests of southern Ontario and ranged northward (Maycock and Matthews, 1966) to study an arctic forest in the tundra of northern Quebec. There are many other studies of this region by both botanists and forestry scientists which are worthy of note, but these will suffice to serve as an introduction to those who wish to pursue the matter in greater detail (Larsen, 1980).

General climatic and habitat relationships of the various species are included in the major taxonomic references on the Canadian and Alaskan flora, for example Porsild and Cody (1980), Scoggan (1978-79), and Hultén (1968). There are, moreover, works on such species groups as the lichens (Thomson, 1984), the genus *Salix* (Argus, 1965, 1973; Raup, 1941, 1943, 1959), as well as species with noteworthy distributions (Porsild, 1958) or those on specific areas (Porsild, 1943, 1957; Raup, 1946, 1947). There are, in addition, a rather large number of other studies on the plant communities of Alaska and northern Canada and these phytogeographical references are for the most part listed in such words as those by Raup (1941), Rowe (1972), Hultén (1968), Larsen (1980), Ritchie (1984, 1988), and Van Cleve et al. (1986).

Other recent studies of the composition and structure of major natural plant communities are now available for northern forest and tundra and provide baseline data for a number of areas; among these are a 1000 km north-south transect in Keewatin (Gubbe, 1976), vegetation and soils studies in central Keewatin (Zoltai and Johnson, 1978), in the region east of Artillery and Great Slave Lake (Bradley et al., 1982), the northern Yukon (Wiken et al., 1981), Quebec (Morisset and Payette, 1983), along the Dempster Highway in western Mackenzie (Stanek et al., 1981; Stanek (1982), and Alaska (Van Cleve et al, 1986) as well as a number of other areas (Argus, 1964; Bliss, 1977a,b,

1981, 1985; Brown et al., 1980a,b; Thompson, 1980; Viereck, 1973, 1975; Carleton and Maycock, 1981; Chapin et al., 1985; Hopkins and Sigafoos, 1951; Johnson et al., 1966; Oechel and Lawrence, 1985; Raup and Denny, 1950; See and Bliss, 1980; Petzold and Mulhern, 1987; Pomeroy, 1985; Harms, 1974; Van Cleve and Viereck, 1981; Tukhanen, 1984; Flinn and Wein, 1988).

Since the early papers appeared, great advances in transportation have made it much easier to conduct studies in such relatively wild and unsettled country as the northern tree-line areas, and additional studies have been appearing with regularity. These are filling what not long ago was a void excepting for the reports indicated above. Moreover, there are now virtually countless special reports by members of the Canadian Forest Service, and many publications on the physiology, taxonomy, ecology, genetics, and biochemistry of individual species making up the flora of the vegetational communities of Canada. The volume of the research conducted in recent years is evidenced by the content of a number of synthesis and symposium volumes appearing. These include those by Tamm (1976), Pruit (1978), Persson (1980), Larsen (1980), Nilsson and Kullman (1981), Wein, et al., (1983), and Olson et al., (1984), as well as Van Cleve et al., (1986) for Alaska.

Large portions of these volumes are directly aimed at readers with a professional interest in forestry applications and forest productivity, less on the structure and function of plant communities and theoretical aspects thereof, or on the broader biogeographical aspects of the species that comprise the plant communities of the boreal forest and tundra regions. The latter topics are often restricted to journal papers or publications of somewhat more restricted interest. Some of the recent ones of pertinence to the study to be reported in the following pages would certainly include, but not be restricted to those of the following: Ritchie (1984, 1977), Black and Bliss (1978, 1980), Zoltai (1975), Corns and Pluth (1984), Stanek et al., (1981), Corns (1983) and Van Cleve et al., (1986). All of these are for the western portions of Canada (and Alaska). For the eastern parts of northern Canada we have such authors as Meredith and Müller-Wille (1982), Moore (1982), Cowles (1982), Légère and Payette (1981), Rencz and Auclair (1978), and Morisset and Payette (1983). Moreover, Cody and Porsild (1980) have carried out a definitive compilation of the species of plants present in the northern parts of Canada.

It is not the purpose here to review, or to even mention, all of the work now published relating to studies of the northern vegetational communities in Canada. The above brief summary is presented in part for the purpose of indicating that during the course of these studies the variety and number of plant communities recognized by these many observers has proliferated rapidly. It is not the intent either to add or subtract ("split" or "lump") these communities, but rather to discern some broadly regional similarities and dissimilarities within a few representative communities over distance, that is to discern relationships among communities on similar topographic sites in different parts of the North American continent, east to west.

We will thus deal with these matters only in passing in the following discussions. It might be noted, however, that there obviously exists an undeniable urge to classify the vegetational community continuum into vegetational types. Whether or not this corresponds to the reality that exists on the natural landscape is, generally, disregarded in favor of the more conceptually manageable division of the continuum into "nodes" or "types." There are pragmatic purposes to be served by so classifying or delineating segments of the continuum, and these are not to be denigrated, but it should also be accepted that these are conceptual accommodations. As more information becomes available on the interaction between environment or habitat and the responses of the individualistic species, perhaps there will be developed a somewhat more realistic approach to plant community structure and function, and it is hoped new statistical and mathematical techniques will also be developed to enhance comprehension of the relationships existing in the vegetational communities. It is obviously not a simple matter to develop the techniques required for such a more accurate representation of the species/environment complex; it is perhaps enough that progress has taken us to the point at which we find ourselves today.

4. Physiography of Northern Study Areas

The position of the continental forest border from western to eastern Canada has been known at least since the time of Louis Agassiz, who presents a map of the natural provinces of North America (Agassiz, 1845), and, in a somewhat more detailed presentation, Cooper (1859) shows "Esquimaux," "Athabascan," and "Algonquin" regions. Later maps become increasingly accurate and treat vegetational zonation on a smaller scale (references in Hustich, 1949, 1966, 1970, 1979; Larsen, 1980). As pointed out in a previous chapter, many early journals of explorers and travelers indicate the location of the forest border in the time before settlement. Palynological studies have furnished quite a detailed history of the vegetation since disappearance of the continental glaciations existing during Pleistocene time. The region was still sufficiently new to science in the early decades of the present century that even simple descriptions are of value. The following paragraphs on the Thelon Game Sanctuary by Clarke (1940) might be taken as an example:

> The timberline is by far the most impressive faunal and floral boundary in Canada. East of Great Slave Lake it is a real line; the forest marches up to Timber Bay on Artillery Lake and halts. Because of the sudden change in altitude from Great Slave Lake to Timberline on Artillery Lake, one passes his last poplar, last jack pine, last tamarack, etc., one after another inside a few miles .... The timberline is reached where the trees are unable to establish themselves under fairly exposed conditions. Beyond that line the spruce is most likely to be found in some sheltered spot where conditions unfavorable for the dominant arctic vegetation and favorable to spruce prevail. Thus, on the loose sand of eskers in the

Hanbury region, small, stunted clumps of spruce are frequently found, both black and white spruce, in tight little clumps a few yards square and three or four feet high, or even higher at times. The best clump on the Thelon is in a place where springs emerge from sandstone and wash down to mineral soil.

A study in the Ennadai Lake area and northward (Larsen, 1965, 1980) revealed that small, isolated clumps of spruce are found as far northward as Yathkyed Lake where they have been observed at 62° 44′N, 98° 38′W, although only small outliers of spruce exist along the Kazan River between Ennadai and Yathkyed (see Appendix H). Isolated clumps also exist at Downer Lake north of the Thlewiaza River, at Roseblade Lake near the Tha-anne River, and at Ameta Lake (Zoltai and Jackson, 1978) as well as some miles south of Heninga Lake (personal observation), and one is located at Padlei. There are clumps on the lower Dubawnt, on the south shore of Beverly Lake, the north end of Sifton Lake, and along the Hanbury River between Ptarmigan Lake and the Thelon River. Clarke indicates that no mention was apparently ever made of spruce near Aylmer or Clinton-Colden lakes, but during field work in this area a number of small clumps of black spruce were found at the west end of Aylmer Lake and at the end of the north arm of Clinton-Colden Lake (Larsen, 1971a,b). The maximum range of spruce northward is thus extended considerably due to the presence of the rather special environment along rivers and shorelines of some lakes. The general deterioration of conditions for spruce growth northward can be seen in a comparison of growth rates and the reproductive capability of spruce in study sites at different latitudes (Larsen, 1965, 1980; Black and Bliss, 1980; Elliott, 1979, 1980; Johnson, 1981; Payette, 1983; Payette and Filion, 1984).

Numerous vegetational studies of the forest-tundra ecotone of the Northwest Territories are now available (Larsen, 1965, 1971a,b, 1972a,b; Maini, 1966; Gubbe, 1976; Ritchie, 1960a,b,c, 1962, 1984; Johnson, 1981; Bradley et al., 1982; Rowe, 1984), and for Labrador-Ungava (Fraser, 1956; Hustich, 1949, 1950, 1951, 1962, 1965, 1966, 1979; Maycock and Matthews, 1966; Morisset and Payette, 1983; Payette and Filion, 1984), although there is great variation in the sampling techniques employed and comparative analyses are difficult. The importance of continued ecological studies in these areas is clearly apparent: knowledge of baseline conditions—those existing in at least relatively undisturbed areas—are now required so that the effects of future environmental changes, especially those induced by climatic change and an expansion of transport, mining, recreational, and other activities, can be assessed and, it is to be hoped, minimized by appropriate protective and management techniques and policies.

Studies cited in previous paragraphs furnish, in varying degrees, statistical data on plant communities in various areas of the forest-tundra transition zone including the Ennadai Lake area in southwestern Keewatin and the area around and between Fort Reliance and Aylmer Lake north of the East Arm of Great Slave Lake (Larsen, 1965, 1971a,b, 1972a,b,c, 1973, 1980). Data from other study areas have subsequently become available, and, in addition, some analyses of these data have revealed dominance and diversity characteristics of the vegetational communities of the various regions. Before elaborating upon the results of these analyses, however, brief descriptions of the major forest-border ecotone study areas will be presented in the remainder of this chapter. Descriptions of areas published in the above-listed journal articles will not be repeated, but will be briefly summarized for purposes of comparison with the added study areas, and, in addition, brief descriptions of regions studied by others will be included for the same purpose, starting with descriptive material condensed from authors studying areas in Alaska and, in particular, the Firth River area in northern Yukon and areas in far northwestern Mackenzie.

Alaska, Alaska-Yukon

In Alaska the forest border extends along the foothills of the southern portions of the Brooks Range and drops southward from Kotzebue to Kodiak Island, thus leaving large areas inland from the western coastal regions barren of trees. The latter is presumably an effect of the cold waters and the winds off the Bering Sea. As Viereck (1979) points out, there is great difficulty defining vegetational zones in Alaska as a consequence of the mountainous terrain, the influence of forest fires, the complex physiographic patterning, and the presence of permafrost as well as the oceanic effect along the western coastal region.

Viereck affirms the observation of Hare and Ritchie (1972) that the high mountains, located precisely where the northern forest ecotone would otherwise exist, eliminate the gradual transition from forest to tundra as found in Canada, Scandinavia, and the Soviet Union, creating instead an abrupt northern tree-line at low elevations on the southern slopes of the Brooks Range. The effect of the Brooks Range and the Alaska Range create an extreme continental climate in the Tanana and Yukon basins, although the Alaskan river systems create a somewhat moderated environment in the vicinity of the rivers and on the floodplains, where trees grow well; on surrounding uplands are found only scattered trees or none at all. The result, Viereck writes, is that it is possible to map a mosaic of vegetation types but difficult or impossible to group these into distinct vegetational zones.

Viereck adds that the tree-line is usually mapped in Alaska by the limit of white spruce in the taiga areas and Sitka spruce (*Picea sitchensis*) in southwestern Alaska. Black spruce (*Picea mariana*) is more limited in distribution than white spruce, but joins white spruce as a tree-line species along the south slopes of the Brooks Range, in the mountains of the Alaska Range, and the Chugach Mountains. Mountain hemlock (*Tsuga mertensiana*) usually is the tree-line species in the alpine areas of southeastern Alaska, and found in places with subalpine fir (*Abies lasiocarpa*, also called alpine fir). White birch (*Betula papyrifera*, also called paper birch) has a distribution similar to white spruce. Balsam poplar (*Populus balsamifera*), surprisingly, is found north and west of the limits even of white spruce. Both *Pinus* and *Abies* are missing in the arctic and western tree-line areas and *Larix* shows a limited range (Viereck, 1979; Viereck and Dyrness, 1980; Van Cleve et al., 1983).

One of the more interesting of the northern extensions of the range of trees in northern North America is that into the Firth River Basin in the Yukon. The Firth River flows northeast from its headwaters in the Davidson Mountains of northeastern Alaska, entering the Arctic Ocean near the far northwestern tip of continental Canada. Studies of the vegetation in a high interior valley basin at the border of Alaska and Canada, some 70 miles south of the Arctic Ocean, show tundra plant communities covering mountain summits and broad terraces within the Firth River Basin, with scattered white spruce and open spruce woodlands on south-facing slopes and certain low terrace edges (Drew and Shanks, 1965). The tree stands give the impression of stability, with no indication that regeneration is presently more vigorous than it has been during the past 250 years, nor were there any indications that the stands were degenerating. Significantly, prevailing southerly winds may allow warmer summer air masses from the interior of Alaska and Canada to prevail down the valley, hence perpetuating conditions that allow the survival of the spruce. Here, too, the spruce occurs on calcareous soils, reported elsewhere to be favorable for growth of white spruce at its northern limits. Moreover, the Firth River Valley was unglaciated during Pleistocene time and may have constituted a refugium for the species during the long glacial epoch.

It should be noted that the results of comprehensive ecosystem studies of forest communities dominated by black spruce in the central interior of Alaska have become available, and much of the information therein yields hints as to the habitat influences that affect spruce at tree-line, in terms of structure and function of the communities, in the Alaska–Yukon region. Interested individuals are referred to Van Cleve and Dyrness; Viereck et al.; Yarie et al.; Chapin; Billington and Alexander; and other contributors to a group of 22 articles in the *Canadian Journal of Forest Research* **13(5):** 695–916, as well as a summary volume (Van Cleve et al., 1986).

It is evident that high-latitude ecotones are very susceptible to environmental change and that plant species, and hence communities that are restricted to small areas with special micro-habitats such as steep north or south facing slopes, will retreat rapidly during periods of unfavorable climatic conditions and will advance, more slowly perhaps, during favorable climatic regimes. Thus, in such regions, variations in coverage by the various species and communities will be related directly to climate. Disturbance of natural habitats in northern regions is increasing, and probably the most extensive surficial modification resulting from development is the construction and maintenance of roads for whatever purpose; changes in surface drainage, local vegetation, soil thermal regions, soil composition, chemical content of surface waters, permafrost, and wildlife. There are numerous studies of these effects, as well as suggestions regarding the techniques most effective for monitoring activities (see, for example, Pomeroy, 1985; Claridge and Mirza, 1981; Spatt and Miller, 1981; Petzold and Mulhern, 1987).

In the Brooks Range, white spruce outliers are found as sporadic occurrences near the continental divide beyond the altitudinal tree-line, particularly on south-facing slopes and in protected habitats. It is apparent that there is here a source of white spruce seed that, if present climate ameliorates, would be available for expansion of forests, presumable to favored sites northward of the present range of the species (Cooper, 1986). This furnishes an example of the potential always present for species to move into habitats newly opened for ecesis. In the Alaskan region, there have been a number of studies of the structure and composition of plant communities now in place, with, thus, the potential of migration if and when the opportunity arises (Pickett and White, 1985; Auerbach and Shmida, 1987; Spetzman, 1959; Argus, 1973; West and DeWolfe, 1974 (bird species); Wiggins, 1962; Wiken et al., 1981; Van Cleve and Viereck, 1981; Hopkins and Sigafoos, 1951; Johnson et al., 1966; Raup and Denny, 1950; Viereck, 1973; Viereck and Dyrness, 1979; Viereck and Schandelmeier, 1980) and also studies along the Dempster Highway (the highway from Dawson to Inuvik) and the far northwest of Canada as well (Stanek et al., 1981; Stanek, 1982; Ritchie, 1984).

Western Arctic and Montane Areas

Canoe Lake Area (68°14′N, 135°52′W)
Airphoto No. A 14363–27 (Figs. 4.1, 4.2),
Trout Lake Area (68°50′N, 138°45′W)
Airphoto No. A 13383–162 (Figs. 4.3, 4.4)*

These areas are in the foothills of the Richardson and British Mountains in the far northwestern corner of continental Canada. They are included here for comparison with other study areas for which more detailed vegetational community data is available, and for comparison with the vegetation of the valley of the Firth River as described above and the delta of the Mackenzie River to be discussed below. The Canoe Lake area is about 20 miles to the west of Aklavik, the latter located in the Mackenzie River valley, and about 135 southeast of the mouth of the Firth River as it enters the Arctic Ocean south of Herschel Island. Trout Lake is located roughly 50 miles southeast of the delta of the Firth River. Both Canoe Lake and Trout Lake are in the relatively low foothills of the mountains; Canoe Lake on the east slope at about 1200 feet elevation and Trout Lake on the northeast slope about 500 feet above sea level. Hills rise 1000 feet or more within a mile of both lakes. Trout Lake is included here primarily for comparison of its aerial photography with that of other study areas. Although the vegetational communities of the area have been studied in detail by Lambert (1968), the data obtained for the purposes of the study reported herein is limited and the description here will be concerned primarily with the Canoe Lake area where more complete data were obtained.

Geological reports indicate that the Richardson Mountains are a northerly trending folded range, about 50 miles in width at the latitude discussed here. The topography is relatively mature, with many broad open valleys. Much of the area is underlain by sandstone. The vegetation of the Canoe Lake area is apparently representative of tundra found at relatively high latitudes and at moderately high elevations, possessing certain characteristics that render it distinct from the tundra of the Caribou Hills to the northeast across the Mackenzie Delta. The general roughness of the topography and the relief in the Canoe Lake area are more extreme than in the Caribou Hills. Climatic conditions also differ from the delta or the Caribou Hills: there is a much higher frequency of precipitation, often in the form of mist and fog, and a much greater probability of snow later in the spring and earlier in the fall. The growing season is noticeably shorter.

The generally more severe conditions in the mountains were well illustrated by the experience of the field party in the Canoe Lake area as late as June 20–27, when a heavy snowfall disrupted field work around Canoe Lake. Some 20 miles westward in the mountains (at an elevation of about 3500 feet) the snow fell to a depth of 2–3 feet, leaving a

*See Appendix for photographic data and acknowledgments.

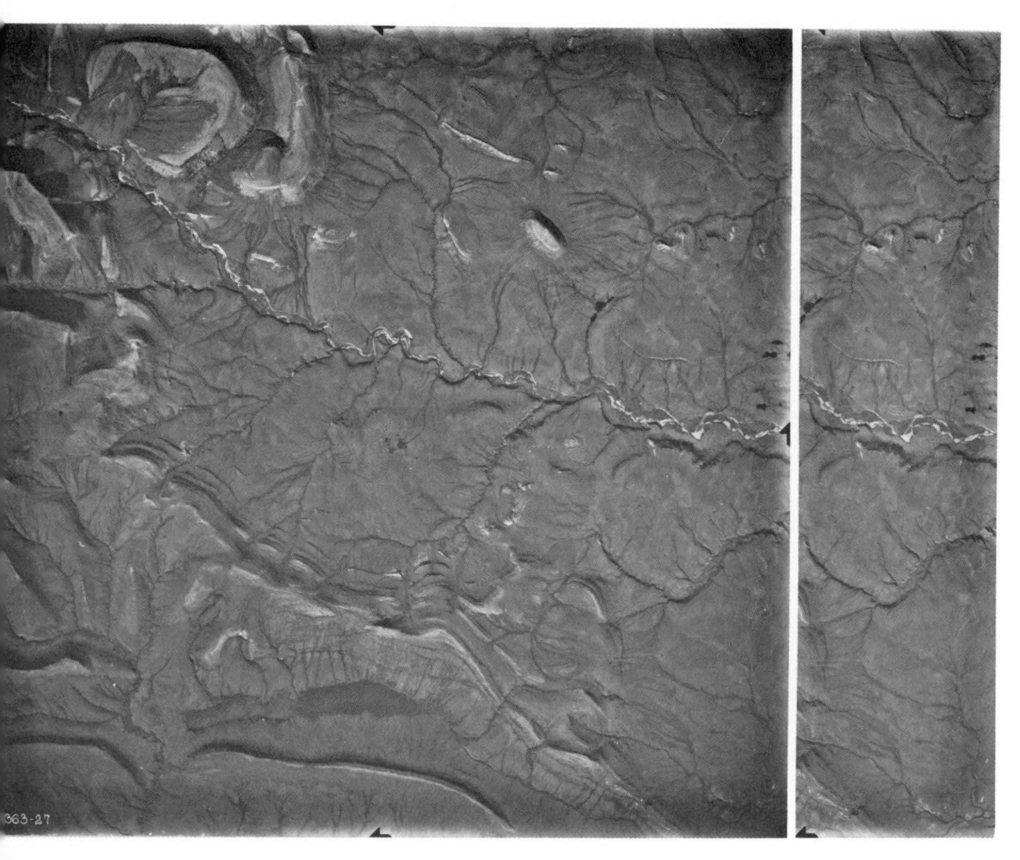

Figure 4.1. Airphoto of Canoe Lake area. Canadian National Air Photo Library No. A 14363-27. (For copyright information, see page 44 and Appendix E: © Her Majesty the Queen in Right of Canada.)

geological team isolated until weather again permitted helicopter flights a week later. At Canoe Lake during this time, ground surface temperatures did not drop far below freezing and the vegetation continued growth apparently unharmed when temperatures returned to normal. Skim ice formed on open water during the nights. On June 16, only a narrow shore lead existed on Canoe Lake, just sufficient to permit operation of a small float aircraft. The lead had not widened appreciably a week later. Only brief visits of aircraft of this type were possible since a wind shift would move the ice and close the lead, opening it on the opposite shore. The air temperatures were consistently in the range 30°–45°F, often as much as 20° colder than those reported by radio from Inuvik at midday. Annual precipitation averages about 9 inches at Aklavik. It is apparent that frequent light rain, mist, and fog at even the relatively low elevations of Canoe Lake result in actual moisture supplies to vegetation at least half again this amount.

Severe climatic conditions are in considerable degree responsible for the vegetational characteristics of the Canoe Lake area and the Richardson Mountains in general. Because of the nature of the topography, there are few extensive areas of low meadow.

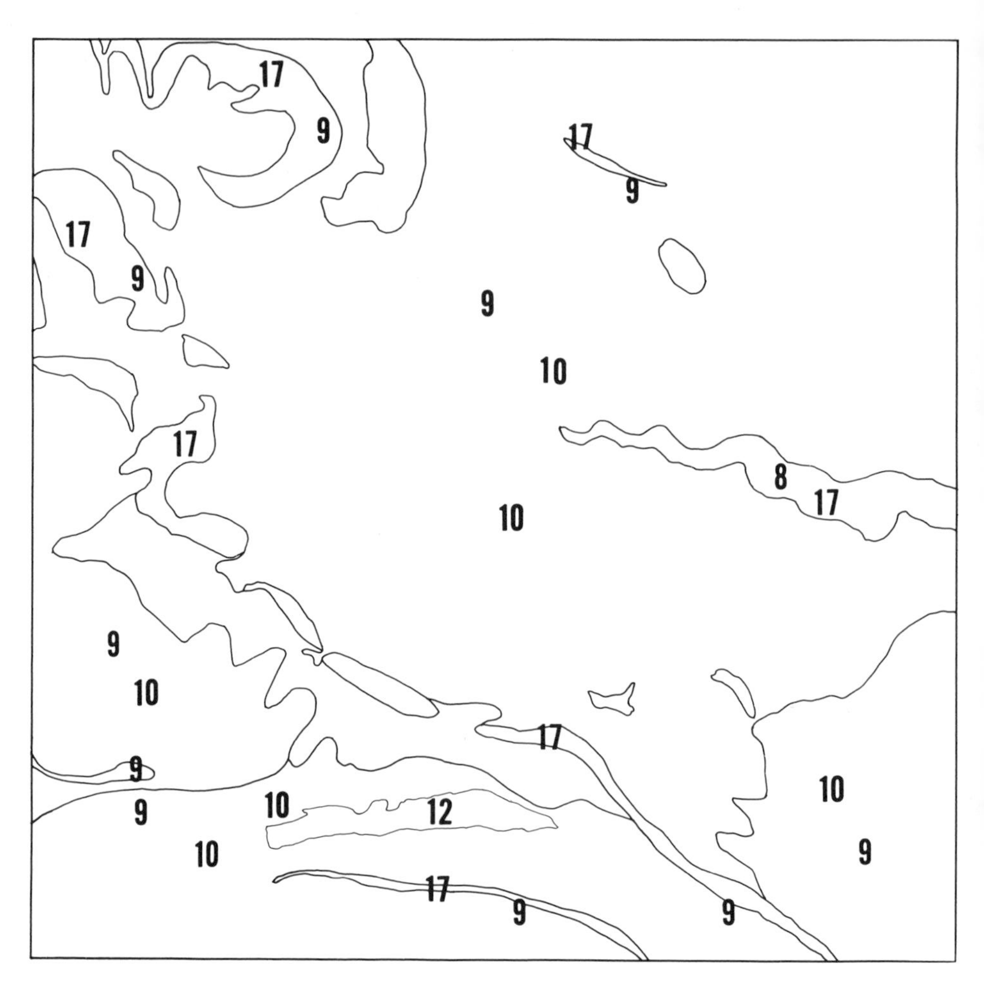

Figure 4.2. Key to Canoe Lake area airphoto. For explanation of numbers on this and all other air photo keys, see facing page.
For information regarding scale and other photographic details, as well as copyright information (© Her Majesty the Queen in Right of Canada), for this and other aerial photographs on subsequent pages, see Appendix E.

Keys to the aerial photographs employ 17 identifying numbers; each number refers to a general vegetational community according to the following categories:

1. Spruce forest; closed-canopied over much of the area.
2. Spruce forest; open-canopied (spruce woodland, lichen woodland).
3. Treed muskeg; low topography where the water table is high and deep accumulations of peat are usually present. Tree densities may vary greatly.
4. Open muskeg or sedge marsh; treeless vegetational communities, found often in association with treed muskeg particularly where accumulated peat has not attained sufficient height to bring the upper surface sufficiently far above the water level to support tree growth. Sedge meadow (sedge marsh) is an open wet meadow with emergent aquatic vegetation, predominantly sedges. Open muskeg and sedge marsh are difficult to differentiate on the aerial photographs.
5. Mixedwood forest; forest with both conifers and deciduous tree species, found commonly south of the northern boreal conifer forest (see Larsen, 1980).
6. Aspen forest; found commonly southward of northern boreal forest (see Larsen, 1980).
7. Jack pine forest; found infrequently in vegetational communities at or near the northern forest border (see Larsen, 1980).
8. Alder-willow-dwarf birch; found most often along the edges of waterways and drainage lines.
9. Rock field community; tundra vegetational community found on uplands, well-drained, where the substrate is often coarse sand and gravel with rocks of varying size.
10. Tussock muskeg community; tundra vegetational community of lower slopes or moderately well-drained flatter areas where a well-developed heath vegetation is often dominant, particularly on drier sites. Sedges intermixed with mosses usually form the individual tussocks, on the summits of which are found an aggregation of lichens and the common Ericaceous species. Tussocks may average a foot or more in height with perhaps an average of a foot between their outer rims. Low areas between tussocks are dominated by sedges, often growing in shallow water.
11. Low meadows; tundra community characterized principally by sedges (including *Eriophorum*) growing (at least in spring) in a shallow accumulation of water, underlain by sedge peat.

Other landscape features indicated by the numbers:

12. Water; 12a represents an ice-covered body of water.
13. Snow and/or ice on ground surface.
14. Disturbance; 14 represents essentially a natural disturbance; 14a specifies disturbance by fire, and 14b disturbance by human activities.
15. Unusual vegetational or topographic feature mentioned in text.
16. Polygonal ground.
17. Vegetation thin or absent because of natural adverse growing conditions; bare rock, bare sand, steep slopes, or other features usually readily identifiable.

Eskers. Eskers are shown as single dashed lines.

Data relating to the scale of the aerial photographs and the date of the flight are included in a table in the Appendix. General descriptions of the regions are included in the text.

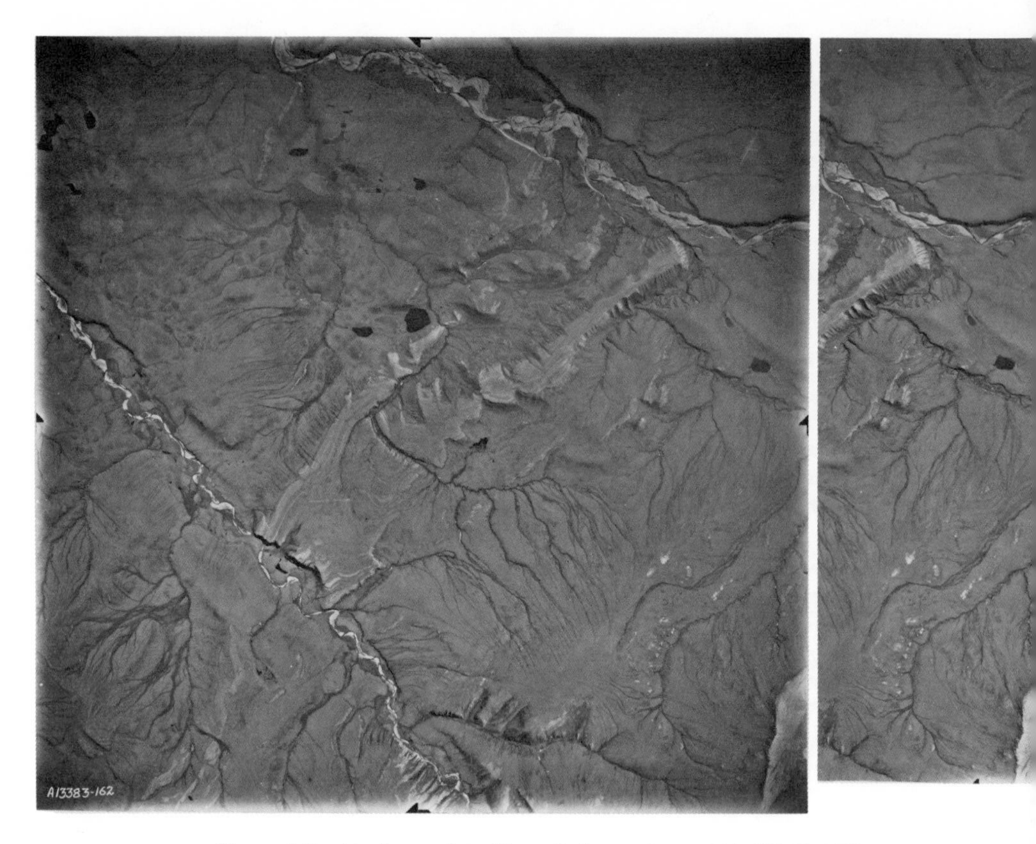

Figure 4.3. Airphoto of the Trout Lake study area (A 13383-162).

However, the generally moist conditions are responsible for the high frequency of sedges (*Carex* spp.) and of the arctic cotton (*Eriophorum* spp.) in tussock muskeg communities*. Even areas characterized by a 3–5% or even 5–8% slope are dominated by a tussock muskeg community, with very pronounced tussock development and a high frequency of *Carex* and *Eriophorum*. Walking is often quite difficult over much of at least the lower slopes.

Polygonal ground, often poorly developed, is to be found throughout the areas occupied by tussock muskeg. Ice wedges are seen frequently where exposed by slumping or cracking of deeper peat deposits. In the areas dominated by tussock muskeg, there is usually abundant evidence of a variety of surficial frost effects. On many small

*The term "muskeg" has been used in various poorly defined ways, but is generally understood to mean a vegetational community underlain by a peat substrate and with the water table near the surface. Ritchie (1956) defines "muskeg" by referring to Hustich (1949, 1955) who "describes a bog forest which has a single tree stratum of *Picea mariana*, and a hummocked ground vegetation of mosses – chiefly species of *Sphagnum* – which bears a varied shrub and herb community dominated by *Ledum groenlandicum*." For an open (treeless) muskeg, the term "peat

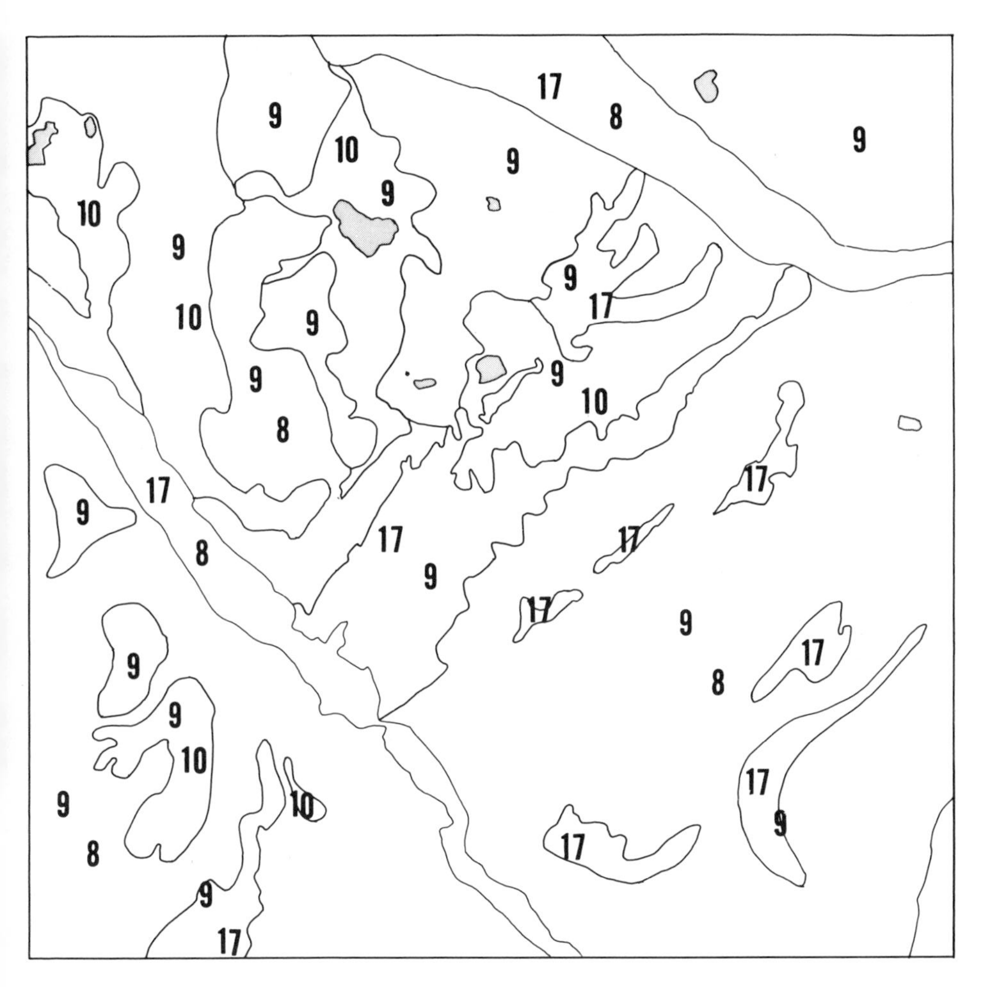

Figure 4.4. Key to the airphoto of the Trout Lake area. For descriptions of features indicated by identifying numbers, see Fig. 4.2.

---

bog" is used by Hustich and Ritchie to differentiate this community from "muskeg," and the primary distinction is that there is no tree stratum in the former. Since the word "bog" is also ambiguous, it is my preference to use "muskeg" for both, employing descriptive antecedents such as "black spruce," "tussock," "open," and so on, before the word muskeg to describe the community on a usually moist or wet peat substrate. This has some degree of validity because it is employed rather widely, even by Hustich (1949, 1954, 1955, 1957), as well as others working elsewhere in Canada, i.e, Argus (1966) and Scotter (1966). My use of "sedge meadow," "tussock muskeg," and "rock field," as well as other terms, can be illustrated by referring to the photographs taken during studies in the Ennadai Lake area (Larsen, 1965) and elsewhere (see refs.), as well as those in Scotter (1966) for an illustration of a "sedge bog" (sedge meadow in my terminology), and in Johnson et al. (1966) for illustrations taken in Alaska of several "sedge meadow" communities and a "tussock field" (tussock muskeg), the latter much more extensive than most sampled in my own studies but quite representative.

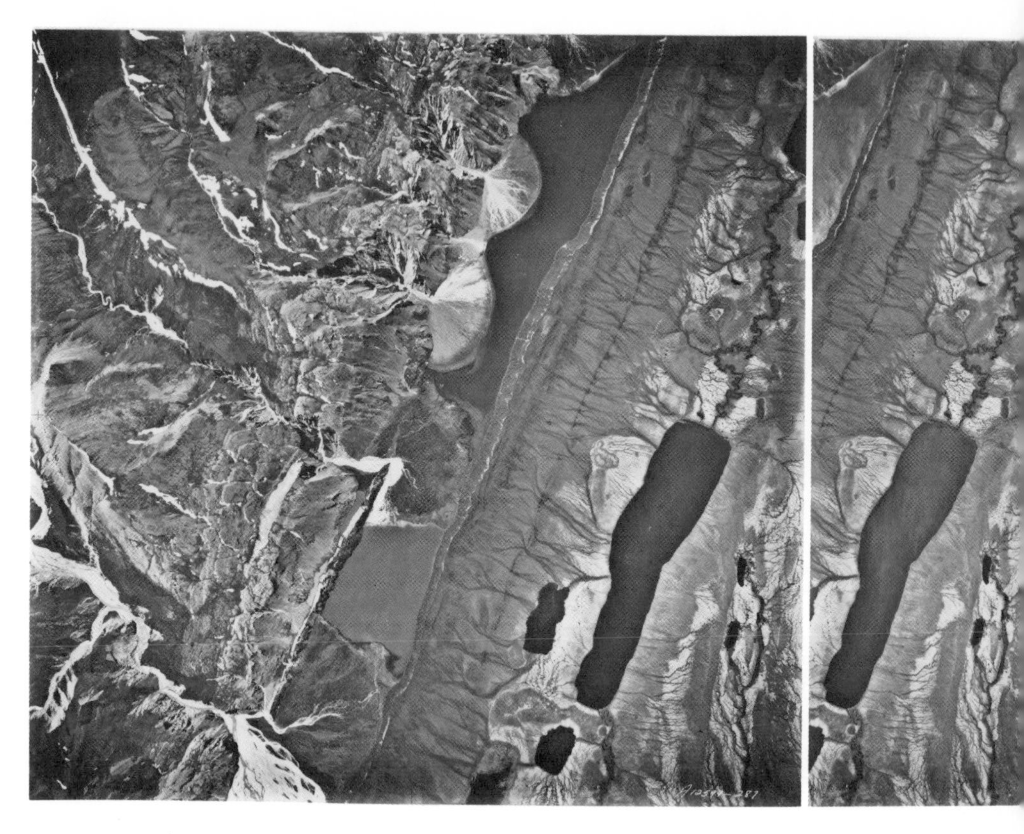

Figure 4.5. Airphoto of the Florence Lake study area (A 12599-287).

spots, varying from a few inches to a few feet in diameter, the vegetation has been markedly disturbed and often the subsoil exposed by the frost activity, an event also noted by Lambert (1972). It is apparent that the soil active layer is rarely deeper than 12 inches on the average and rarely as deep as 18 inches maximum by the end of summer on the tussock muskeg sites.

Shown in the airphoto is a small portion of the Cache Creek drainage basin. Although there are gentle slopes of 0–4% adjacent to the main stream, the entire drainage basin is enclosed in most places by low vertical bluffs, with steeper slopes in the range of

Arctous = Arctostaphylos (see Polunin, 1959; Cody and Chillcott, 1955; Ritchie, 1956). As used here, *Arctous* includes both *A. alpina* and *A. rubra*, as described by Polunin. Ritchie points out: "At Churchill it is possible to distinguish this species from *A. rubra* only when ripe fruits are available; the distinguishing vegetative characters which Fernald describes were found to be unreliable with much of the material." The difficulties of field identification are insurmountable, and since the species occurs often in quadrat sampling, the collection of doubtful specimens would be impossibly burdensome not to mention the time required later for more accurate identification in laboratory or herbarium. Thus, in sampling, *A. alpina* includes *A. rubra* whenever it was encountered. All subsequent reference will, however, be in terms of *Arctostaphylos alpina*, preferred usage at present, without differentiating *A. rubra*.

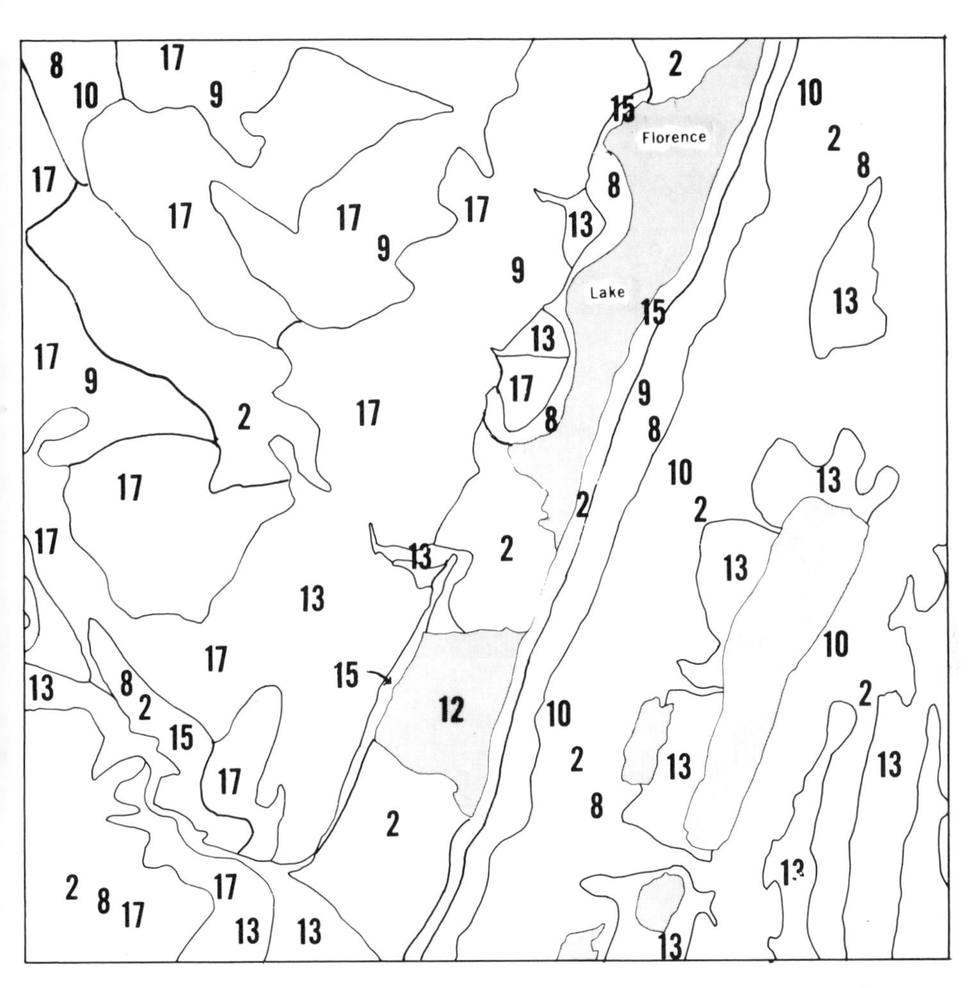

Figure 4.6. Key to the airphoto of the Florence Lake area. For descriptions of features indicated by identifying numbers, see Fig. 4.2.

35–40% or even to 45–50% and greater. The general drainage pattern is dendritic, with examples of radial drainage descending from the rounded sedimentary outliers.

Upper hill slopes and summits are occupied by rock field plant communities; uplands supported relatively higher frequencies of *Arctous alpina**, *Hierochloe alpina*, *Vaccinium uliginosum*, *Saxifraga tricuspidata*, *Phyllodoce caerulea*, *Luzula* spp., and *Oxytropis* spp. than the tussock muskeg communities on the lower slopes.

Florence Lake Area (65°08′N, 128°03′W)
    Airphoto No. A 12599–287 (Figs, 4.5, 4.6)
Carcajou Lake Area (65°39′N, 127°52′W)
    Airphoto No. A 12148–467 (Figs.4.7, 4.8)

---

*See page 48.

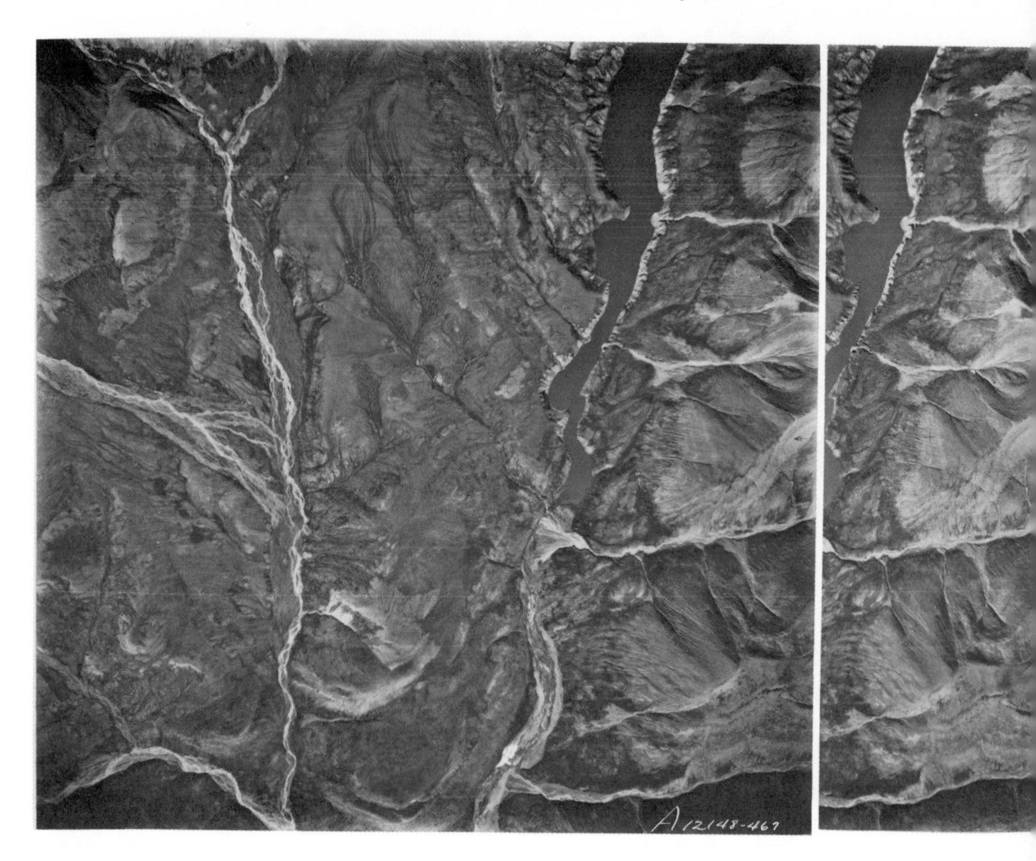

Figure 4.7. Airphoto of the Carcajou Lake study area (A 12148-467).

To the west of the Mackenzie River in the vicinity of Norman Wells the very gently rolling alluvial plain is about 20 miles wide, and beyond it rise rather sharply the eastern foothills of the Mackenzie Mountains. Across the river from Norman Wells is a small cluster of abandoned buildings, the eastern terminus of the legendary Canol Road, built in the early years of World War II to service construction of the oil pipeline between Norman Wells and Whitehorse (Finnie, 1959). To illustrate the general rise of the east slope of the mountains in this area, the old Canol Road reaches the highest point of its 513 miles only 80 miles west of Camp Canol. Here it crosses the broad, flat-topped range known as the Plains of Abraham, where the road attains an elevation of 5700 feet. Elsewhere along the road, peaks rise to considerably greater heights than this. The road is no longer passable for conventional vehicles and is seldom visited.

The botany of the region was surveyed rather extensively in 1944 by Porsild, whose subsequent publications (1945, 1955, 1974) provide good general descriptions of the area and the history of early exploration as well as lists of the plants collected. Porsild writes: "This alpine plateau is covered by a thick mantle of weathered limestone rubble through which protrude ledges or cliffs of solid rock .... Everywhere on sloping ground solifluction and live stone creeps can be observed, and on level ground polygon

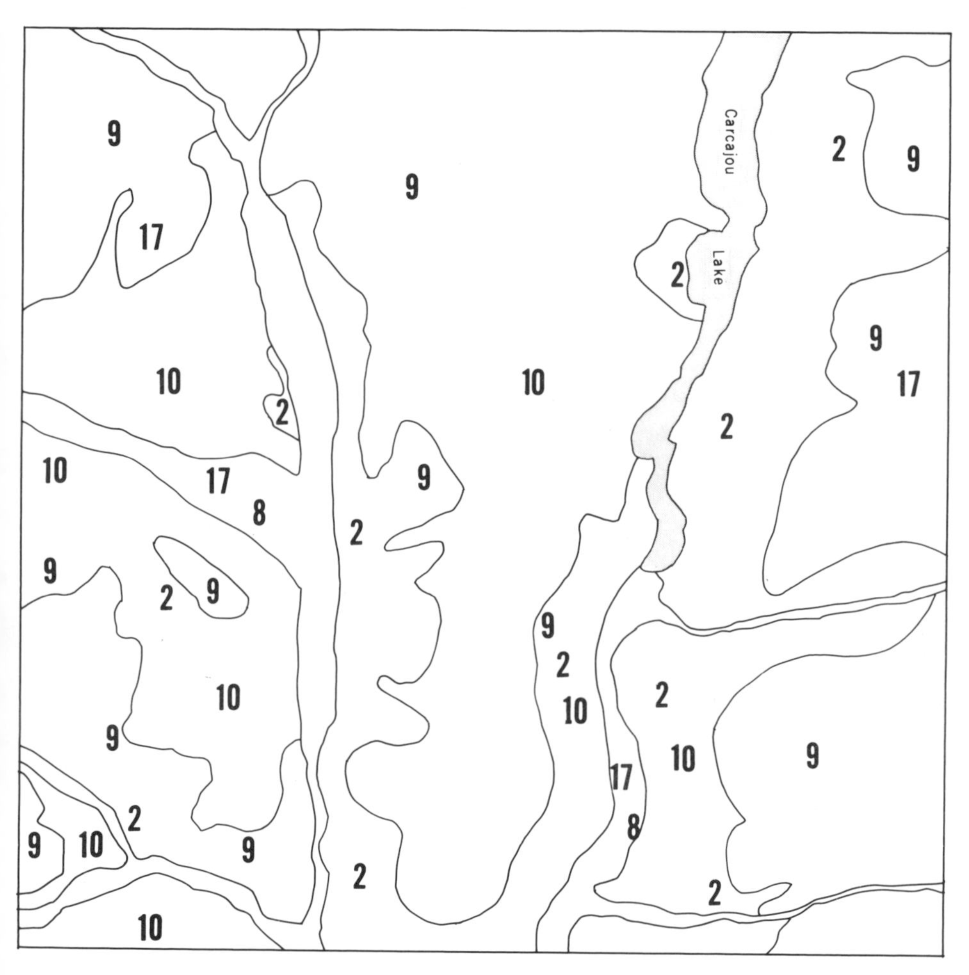

Figure 4.8. Key to the airphoto of the Carcajou study area. For descriptions of features indicated by identifying numbers, see Fig. 4.2.

formation is active. Outcrops of rock everywhere show evidence of rapid weathering by frost action. From the abundance of water in the soil and the seepages noticeable at the bottom of all hills and slopes precipitation appears to be high."

At Florence Lake the floor of the valley is forested with relatively large white spruce. The lower slope levels have scattered stunted white spruce as well as black spruce. The upper slopes and summits are devoid of trees and a carpet of tundra vegetation prevails over much of the surface. In contrast, at Carcajou Lake, which is both deeper within the mountain system and at higher elevations, a richer arctic component in the vegetation is apparent and the forested areas in the valley tend to be dominated by black spruce.

At higher elevations above Florence Lake, the major proportion of the surface is given over to tundra vegetation, although rather extensive stands of dwarfed spruce with an understory dominated by shrubs are found in the swales or "sheep meadows"

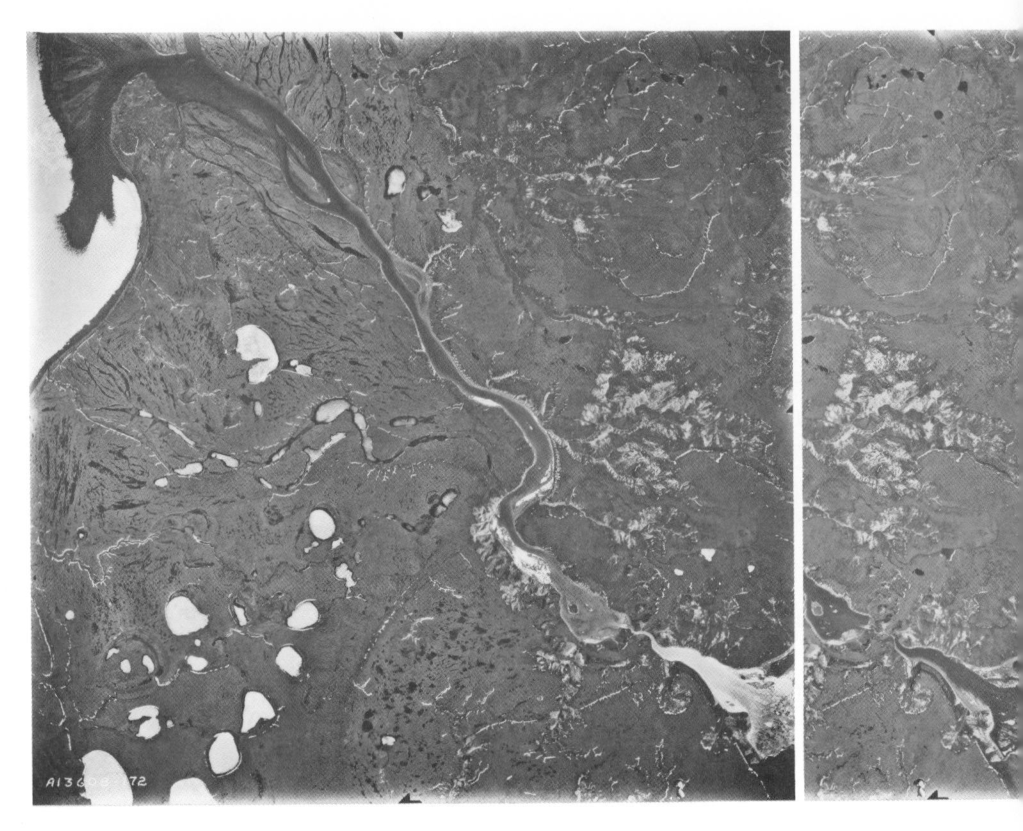

Figure 4.9. Airphoto of the Coppermine study area (A 13608-172).

up to elevations of 3500–4000 feet. At Carcajou Lake the complex topography and irregular patterns of climatic factors—wind, radiation, and evapotranspiration—render the pattern of communities over the landscape too complex for easy interpretation. Spruce stands of highest density are found on southeast-facing slopes, on flat areas along streams, at the base of mountain slopes, and in drainage ravines. This is apparently a response to moisture because such slopes will have lower rates of evaporation in late afternoon when temperatures are highest. In winter, snow accumulates on such slopes and in such depressions, adding to protection of trees from desiccating effects of the mountain winds (Tranquillini, 1979). On low and relatively moist north-facing slopes, spruce is present only as scattered individuals, dwarfed or decumbent, if any are present at all.

The tussock muskeg community is by far the most widespread in the Carcajou Lake area, occupying large portions of the total land surface and attaining full development even on the upland areas, wherever drainage is sufficiently impeded to allow the lush, turfy growth characteristic of the tussock community. The tussock communities on the upper slopes must be considered to be stable and of long duration. In many there exists little evidence of extensive frost action and where it has occurred those species com-

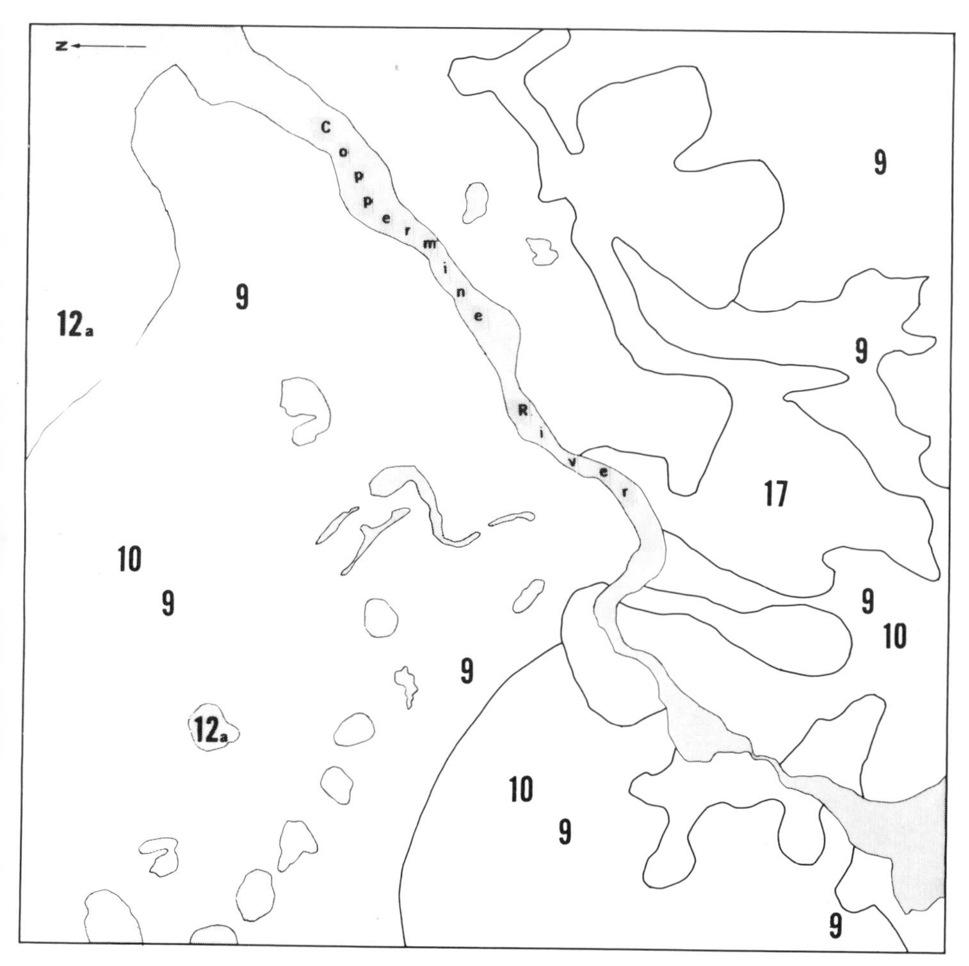

Figure 4.10. Key to airphoto of the Coppermine study area. For descriptions of features indicated by identifying numbers, see Fig. 4.2.

monly found in the adjacent undisturbed areas appear to rapidly recolonize the disturbed site.

It is apparent that the tussock structure serves to retain water which accumulates, either as snow in winter or as rain in summer, and despite the degree of slope the tussock structure tends to produce a network of small pools and channels which impede drainage and produce conditions affording ample moisture over areas with a relatively wide variation in degree of slope.

The tussocks of *Eriophorum* persist because of the continued availability of water; on the tussock summits such species as *Dryas*, *Ledum*, *Betula*, and *Empetrum*, along with other characteristic species and a rather uniform assemblage of lichens and mosses grow. Important studies of the vegetation of these and nearby areas are now becoming increasingly numerous (See and Bliss, 1980; Bird, 1974a,b; Kershaw, 1984; Bird et al., 1980; references cited in these publications).

Figure 4.11. Tundra in the Coppermine study area. The sharp hills rise behind the village of Coppermine, located along the coast of the Arctic Ocean in Coronation Gulf.

Coppermine Area (67°47′N, 115°15′W)
Airphoto No. A 13608–172 (Figs. 4.9, 4.10)

The small settlement, Coppermine, lies on the arctic shores at the mouth of the Coppermine River. Near here, Samuel Hearne and his band of Indians reached the area during 1771 in search of a reported source of copper. In his paper on the natural history of the Coppermine area, Hicks (1955) affirms that "the northern limit of trees is approximately 20 miles south of the settlement." His description of the area affords a good general impression of the tundra plant communities:

> The tundra is comparatively flat and composed of caribou mosses, grasses, sedges, and scattered showy-flowering plants. The low-lying area in the settlement is marshy and produces many species of perennials. *Carex* meadows are noticeable as one moves up the valleys toward the three rock cliffs on the south. There are mats of *Ledum*, *Rhododendron*, *Vaccinium*, and occasional patches of dwarf Betula in many places on rock or soil.

The surficial geology of the region has been described by Craig (1960), who points out that the Wisconsin Laurentide ice created a "profusion of glacial landforms" in the region between Great Bear Lake and the coast, with numerous moraine ridges and "vast areas of hummocky moraine." On the area in the immediate vicinity and south of the village, outcroppings of basalt are in evidence, and according to Hicks a "broad valley directly back of the settlement fans out into smaller valleys between high cliffs, where it meets the tundra country . . . . Three basaltic cliffs stand in bold relief a mile to the south of the settlement. The highest of the cliffs . . . has an abrupt drop of approximately 150 feet, a characteristic of all the cliffs in this region."

Large areas in the vicinity of the village are either relatively flat or very gently sloping, with a vegetation that might be categorized as either—or intermediate between—*Carex* meadow or tussock muskeg, the latter more widespread on the gentle slopes of the lower portions of the hills, the flat terraces which appear to once have been floodplains of the river, and the older beach ridges which can be seen to parallel the contours of the hills at some considerable height above the present day shoreline. On many of the gentle slopes, the tussock muskeg community gives the impression of age and stability, the branches and twigs of individual plants tightly intermeshed and with an unusually high density of well-formed lichens on the ground. Where the tussock muskeg community is well-developed, the originally lower and wetter intervening areas between tussocks have filled with a dense moss, lichen, and shrub mat (Fig. 4.11; see also Cody, 1954).

At the end of the first week in August, the following temperature profiles were obtained (on different days) in tussock muskeg communities in the Coppermine area:

Depth (inches)	Tussock Muskeg Temperature Profile	
1	14°C	18
2	12	10
3	9	9
5	7	
6	6	3
8	5	
12	3	2

Depth to the upper surface of the permafrost in a rock field community, measured as the average of 20 observations, was 18 inches, with a range between 12 and 24 inches. In a *Carex* meadow, the average of four observations was 15 inches, and the average of 34 observations in a tussock muskeg was 18 inches with a range of 12 to 24 inches.

Large areas of terrain in the vicinity of the village are either relatively flat or very gently sloping upland, the surface of which bears the vegetational communities that range between *Carex* meadow and tussock muskeg. The summits of the highest hills are covered with sand, gravel, and detritus and support a rather dense vegetational cover. This, as in other sections, has been termed the rock field community (though in many areas exposed rocks represent only a small percentage of the ground surface; the term is used in such instances for convenience). The latter represents the most xeric of the three environmental nodes.

The tussock muskeg community is found over a relatively large proportion of the land surface, occupying the gentle slopes of the hills, flat terraces, and old beach ridges. A tussock community dominated by *Cassiope tetragona* occupies terrain that is slightly more elevated than the muskeg dominated by *Salix* species. Lichens are abundant in the tussock muskeg community, particularly the *Cornicularia* and the common *Cetrarias*

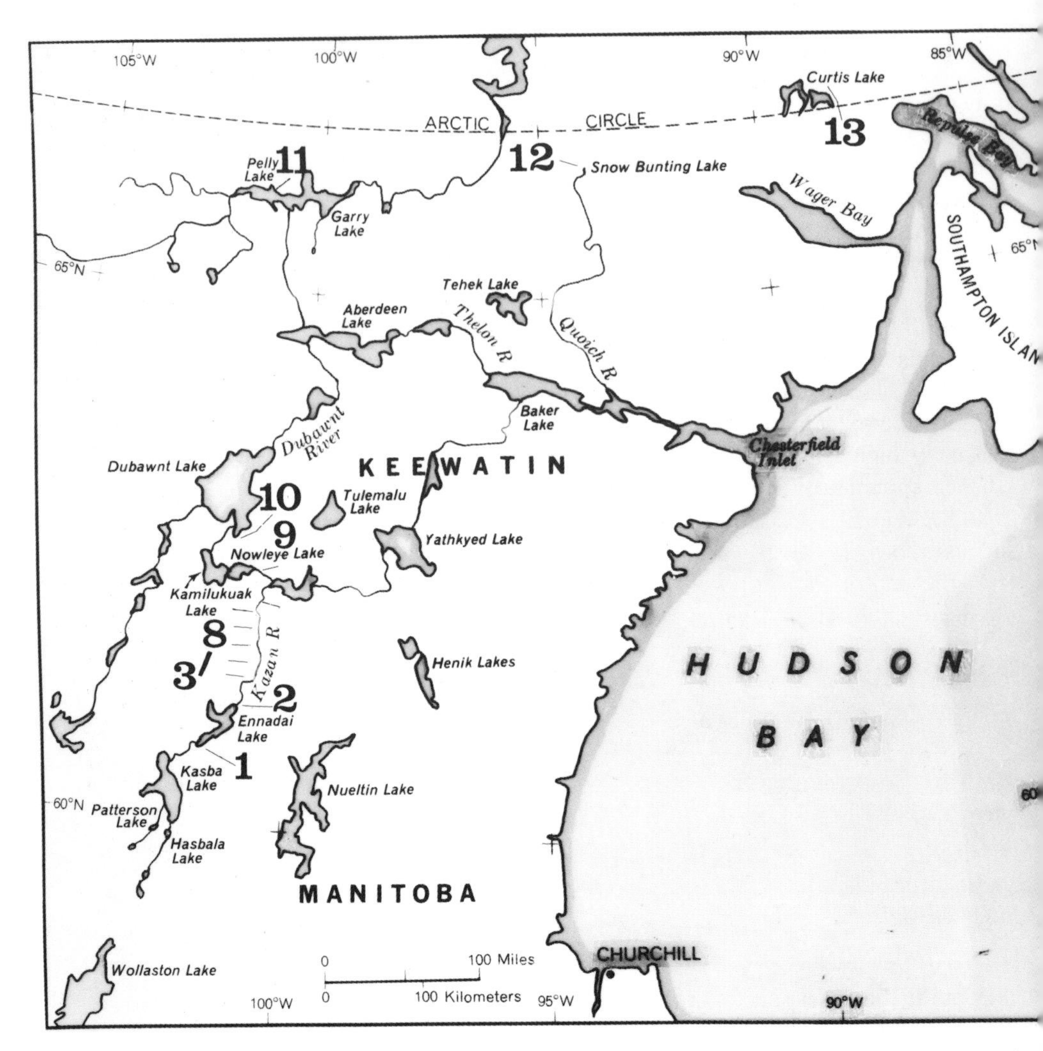

Figure 4.12. Map of study areas in Keewatin. 1—south end of Ennadai Lake; 2—north end of Ennadai Lake; 3–8—Kazan River study areas; 9—Kazan River tributary; 10—Dubawnt Lake; 11—Pelly Lake; 12—Snow Bunting Lake; 13—Curtis Lake. The numbers assigned to the areas are those employed in the summary table (Table 5.1). For detailed coordinates of study area, see Appendix F.

as well as the somewhat more rare *Cetraria richardsonii*. *Betula*, *Ledum*, and *Vaccinium uliginosum* are common associates.

A limited number of botanical collections were made in the area during the early years of this century, and one of the more extensive early collections was that of Findlay in 1951, the specimens of which were identified by Cody (1954) who states that the collection is small but of interest because it established the eastern limit of

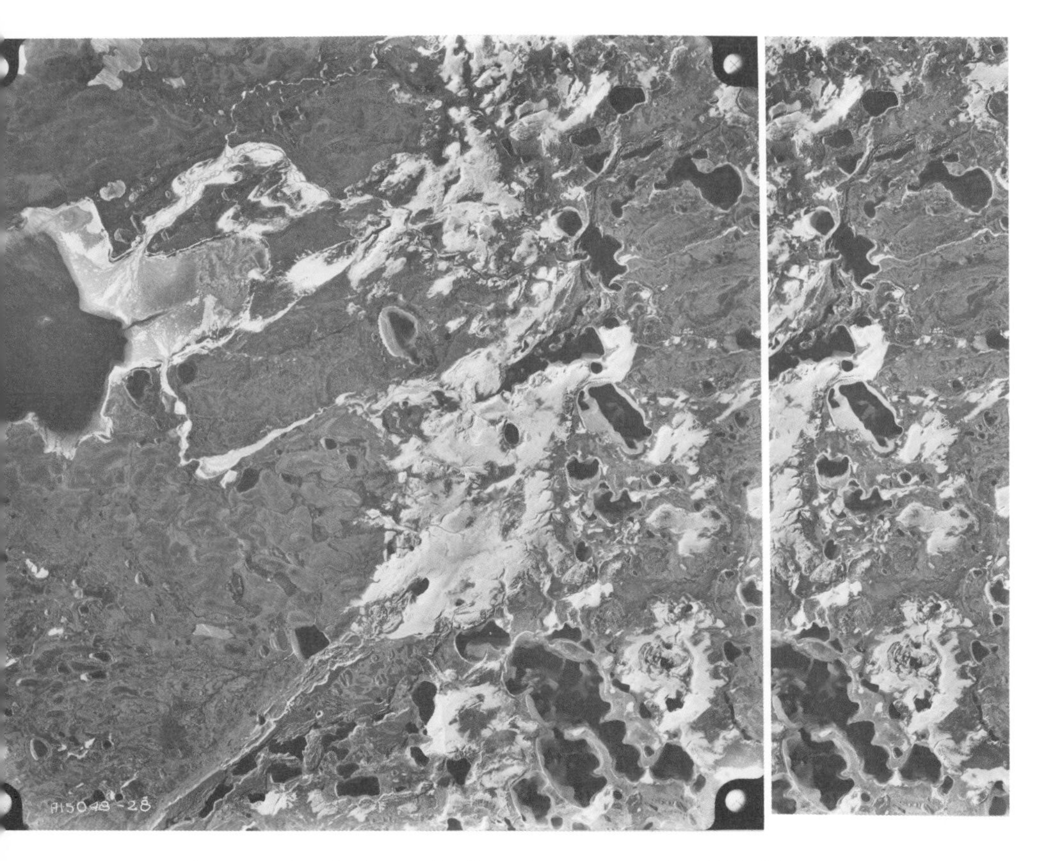

Figure 4.13. Airphoto of Pelly Lake study area (A 15048-28).

the range of many species along the arctic coast as well as the northern limit for a number of other species.

Northern Keewatin

The study areas in northern Keewatin lie very close to the Arctic Circle, extending from the Pelly Lake area on the west to Repulse Bay on the east, and they were selected primarily because the climate of the region is apparently the most severe in both winter and summer of any area in the Canadian northern interior. Permanent snow banks persist in most of the areas at least throughout most summers, and from their size one suspects that this must be true of all summers under present climatic regimes. The areas were selected for their intrinsic interest as well as for the purpose of comparison with vegetational communities sampled elsewhere in Canada (Fig. 4.12).

Vegetational community sampling in the northern areas during the course of this study affirm the statement (Drury, 1962) that there "are no special vegetation types

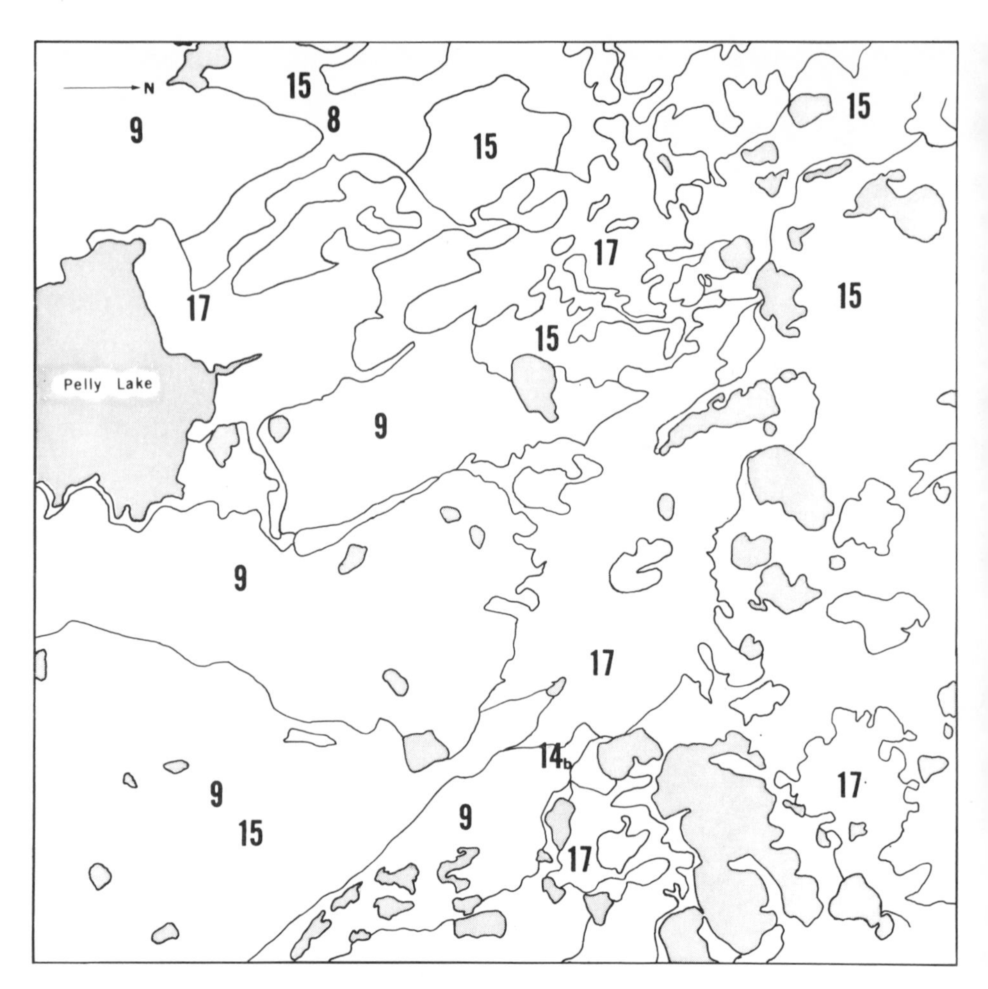

Figure 4.14. Key to airphoto of Pelly Lake study area. For descriptions of features indicated by identifying numbers, see Fig. 4.2.

that occur on slopes, or on hilltops, or on valley bottoms" in the sense that all species found in the area are very apt to have individual representatives on any or all of these topographic positions, on at least small and locally favorable sites. Thus *Andromeda polifolia* appears with greatest frequency in wet lowlands but will also be found—albeit less often—on the summits of the rock fields. The sites available to this species will be fewer on the rock fields than in the lowlands, and so it is less abundant on the latter but seldom does it appear to be excluded altogether. The behavior of species, alone and in association with one another in the various communities, appears to be determined by raw physical forces to a greater degree than is the case in the plant communities of temperate and tropical regions.

Frost action is the chief disturbing influence upon the vegetation. Following upheaval of the soil surface by frost, the species making up the community existing on the site

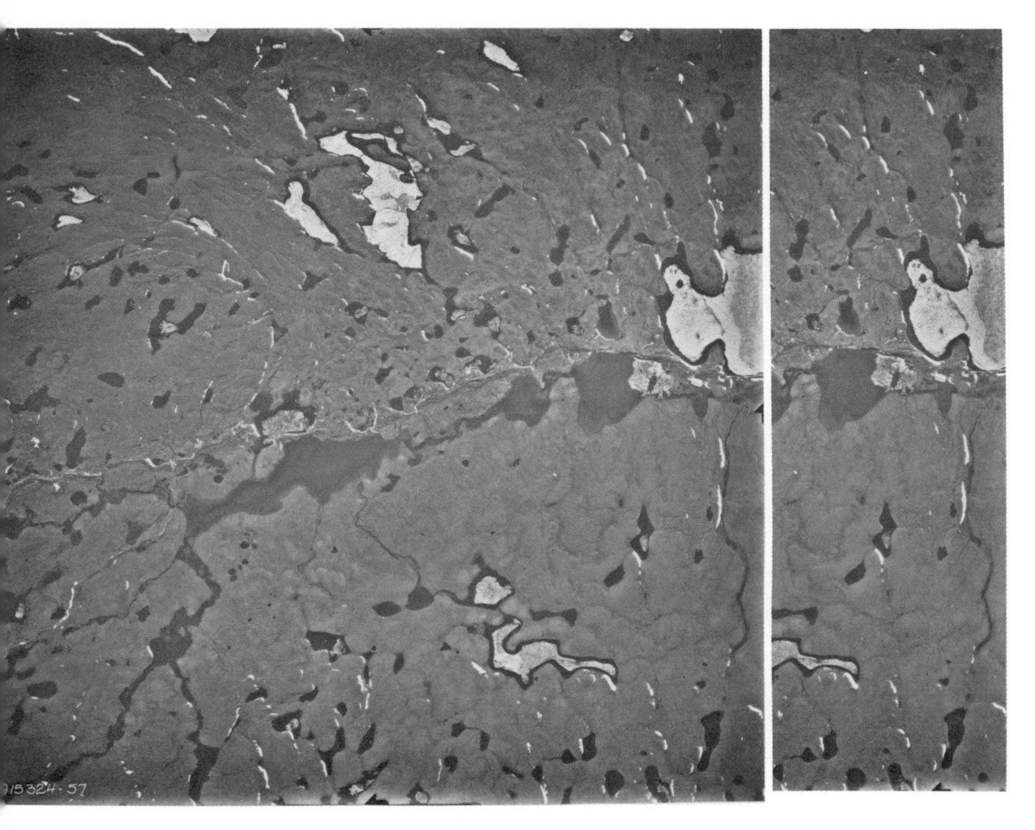

Figure 4.15. Airphoto of Snow Bunting Lake study area (A 15324-57).

before disturbance will usually recolonize the site. Some species appear, however, to have preferences for recolonization of polygon centers, some for rims, and so on, but elucidation of these characteristics is not the purpose of the study here. Investigations of pattern on such areas is nevertheless of considerable interest, and further efforts to designate the "micro" associations on these smaller terrain features must await the investigations of others. Since 1 m^2 (one square meter) quadrats were employed throughout the studies reported in this and other volumes, taking in more than 40 study areas in forest and tundra, it was necessary to resist the temptation to change quadrat size so that community composition for the smaller physiographic features might better be revealed. For a more complete review of the sampling techniques, reference is made to a previous publication (Larsen, 1980); for a detailed description of the studies conducted in northern Keewatin, reference should be made to the report on this work (Larsen, 1972c).

Pelly Lake Area (66°00′N, 101°15′W)
Airphoto No. A 15048–28 (Figs. 4.13, 4.14)

The area has been described as underlain by gneissic granitic rocks, including some schist, and a zone of pitted outwash many square miles in extent was apparently

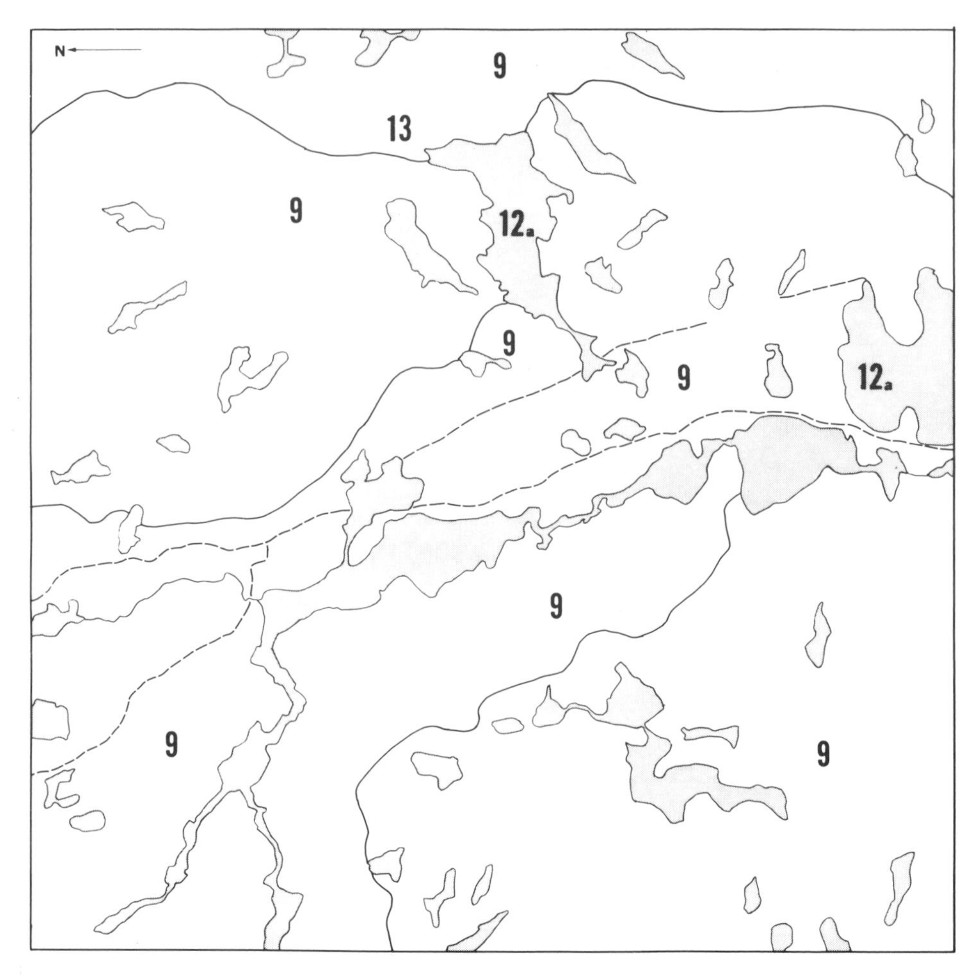

Figure 4.16. Key to airphoto of Snow Bunting Lake study area. For descriptions of features indicated by identifying numbers, see Fig. 4.2.

deposited during a halt in the retreat of the ice front (Craig, 1961, see map; 1964, p. 18). The southern edge of the outwash plain is marked by an abrupt termination of the sand and gravel outwash deposit, with barren sand slopes descending 50 to 100 feet or more to the surface of coarse boulder and gravel drift possessing familiar surface characteristics, indicating extensive former glaciation.

The large area of outwash sand is clearly visible on the Pelly Lake area airphotograph. On areas covered with vegetation, designated number 15, the plant species making up the vegetational mat are a combination principally of dark-colored lichen species, grasses, with scattered dwarf shrubs and herbaceous species forming a loose turf easily broken away from the ground surface. This unusual community has been described in detail (Larsen, 1972). The substrate of the area designated by the number

Figure 4.17. A conspicuous feature of the vegetation of the Snow Bunting Lake and other areas in northern Keewatin is the dense carpet of lichens covering the ground on such favorable sites as rock fields, esker summits, and esker slopes with a south-facing exposure. The lichens are principally *Alectoria nitidula* and *A. ochroleuca*, shown here with tufts of the grass *Hierochloe alpina*. Lower areas with moderate drainage, however, have a lichen carpet in which these *Alectoria* species are minor components and *Cladonia* (*Cladina*) species attain dominance. For a detailed discussion see Larsen (1972).

9, to the lower left on the photograph, is glacial till. This area is beyond the edge of the outwash plain which extends largely throughout the upper portion of the photograph and for a number of miles north from the area shown. A flat-topped ridge of coarse sand and gravel extends from the outwash plain (at 14b) toward the lower left. This is covered with a rock field vegetational community. The point 14b shows several abandoned buildings that were part of a field station for aircraft crews during the aerial photographic survey of the region. The airstrip extends roughly toward the upper right from these structures.

The large areas of outwash plain are relatively flat and are covered for the most part with a thin vegetation, principally tufts of *Hierochloe alpina* in a continuous carpet of *Alectoria nitidula* and *A. ochroleuca.* Slopes of depressions often lack vegetation, particularly those with a southwest exposure, while slopes with a northern exposure are often covered with a continuous community that includes *Ledum decumbens*, *Cassiope tetragona*, *Vaccinium vitis-idaea*, and *Hierochloe alpina*. There is here a lichen cover of *Alectoria ochroleuca*, *A. nitidula*, *Cetraria cucullata*, *C. nivalis*, and *Cladonia alpestris (Cladina stellaris)*. Some degree of solifluction activity usually is in evidence

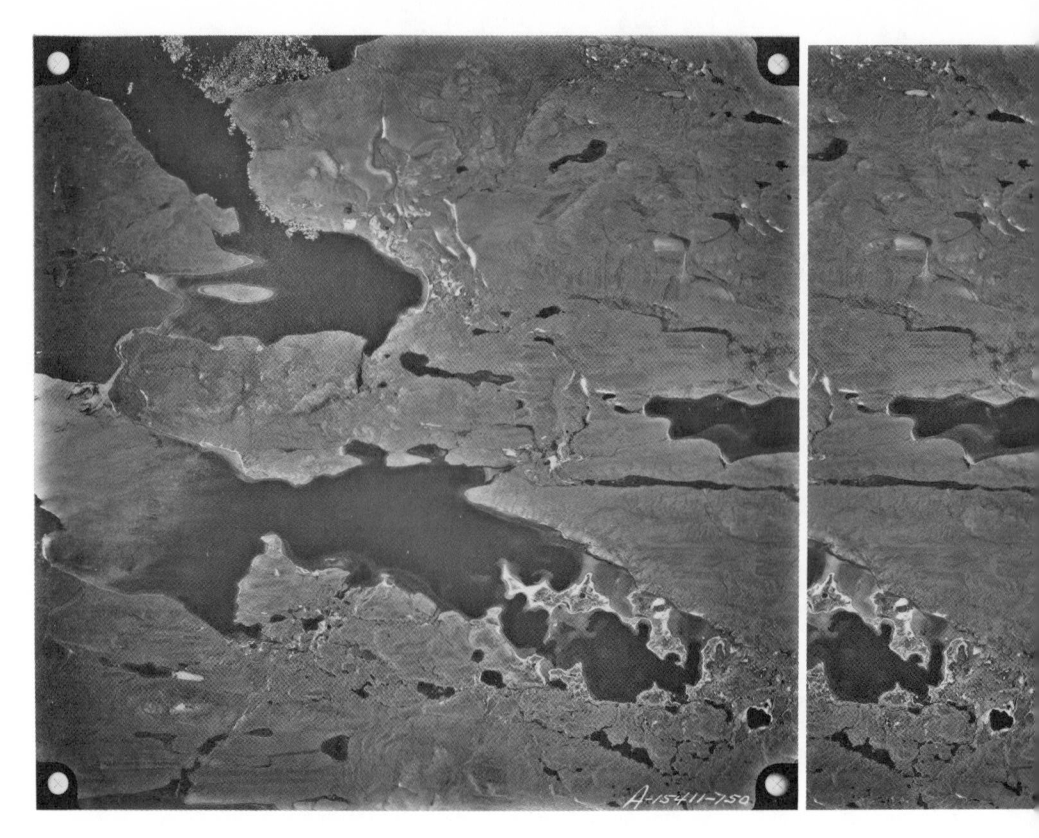

Figure 4.18. Airphoto of the Curtis Lake study area (A 15411-150).

near the summits of the slopes. The exposed surface material, often lag gravel and coarse sand, is frequently undergoing colonization or recolonization by the lichen components of the community, chiefly the *Alectorias*, with *Carex* spp., *Silene acaulis*, *Luzula confusa*, *Saxifraga tricuspidata*, and *Hierochloe* somewhat less frequent.

Low areas with gentle slopes, and the margins of ponds, are often covered with a continuous mat of *Carex* and *Eriophorum* species. Areas slightly elevated above generally low surroundings are colonized by *Ledum decumbens*, *Vaccinium uliginosum*, and *Andromeda polifolia*. Along drainage lines from upland slopes to the meadows, *Cassiope tetragona* is much in evidence, along with *Ledum decumbens*, *Luzula spadicea*, and *Vaccinium vitis-idaea*.

It appears that tussock muskeg communities are poorly developed in the Pelly Lake area, even on sites transitional between low meadows and rock fields, where they attain their most advanced development elsewhere to the south.

The *Alectoria-Hierochloe* vegetational mat is a most interesting community. *Alectoria nitidula* is a dark brown, nearly black, species, giving the ground a very dark aspect wherever it is abundant. The habitat where it is very abundant is often extremely

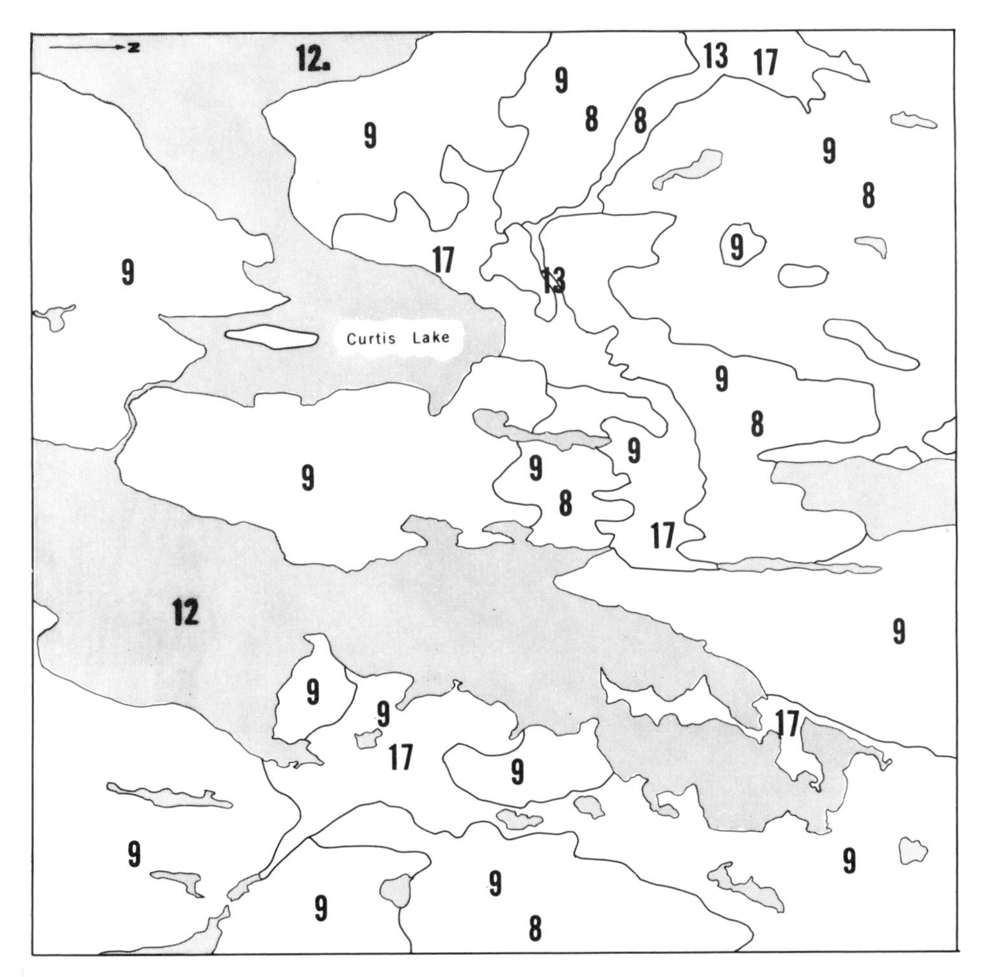

Figure 4.19. Key to the airphoto of the Curtis Lake study area. For descriptions of features indicated by the identifying numbers see Fig. 4.2.

dry, and at most times the *Alectoria* cover is so desiccated as to crunch audibly under foot. During early morning hours, however, the *Alectoria* mat often has absorbed sufficient moisture to become soft and pliable. Even under the early morning reduced light conditions and cool air temperatures it seems reasonable that the dark thalli absorb sufficient solar radiation to attain temperatures permitting photosynthesis before they become desiccated and inactive once again.

The rock field community in this area is found on the drumlinoid forms composed of glacial till and is represented often by a relatively long species list, including principally the *Ericaceae*, *Carices*, and *Gramineae*. Higher densities of *Ledum* and *Cassiope* occur around the periphery of exposed rocks and in shallow drainage lines. *Salix rotundifolia* (or *S. phlebophylla*) and *Oxytropis arctica* are found on some sites, giving

Figure 4.20. A rocky hill summit in the Curtis Lake study area. The eastern end of Curtis Lake is shown in the background. For a detailed discussion of vegetational communities in the area see Larsen (1972).

the community a certain resemblance to the rock field communities at Canoe Lake at considerably higher elevations of the Richardson Mountains in the Yukon. A more detailed description of the area is given in Larsen (1972c).

Snow Bunting Lake Area (66°10′N, 94°25′W)
Airphoto No. A 15324–57 (Figs. 4.15, 4.16)

The surficial geology is evidently typical of large areas of the central Canadian and especially the northern Keewatin region, with an abundance of glacial landforms (Craig, 1961), particularly sandy eskers and rock fields formed of large, coarse, generally angular boulders. Most of the area is covered with glacial drift, as described by Craig, ranging in texture and size from sand to gravel to large boulders, and a few outcrops appear to the northeast of the lake. In general appearance, the landscape is gently rolling, with rugged rock fields and rock outcroppings to break the surface. Snow Bunting Lake forms the headwaters of the Quoich River, and is located roughly on the height of land between the watersheds flowing south into Baker Lake and north or northeast into the Back River and Chantry Inlet. Lakes and ponds are common, often linked by small streams flowing through shallow immature channels.

The conspicuous vegetational feature is the dense carpet of lichens over the rock fields, esker summits and slopes, dominated by *Alectoria nitidula* and *A. ochroleuca*

Figure 4.21. Mackenzie Delta area viewed from atop the shoreline slopes in the vicinity of Reindeer Station. Trees are confined to the delta deposits and the shoreline slopes; uplands in the area, extending to the right beyond the area of the photograph, are tundra.

(Fig. 4.17), similar in many respects to the community at Pelly Lake. The rock fields in the lower areas, however, have a dense lichen carpet in which these *Alectoria* species are minor components and where *Cladonia* and *Cladina* species attain dominance. The moss *Rhacomitrium lanuginosum* is frequent, at times occupying areas in nearly pure stand. The low meadows, dominated by *Eriophorum* with *Carex* spp. of less significance, occupy the extreme lowland areas. *Carex stans* with occasionally *Saxifraga foliolosa* and *S. cernua*, along with grasses, line the turfy shores of ponds in the meadows.

Curtis/Stewart Lakes Area (66°50′N, 88°55′W)
Airphoto No. A 15411–150 (Figs. 4.18, 4.19)

The Curtis and Stewart Lakes area is located about 75 miles west-northwest of Repulse Bay and about an equal distance northeast of the west end of Wager Bay, on terrain that

Figure 4.22. Deep deposit of peat along the eastern shore of Colville Lake; a short stretch of sand beach extends from the base of the peat deposit to the lake edge where the photographer stood.

might be described as rolling uplands, principally drift-covered, with a relatively large proportion of the surface given over to outcropping rocks and a variety of glacially-formed ridges and hills. The highest hills, some of which are steep and craggy, rise to elevations of 500 feet above the lake. Large numbers of ice-rafted boulders now rest randomly over the surface of the terrain, attesting to submergence during a post-glacial period. Excepting for the more rugged appearance resulting from the jagged outcropping hills, the region resembles the Snow Bunting Lake study area in its variety of glacial landforms (Craig, 1961). In contrast to the latter area, however, the Curtis and Stewart Lakes area contains a number of permanent snow banks, resting in the deeper declivities of drainage lines and on beds of boulders from which the finer particles have been removed by the meltwater streams. There were no well-developed eskers in the study area, and the few low sandy hills were covered with rock fields or a virtually bare lag gravel surface (Fig. 4.20). Eskers seen from some 10 or more miles west of the camp appeared largely devoid of vegetation on the summits and upper slopes.

Surface material of rock field summits is primarily sand and gravel with an average of 10–25% bare exposed rock surface. *Saxifraga octopetala* is relatively common in the

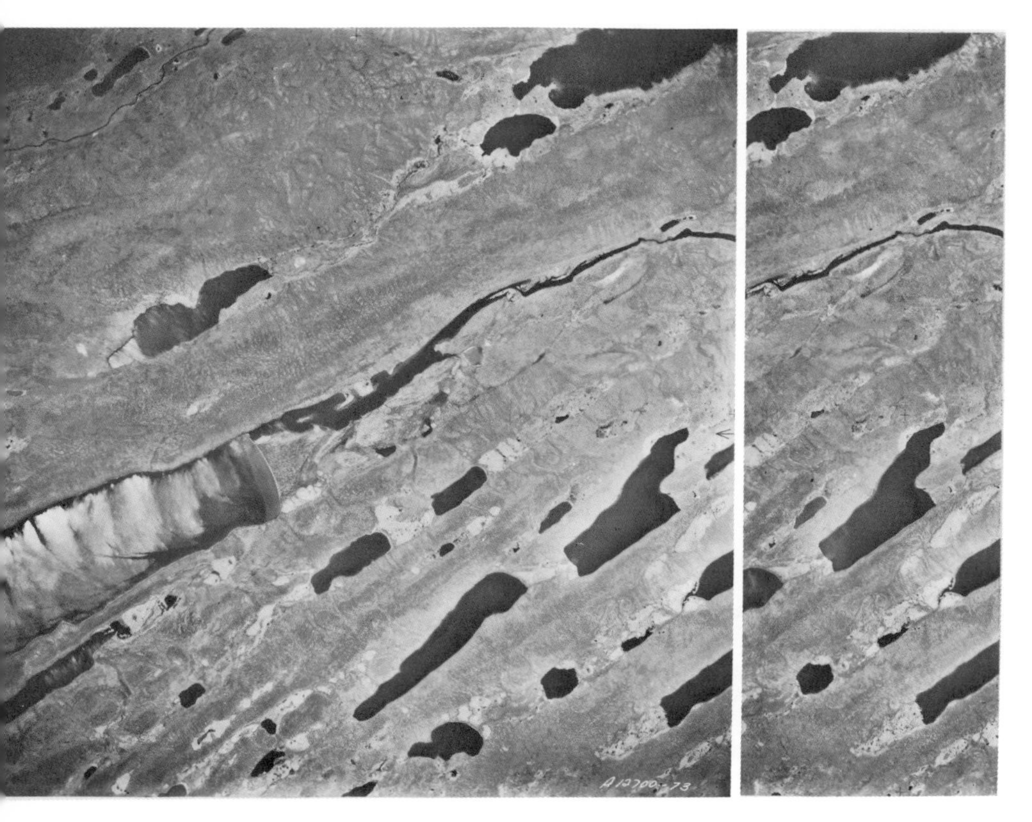

Figure 4.23. Airphoto of the "East Keller" Lake study area (A 12700-73).

central portions of mud boil sites, along with *Silene acaulis* and *Carex stans* and a few other species. *Ledum decumbens* is found in sheltered spots around rocks. *Carex nardina* and *C. bigelowii* are relatively common, along with *Polytrichum* and other mosses and a relatively wide variety of lichen species. *Hierochloe alpina* is distributed widely.

At Repulse Bay, the surface of the rugged terrain is characterized by rocky outcropping hills separated by relatively flat areas of drift, the latter somewhat reduced in total extent over that found in other study areas. The irregular terrain probably accounts for the size and number of late snow patches and permanent snow banks persisting throughout the summer, all of which occupy cliff bases in the lee of winds where snow accumulates to depths. A rich and turfy lichen and forb community is found around such areas, probably the consequence of the continuous supply of moisture.

Northern Forest Limit: Western

It is in the far northwestern tip of Canada that the continuous spruce forest reaches the northernmost extent of its range in North American, occupying the islands and margi-

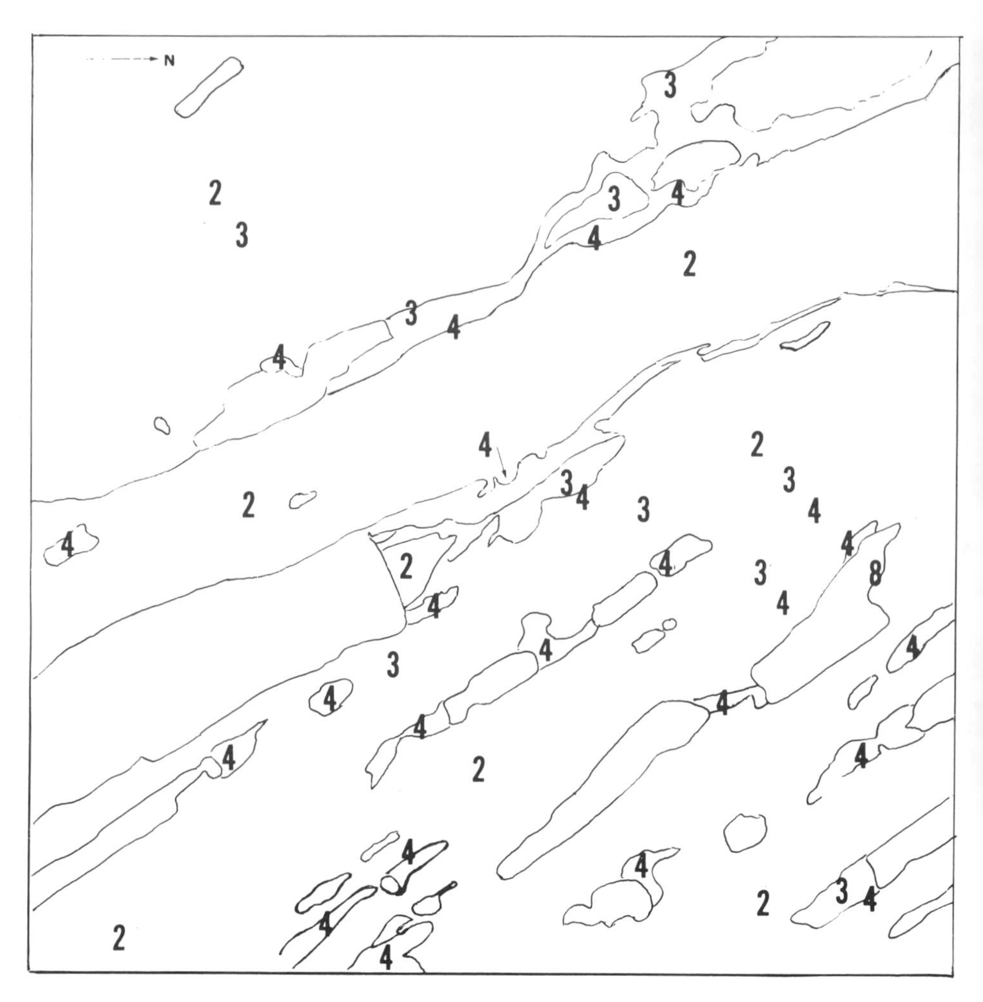

Figure 4.24. Key to airphoto of "East Keller" Lake study area. For descriptions of features indicated by the identifying numbers, see Fig. 4.2.

nal slopes of the Mackenzie River Delta region nearly to the Arctic Ocean. It is apparent that the river and its delta valley ameliorate the environmental conditions over those on surrounding uplands as a habitat for spruce. As one travels northward along the river from Inuvik, uplands gradually become devoid of trees while the islands of the delta remain covered with forest. Northward the trees are often found only in the immediate vicinity of the shorelines of the river or of the innumerable lakes, finally giving way to dense willow thickets and wet meadows a few yards from the water's edge. Muskeg dominated by black spruce growing in association with shrubs and mosses occurs in the southern parts of the delta and occupies uplands south of the forest border eastward across to the Anderson and Horton Rivers.

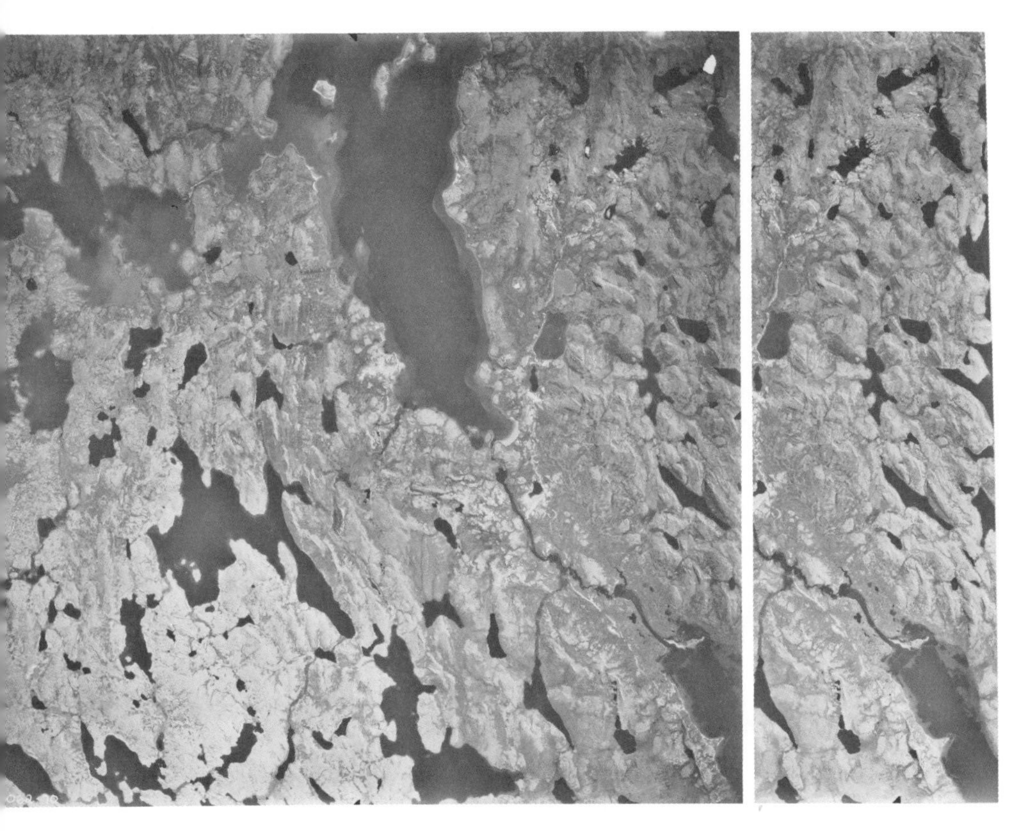

Figure 4.25. Airphoto of Winter Lake study area (A 14062-70).

Along the east side of the river in the vicinity of Reindeer Station, steep slopes rise to elevations of as much as 500 feet, forming the westward boundary of the Caribou Hills. The gently rolling uplands are dominated by tundra while the steep west-facing slopes support many species with a restricted distribution in the area, including *Shepherdia canadensis*, *Selaginella sibirica*, *Myosotis alpestris* spp. *asiatica*, *Lathyrus japonica*, and *Silene repens* spp. *purpurata* (Cody, 1965). On limestone hills near Inuvik, other species include *Linum lewisii*, *Woodsia glabella*, *Cystopteris fragilis*, and *Galium boreale*. West of the delta region, tree-line retreats southward, following the shoreline of the river and the foothills of the mountains; scattered trees are found farther west on the Old Crow flats and northward along the valley of the Firth River (Drew and Shanks, 1963), as noted above. Past and present vegetation of the general region is discussed fully by Ritchie (1984), whose palynological studies have clearly elucidated the post-glacial history of the vegetation in this and other western regions.

In the western forest-tundra transition zone, the forest border is much more irregular than it is in the Artillery Lake and Ennadai Lake areas to be discussed below. Somewhat more extensive areas appear to be given over to lichen woodland in some places, partic-

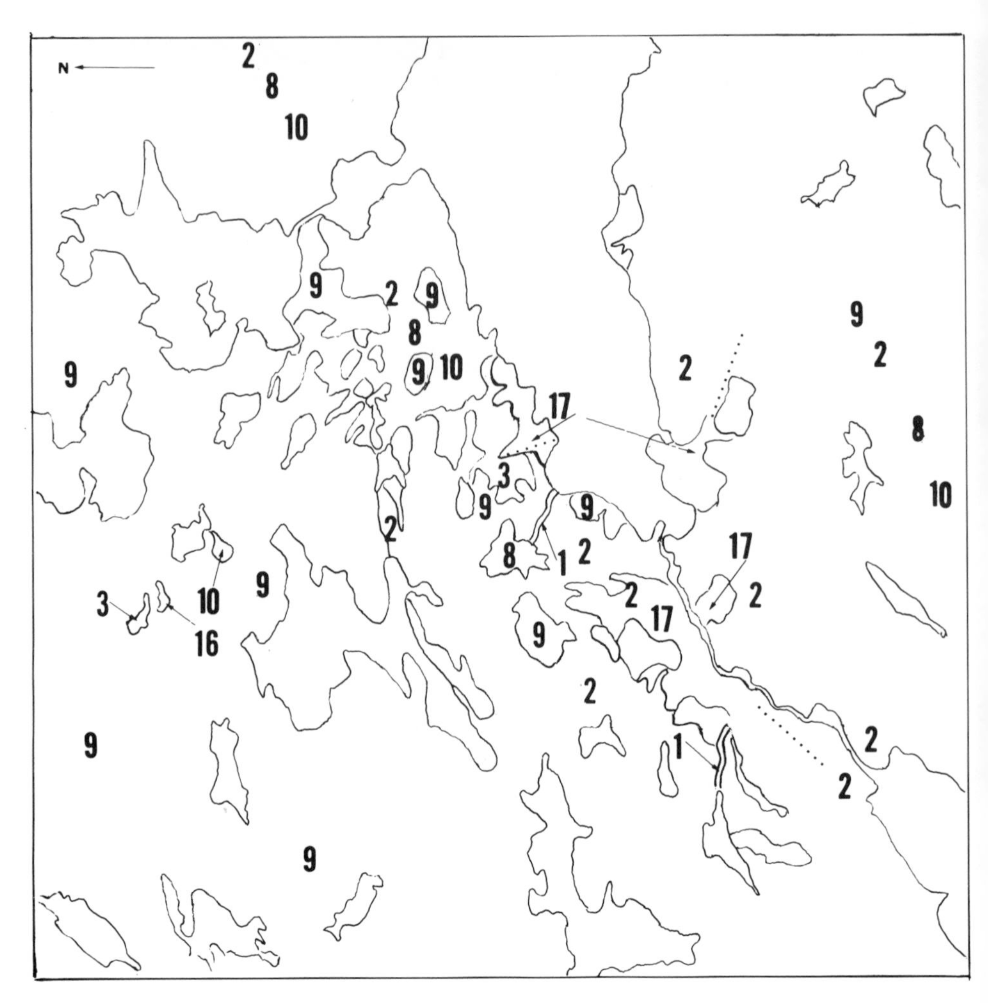

Figure 4.26. Key to airphoto of Winter Lake study area. For descriptions of features indicated by the identifying numbers, see Fig. 4.2.

Figure 4.27. Typical vegetational communities of the Winter Lake study area, showing black spruce stand in protected declivity along the shoreline and an upland rock field in the right foreground. Unidentified grave is shown in center of picture.

ularly in the Colville Lake area, in which trees are widely spaced and with an understory rich in lichens. The characteristics of the ecotone in this region may be the consequence of somewhat greater differences in elevation from place to place than is the case eastward, and also to a maritime influence originating in the Arctic Ocean. In the mountains to the west of the Mackenzie River, at the latitude of Inuvik, the timberline in found at elevations of about 1000 feet above sea level. Southward, approximately west of Norman Wells, the tree-line in the mountains is between 3000 and 4000 feet. In both areas, the ecotone is irregular and patchy due to the effects of slope, topography, and montane climatic influences.

Mackenzie Delta and Caribou Hills

Reindeer Station (68°42′N, 134°07′W)
Inuvik (68°21′N, 133°44′W)

Forest in the delta extends to approximately 68°45′N, farther than in either the uplands known as the Caribou Hills to the east (Fig. 4.21) or the foothills of the Richardson

Figure 4.28. Another view of the Winter Lake study area, showing black spruce community on upland ridge to the right in the distance. A rapids of the Snare River marks the outlet of Winter Lake, beyond the point to the right, and here John Franklin, John Richardson, George Back, Robert Hood, and others of Franklin's expedition established Fort Enterprise, where they wintered 1820–1821.

Mountains to the west. The Caribou Hills rise to elevations of 800 feet within a mile of the east bank of the Mackenzie River, and the montane foothills to the west attain even higher elevations in the same distance. Both east and west of the delta, the limit of forest is not as finely delineated on the land surface as might be inferred from the conventional maps because fingers of forest and tundra extend for many miles north and south of what can be considered the mean position of the forest border, a response to topography, winds, soil moisture, permafrost, and solar radiation.

Coring studies have shown that about 180 feet of stratified silts, fine sand, and organic material overlie 50 feet of dense silty clay in the delta, deposited by the present river in its post-glacial meanderings. The evidence appears to preclude extensive channel shifting during the past several hundred years. In the northeastern lowlands of the delta area, pingoes are the most striking relief features; nearly 1400 of these conical hills have been counted in the area east of the present delta and north of the Eskimo Lakes (Johnston and Brown, 1965; Mackay, 1963; Müller, 1962).

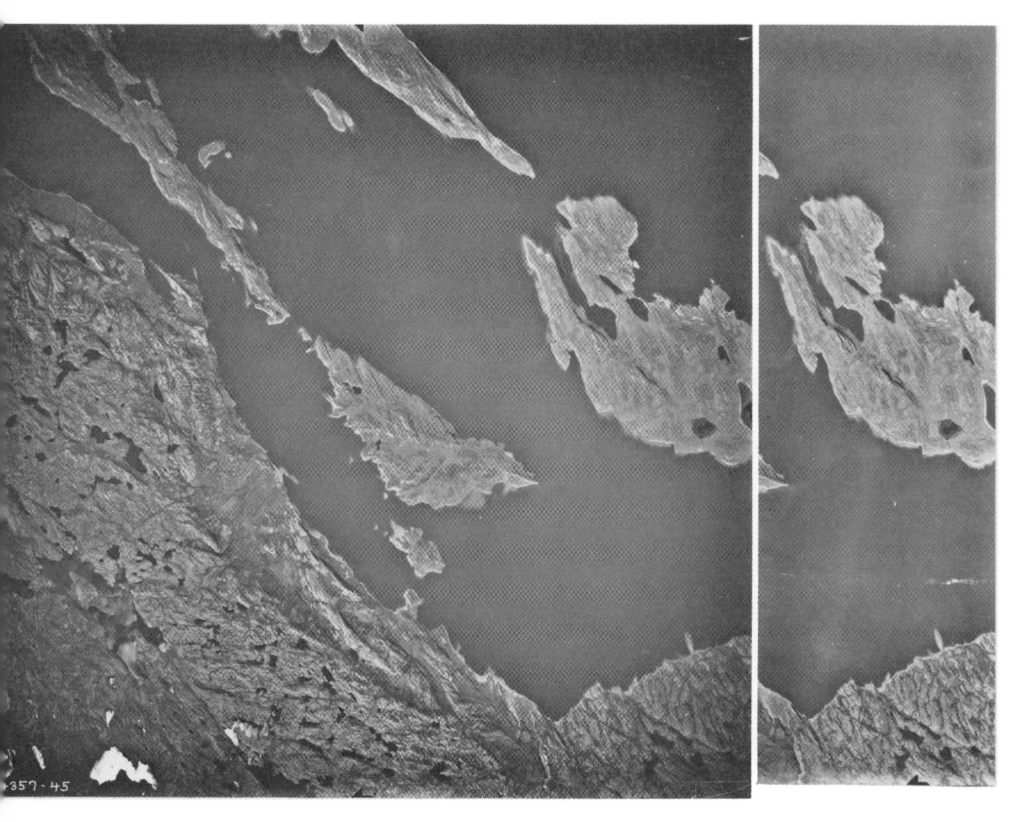

Figure 4.29. Airphoto of the Fort Reliance study area (A 14357-45).

Southward from tundra in the Caribou Hills, the vegetational sequence is, in general, from tundra communities with some willow and ground birch to increasing amounts of the latter, then finally to open woodland and ultimately to continuous woodland. At Reindeer Station, trees are confined to the shoreline bluffs and the forest cover in the delta is relatively thin, with trees lining the channel banks and intermixed with extensive areas of willows, alder, sedges, and muddy lakes. In the delta, white spruce is the more common of the species, often with a dense understory of willows and alders.

The Inuvik area is quite heavily forested, with small but not markedly dwarfed or decumbent trees. Black spruce is dominant in the treed muskegs, with trees averaging 30 or more feet apart and often less than about 4 inches in diameter at breast height. Tamarack, however, reaches fairly large sizes in the muskeg forests, although it is in much lower densities than is the black spruce. White spruce is confined largely to uplands, ridges, and stream banks, and grows to large sizes on such sites. Permafrost is very near the surface in treed muskeg, even toward the end of summer; it ranged between 10–12 inches below the surface of the treed muskeg toward the end of July. Permafrost at Arctic Red River is about 350 feet thick, and numerous other observa-

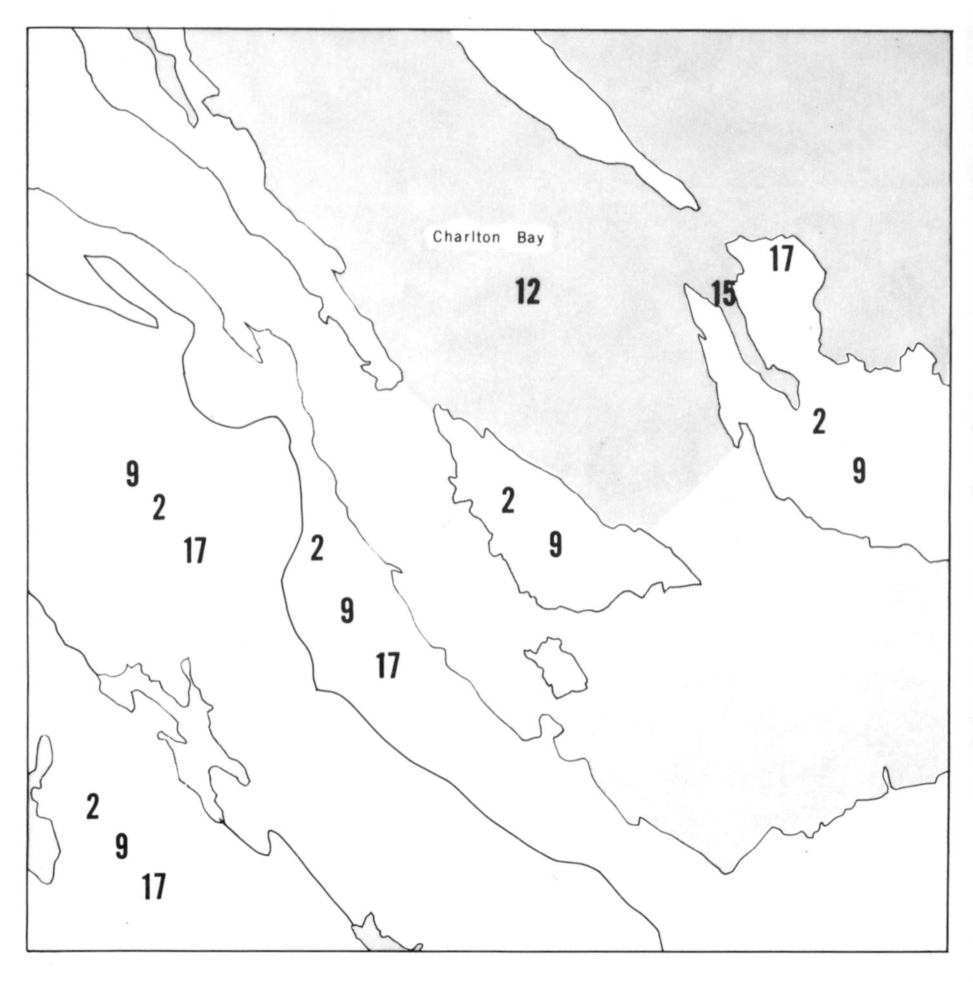

Figure 4.30. Key to airphoto of the Fort Reliance study area. For descriptions of features indicated by the identifying numbers, see Fig. 4.2.

Figure 4.31. Rocky uplands border much of the basin of Great Slave Lake; this view is from the east side of Charlton Bay at the extreme eastern end of the lake. Spruce forest covers much of the land in the area, although the forest grades into tundra some 30 miles north of this point.

tions on permafrost characteristics in the area are given by MacKay (1967). A general description of the vegetation in the Inuvik and Reindeer Station areas is now available in the literature; these deal specifically with the mosses (Holmen and Scotter, 1971) and with the forests (Zoltai, 1975).

Extensive fires have markedly influenced the vegetation in the immediate vicinity of Inuvik in recent years (see Larsen, 1972a). Due to this fact, and due also to the extensive settlement in the area, the photographic coverage of the Inuvik area has been omitted.

Colville Lake Area

(67°15′N, 120°00′W)

Major communities occupying the largest proportion of the undisturbed land surface in the Colville Lake area are white spruce on the rolling well-drained uplands and black spruce in the lower areas. Disturbance by fire and to some extent by cutting for the vil-

Figure 4.32. Ancient beach ridges are now elevated above the level of Great Slave Lake by as much as 200 feet. This ridge is along the shore of Charlton Bay at the east end of the lake. The sand and fine gravel of the ridge is dominated by *Stereocaulon* and *Cladonia* (*Cladina*) lichen species.

lage at the south end of the lake has been extensive in the area studied, and for this reason the aerial photographic coverage and interpretation has been omitted here.

For the most part, the forest is open and park-like, and on undisturbed areas gives every appearance of having existed in its present condition for a long period of time. Notable among the more unusual features of the area is the relatively abundant moss mat of the palsa type, in which a thickening of the mat has evidently been accompanied by a marked reduction in the density of the black spruce, the dominant on such sites. On some of the better developed palsas, dead trees are occasionally to be found in some abundance, and it appears they have suffered disruption of root systems by frost action

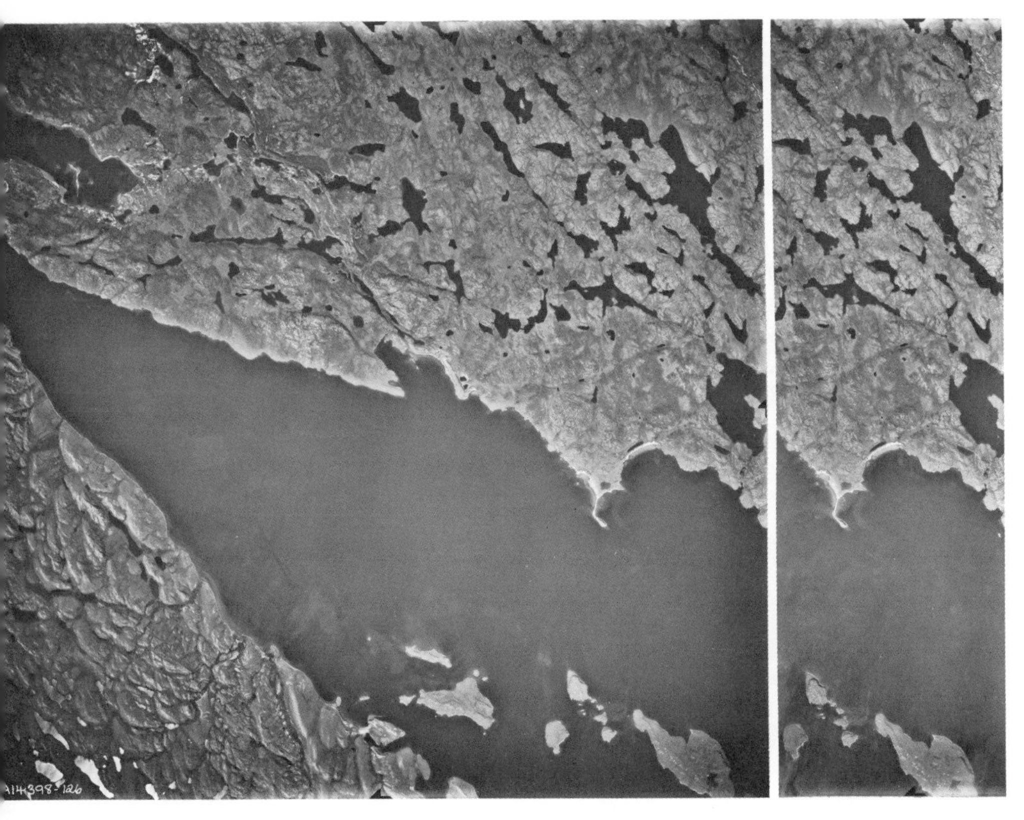

Figure 4.33. Airphoto of the south Artillery Lake study area (A 14398-126).

and by a general elevation of the surface of the moss-covered peat. This latter trend would ultimately result in encasing of root systems by frozen moss peat at a depth that would be in the zone of permanent permafrost. At times, the deep accumulation of peat has occurred at the edge of the lake, resulting in an easily accessible profile of the peat accumulation (Fig. 4.22).

Central Forest Border Study Areas

Cartridge Mountains:
"East Keller" Lake (63°55′N, 120°28′W)
  Airphoto No. A 12700–73 (Figs. 4.23, 4.24)

The study area is around the north end of an apparently unnamed lake, called "East Keller" for purposes of identification here. It is located between Keller Lake and Lac Tache, between 60 and 70 miles south of McVicar Arm of Great Bear Lake, in the northern interior plains, and in roughly the center of the District of Mackenzie. Although the

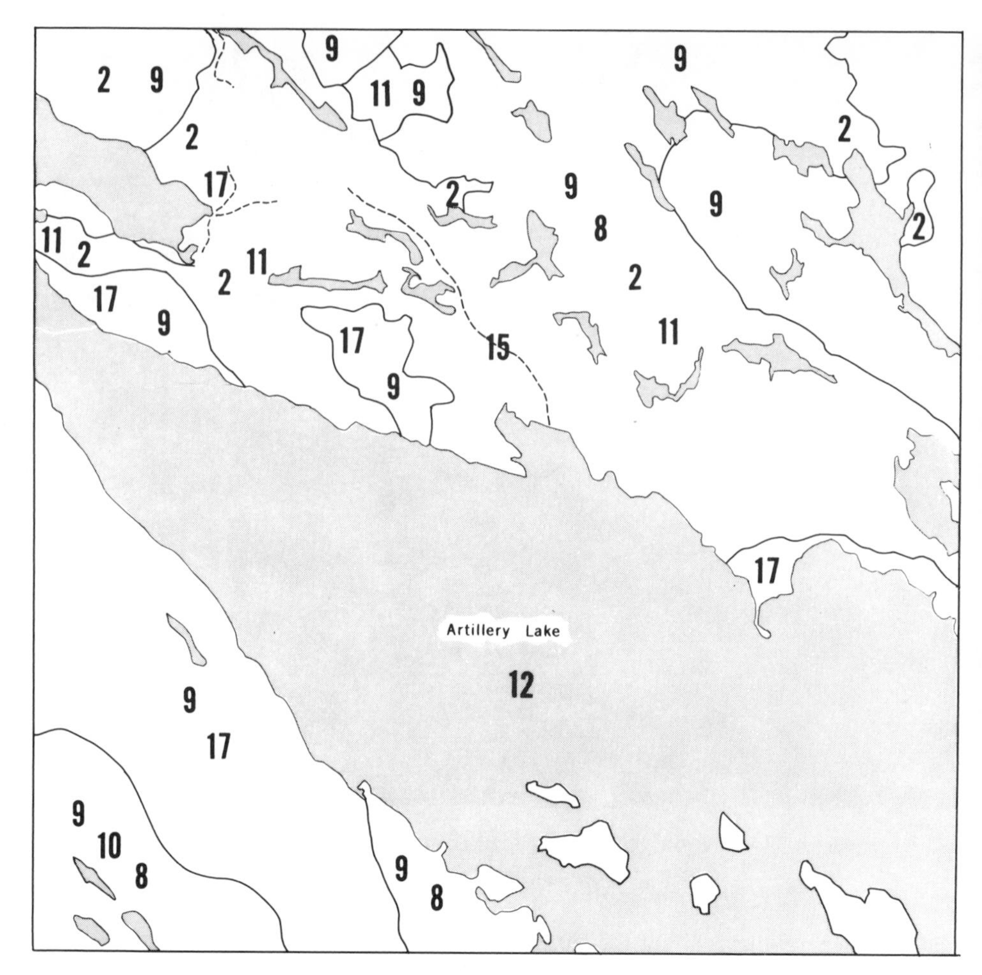

Figure 4.34. Key to airphoto of the south Artillery Lake study area. For description of features indicated by the identifying numbers, see Fig. 4.2.

area is many miles from the northern forest border, this is due largely to the effect of Great Bear Lake, which supports tree growth around its northern shore, probably at least in part as a consequence of the environmental effects of the lake itself. The area is within the forest zone classified as the northwestern transitional boreal forest (Rowe, 1972).

In this area, there is generally low relief, with a relatively shallow depth of glacial till over Palaeozoic and Cretaceous sedimentary bedrock. There are many lakes, with eskers, till ridges, drumlins, and occasional rocky outcrops or knobs of one kind or another. In the area, jack pine reaches near the northern limit of its range. There are stands of white spruce although black spruce appears to be the more abundant species; white birch and tamarack are present although not in great abundance. The stands are

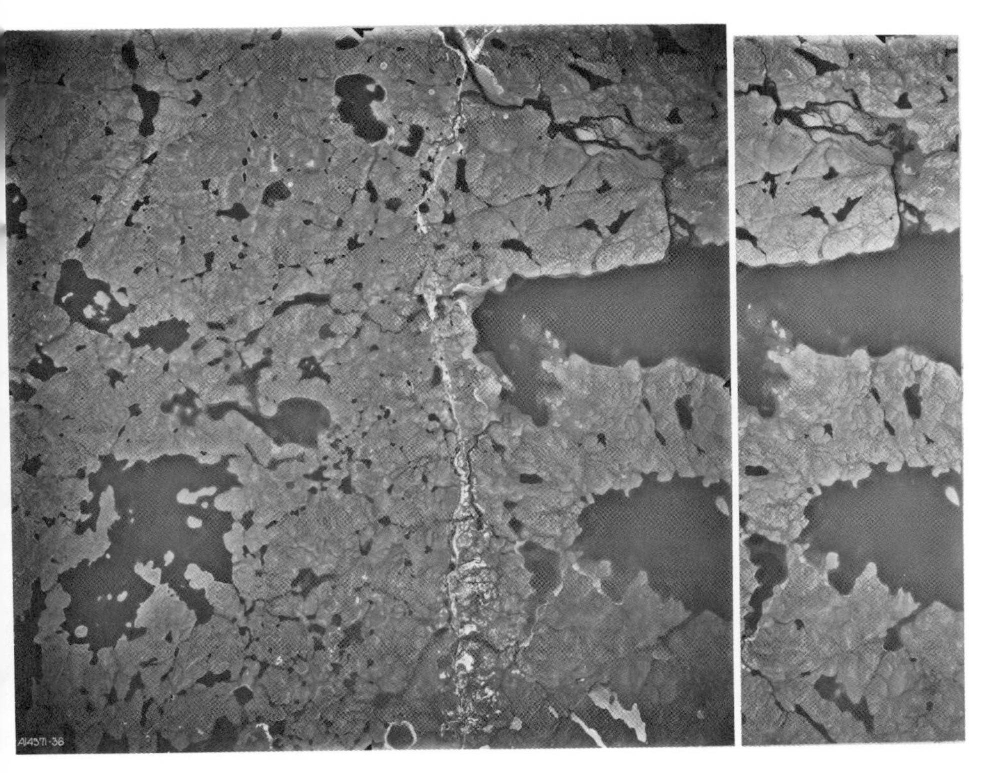

Figure 4.35. Airphoto of north Artillery Lake study area (A 14371-36).

open and park-like with an abundant ground cover of lichens and common shrubs, *Ledum* and *Betula*, attaining great density.

The area of the photograph is densely forested, the relatively dense black spruce stands covering both uplands and lowlands of gently rolling glacial drift. A white spruce woodland is found more rarely, almost invariably on the level areas of sandy outwash. The lowland stands of black spruce are quite dense in places, grading perceptibly into the wetter lowland treed muskegs with widely scattered dwarfed black spruce. The jack pine is rare and found on occasional high sandy uplands.

Forested areas shown in the photograph are principally of the black spruce upland type with the white spruce woodland confined to the sandy shorelines of the lakes. Light areas in contrast to black spruce forest are sparsely treed open muskeg, with a high density of ground lichens, principally *Cladonia* and *Cladina* species, and mosses.

Winter Lake Area (64°30′N, 113°00′W)
Airphoto No. A 14062–70 (Figs. 4.25, 4.26)

Winter Lake is on the Snare River, roughly 40 miles west of the Coppermine River at this latitude, and 50 miles northwest of MacKay Lake. It is about 250 miles due east of

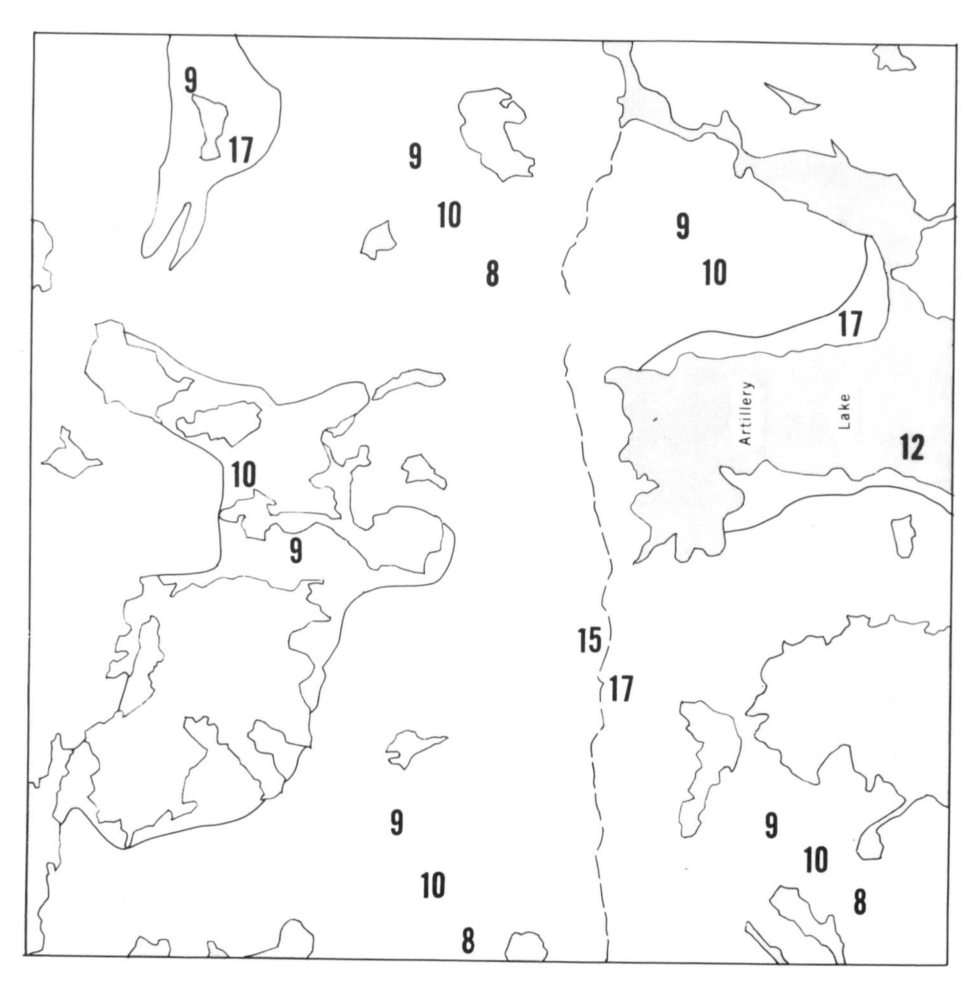

Figure 4.36. Key to airphoto of north Artillery Lake study area. For description of features indicated by the identifying numbers, see Fig. 4.2.

the "East Keller" Lake area described above and is very near, if not actually constituting, the northern limit of spruce at this longitude. The topography is rolling, with deep deposits of till; outcrops occur only rarely. The summits of the highest hills rise some 300–400 feet above the lake. The relief in the area may account for the extension of spruce groves into tundra within the valley of the Snare River system, northeast of the general regional position of the edge of spruce forest. Black spruce stands, composed of dwarfed spruce attaining a maximum height of perhaps 10–12 feet, are found in favorable lowland areas and protected declivities (Figs. 4.27, 4.28). The rough terrain also very likely accounts for sites in which deep snow accumulates in winter, affording protection for the spruce. Although the boles of some individual spruce rise above the general level of the terrain in some instances, snow abrasion and general desiccation is evident at the edges of ridge and esker summits (Fig. 4.28). In the immediate vicinity

Figure 4.37. View of the landscape at the south end of Artillery Lake; this is the Lockhart River at the point where it flows out of the lake. In this area, there are rather large stands of spruce in favorable sites but a few miles north the tundra becomes virtually continuous (Larsen, 1971).

of Winter Lake, spruce groves of more than an acre in extent are found virtually all of the way to the eastern end of the lake. Beyond this, to the northeast, however, tundra dominates, as it does to the southeast as far as Matthews, Courageous, and MacKay Lakes (64°05′N; 111°15′W), where spruce groves are again encountered, also reaching here their northeastern limit in the area. On the west side of Matthews Lake, black spruce is fairly common in sheltered valleys, and may at times attain a height of 20 feet although most specimens are severely twisted and stunted. It is of interest that here there are numerous dead stumps in sheltered valleys, some much larger than the living trees. On the east side of Matthews Lake, the spruce trees are little more than shrubs, rarely over two feet high, growing in compact clumps along the edges of lakes and ponds (Cody and Chillcott, 1955). The latter authors state that a geological fault runs the length of Matthews Lake, with basic volcanic granite on the west side of the lake and slaty sedimentary rock with some intrusions of granite and acid volcanic rock prevalent on the east.

A rather extensive stand of white spruce is to be found along a creek entering Winter Lake on the north shore, about 1.5 miles from the west end of the lake. The trees and willows in dense stands are rooted between heavy boulders forming the bed and shores

Figure 4.38. View of the landscape at the south end of Artillery Lake, with a few scattered spruce around the shoreline. The view is toward the northeast.

of the stream. Trees with a diameter of 8–10 inches at the base are common. Elsewhere, white spruce is confined to sandy alluvial or glacial deposits, along with black spruce in stands with widely spaced individual trees. Black spruce occupies other treed areas in virtually pure stand. Patches of treed muskeg are found in low relatively flat areas and on hillside slopes where snow accumulates to sufficient depths for protection of the trees during winter. In areas occupied by black spruce and with a ground cover of mosses and herbaceous species rooted in deep peat, the upper surface of permafrost on July 22 lay at depths averaging about 9 inches. In tussock muskeg, it was found at depths averaging 14–18 inches.

Rarely do black spruce attain a basal area at breast height of more than 12 square inches, and the trees are rarely more than 10 feet in height. Decumbent individuals are found on windswept areas. Occasional survival of trees on esker summits or hills and ridges of sandy till can probably be explained by the rapid thawing of surface soil horizons early in spring, affording a meltwater supply for roots at a time when desiccation of leaf tissues (winter kill) would appear to be a hazard. The survival of white spruce in the rocky shorelines of the creek mentioned above can probably be explained by the fact that running water is available early in spring to afford roots an opportunity

Figure 4.39. A grove of spruce on Crystal Island in Artillery Lake. The island is largely of dolomitic limestone, and supports tree growth in protected sites where snow accumulates to considerable depth in winter (see Larsen, 1971).

to avoid desiccation (see description of similar site of white spruce stand at Ennadai Lake; Larsen, 1965, and below). The forest at Winter Lake shows no evidence of currently making an expansion into tundra. Many gnarled trees on summit areas appear to be relicts of some previous time when conditions permitted more extensive forest cover in the region. These trees appear not to be capable of regenerating even by layering, and where they have been cut by native hunters on lookout points (for fires) there now remain only dried stumps.

The slopes of the hills are covered over extensive areas by a growth of dwarf birch, waist-high and so dense as to make walking difficult. Just below these patches on the lower slopes are commonly found the areas of tussock muskeg, with moss hummocks colonized by heath species. Between hummocks is found a fairly dense growth of sedge species. The tussock muskeg community is well represented in the Winter Lake area, contrasting to its rarity or absence farther north, but not more than 10% of the total land area is occupied by this community. Flat areas with an accumulation of water at or very near the surface are occupied by meadows with *Eriophorum* and *Carex* species dominant.

Figure 4.40. Tundra in the vicinity of the Hanbury Portage on the east shore of Ptarmigan Lake, northeast of Artillery Lake.

The species in each community are few in number and confined almost entirely to members of that group identified elsewhere (Larsen, 1965, 1971, 1980) as those ubiquitous species found both north and south of the forest border but reaching highest frequencies in the forest/tundra ecotone. The area shows much evidence of former human occupancy in the form of tent-rings and birch bark canoes in final stages of deterioration. A late snow bank was visible from the air on July 26 at the northeast end of the lake about a half mile from the shoreline.

It should be mentioned that Winter Lake is the area where the Franklin expedition of 1819–1822 established their "Fort Enterprise," located on a hilltop exposed to the full force of the winter winds but affording—unnecessarily—an excellent field of fire in all directions, as was the military custom in those days. A small plaque on the hilltop now commemorates the winters spent in that house, which "admitted the wind from every quarter."

The Fort Reliance Area (62°40′N; 109°07′W)
Airphoto No. A 14357–45 (Figs. 4.29, 4.30)

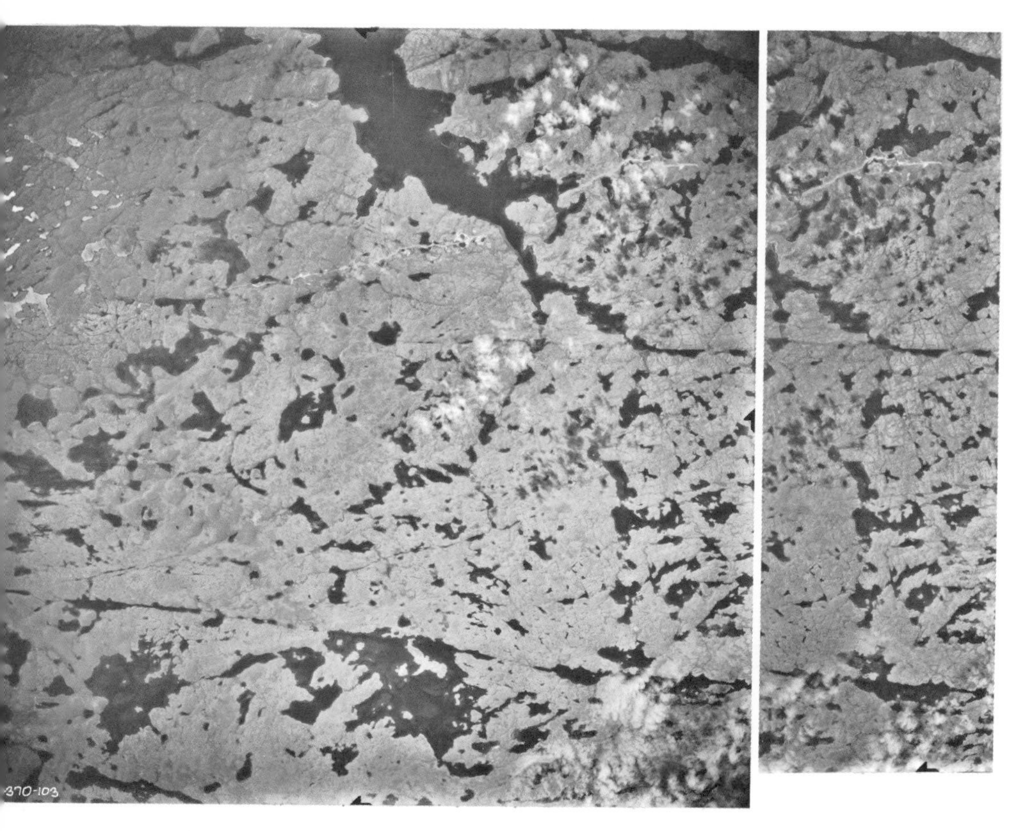

Figure 4.41. Airphoto of the Clinton-Colden Lake study area (A 14370-103).

The Fort Reliance area, located at the extreme east end of Great Slave Lake, is rugged along the lake shores, becoming less so as one moves progressively away from the lake (Fig. 4.31). The region has been described (Wright, 1952) as follows: "The granitic uplands bordering the lake basin present a monotonous succession of low rocky hills and ridges, with local relief rarely exceeding 250 feet . . . . The monotonous aspects of the bordering uplands contrasts sharply with the rugged and picturesque topography within the lake basin. There, vertical cliffs of diabase and limestone in places rise several hundred feet from the water, or form cappings over steep slopes of softer rocks, particularly shale . . . . Although glacial boulders are abundantly distributed over most of the area, thick morainal deposits are essentially restricted to the northwest part of the area, particularly south of Artillery Lake. Bouldery hills, 50 to 100 feet high and composed of unsorted, angular, granitic, and gneissic boulders and coarse gravel, are conspicuously well displayed . . . ."

The geological survey maps of the McLeod Bay area at the east end of Great Slave Lake reveal rather large areas of limestone, dolomite, slate, and shale on the south shore, and sandstone, quartzite, slate, and shale on Fairchild Point, which is the site of

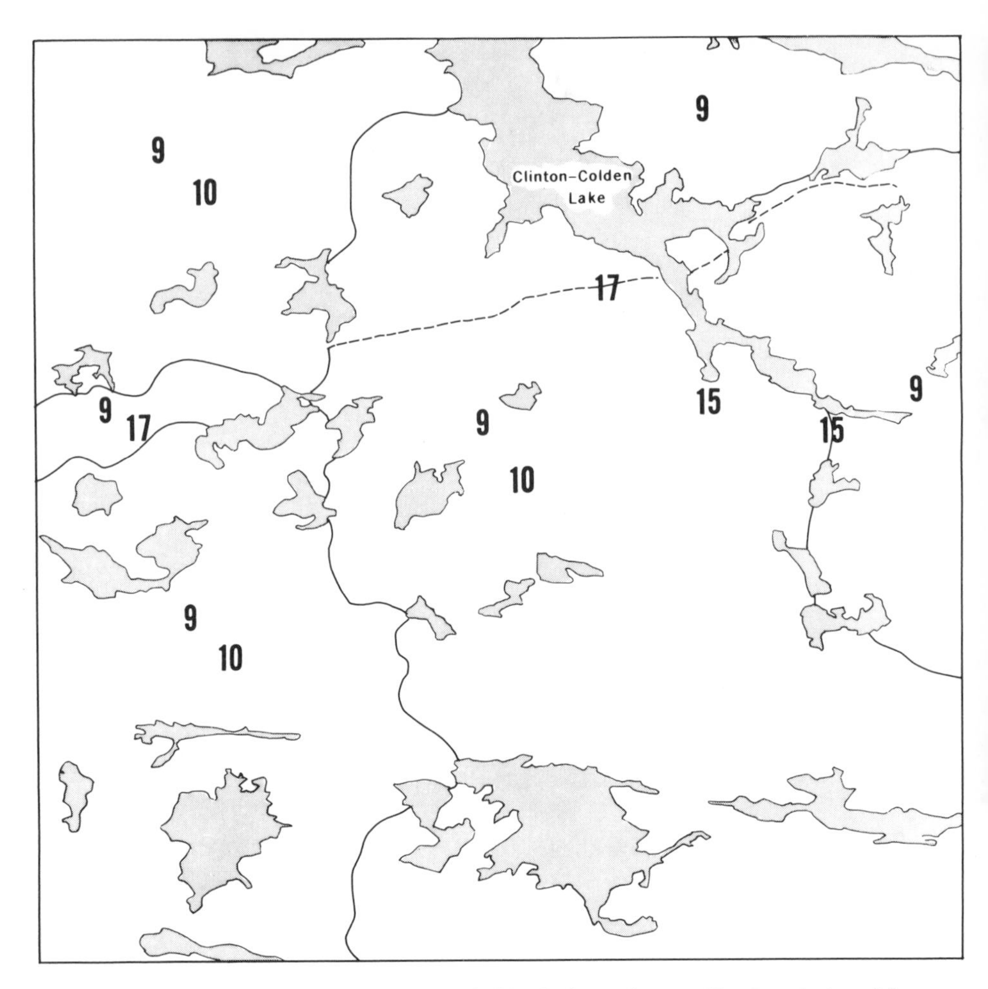

Figure 4.42. Key to airphoto of the Clinton-Colden Lake study area. For description of features indicated by the identifying numbers, see Fig. 4.2.

Fort Reliance. Vast tracts of land to the north are given over to granitic and gneissic rocks, with smaller areas of quartzite and greywacke (Wright, 1967). Ancient beach ridges are a a conspicuous feature of the terrain around the lake (Fig. 4.32), and the vegetation has been described (Lindsey, 1952).

Both black and white spruce are found in the area around Fort Reliance, the white spruce principally on upland soils derived from limestones and along the sandy shorelines, black spruce on both the forested lowlands and the slopes and summits of the hills. The understory vegetation on Fairchild Point is of particular interest since here are found such species as *Dryas integrifolia* and *D. octopetala*, as well as *Rhododendron lapponicum*, species ordinarily found only much farther north. For a more detailed account of the vegetation, see Larsen (1971) and below.

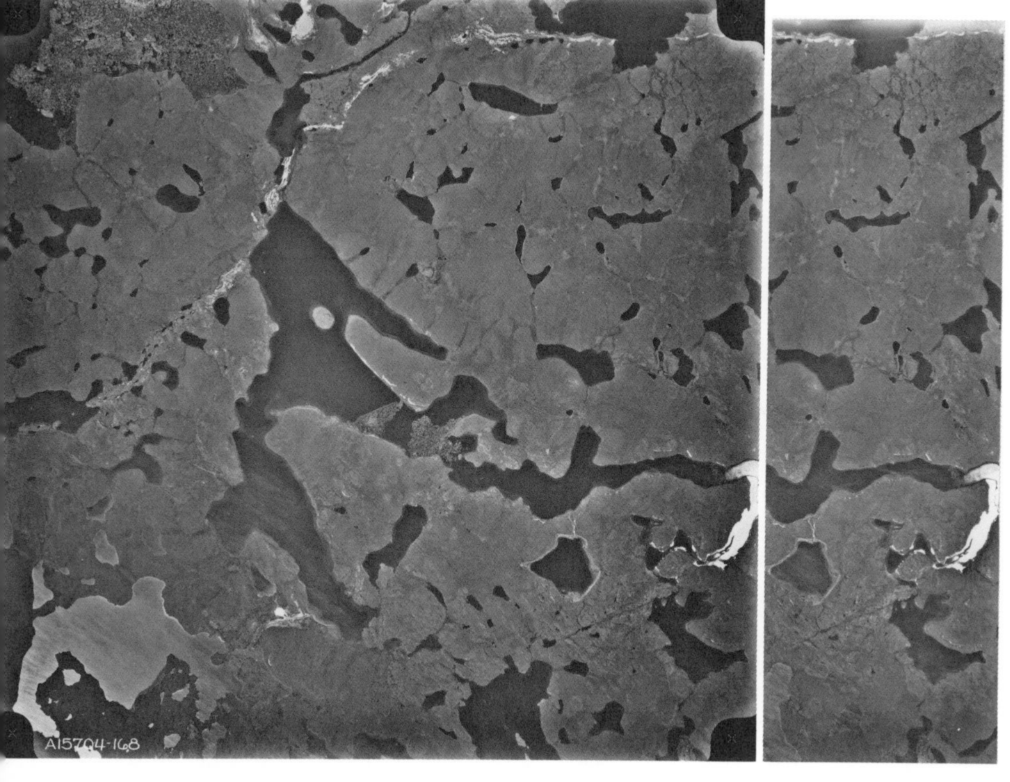

Figure 4.43. Airphoto of the study area at the north end of the north arm of Aylmer Lake (A 15704-168).

Artillery Lake	Airphoto Nos.
South End: (63°05′N: 108°05′W)	A 14398-126 (Figs. 4.33, 4.34)
North End: (63°27′N: 107°40′W)	A 14371-36 (Figs. 4.35, 4.36)

Pike's Portage Route between the east end of Great Slave Lake and the south end of Artillery Lake links a number of long narrow lakes lying along the McDonald Fault, north of which the parent material is principally gneissic and south of which it is largely quartzite or a gneissic, granitic complex. Along this portage, vegetation is principally black spruce forest once Great Slave Lake is some distance behind, grading perceptibly into tundra as one approaches Artillery Lake. A single stand of jack pine was observed by the author along this portage route, near the terminus at the east end of Great Slave Lake. From Artillery Lake to the southeast, tundra extends for many miles, tree-line dipping rather steeply southward from the south end of Artillery Lake (Figs. 4.37–4.39; see also Clarke, 1940).

The south end of Artillery Lake is surrounded by hills with deposits of drift in the depressions between, and as one travels north the deposits of glacial drift appear to

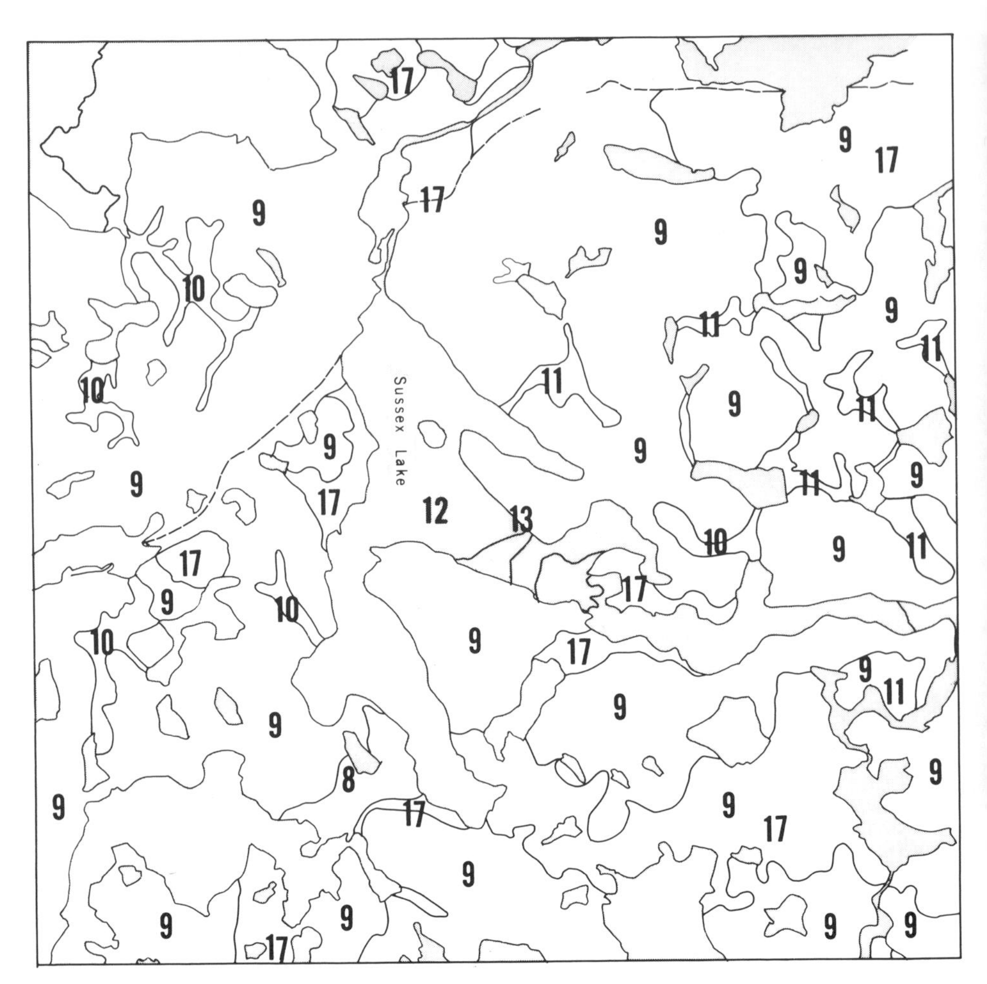

Figure 4.44. Key to airphoto of study area at the north end of the north arm of Aylmer Lake. For description of features indicated by the identifying numbers, see Fig. 4.2.

become deeper and bedrock exposures less frequent. Halfway up Artillery Lake the hills are rolling with gentle slopes, the surface material is gravel and sand intermixed with larger rocks of all shapes and sizes, and this drift is deep and nearly continuous over the surface of the earth. Sandy eskers and gravel deposits of one kind and another are frequent. Essentially this same surface geology is then found for a good distance north, around Ptarmigan, Clinton-Colden, and Alymer Lakes, as well as the region around the Hanbury Portage, all of which were the sites for the vegetational sampling described later.

Black spruce is found in large groves along the west shore and inland at the south end of Artillery Lake, becoming scattered and rare toward the north end. Small clumps of dwarfed black spruce are found nearly to the mouth of the Lockhart River along the

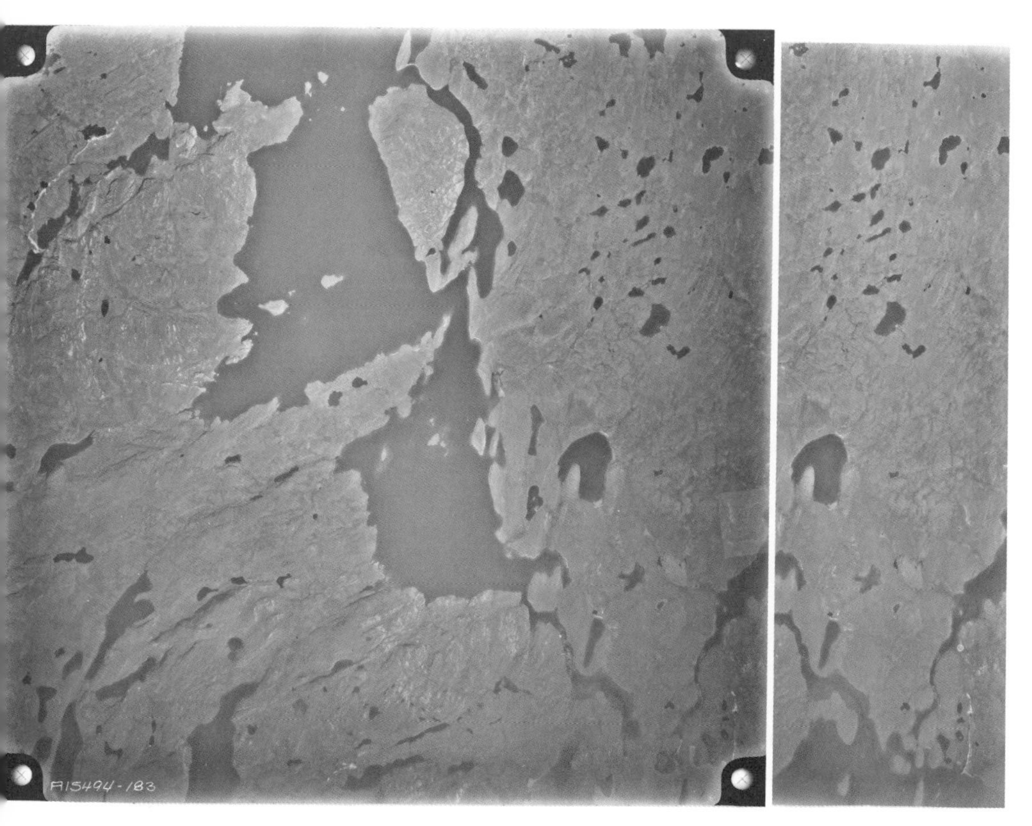

Figure 4.45. Airphoto of study are at the east side of the north arm of Aylmer Lake (A 15494-183).

east shore, although markedly more scattered toward the north. Black spruce is also found in protected sites along the esker which crosses the Lockhart River between Artillery and Ptarmigan Lakes; it then occurs with decreasing frequency northward. These small clumps of dwarfed spruce are located in sites where they are protected from prevailing winds and where snow accumulates to considerable depth in winter.

A lichen woodland forest is found at places along the portage route between Great Slave and Artillery Lakes, with widely spaced spruce trees and an understory dominated by *Cladonia* and *Stereocaulon* lichens, with denser aggregations of *Vaccinium-vitis idaea* and *Empetrum nigrum* directly beneath the individual trees. The substrate is primarily well-drained sand and gravel. The greater proportion of black spruce stands across the portage, however, occupy those areas where till and products of weathering of rocks have accumulated in the declivities between outcropping knolls and hills. The scattered dwarfed spruce seldom exceed 20 feet in height or 12 inches basal area (breast height).

From Artillery Lake northward, a portage route follows the Lockhart to Ptarmigan Lake, where it is possible to attain access to Clinton-Colden and Aylmer Lakes by

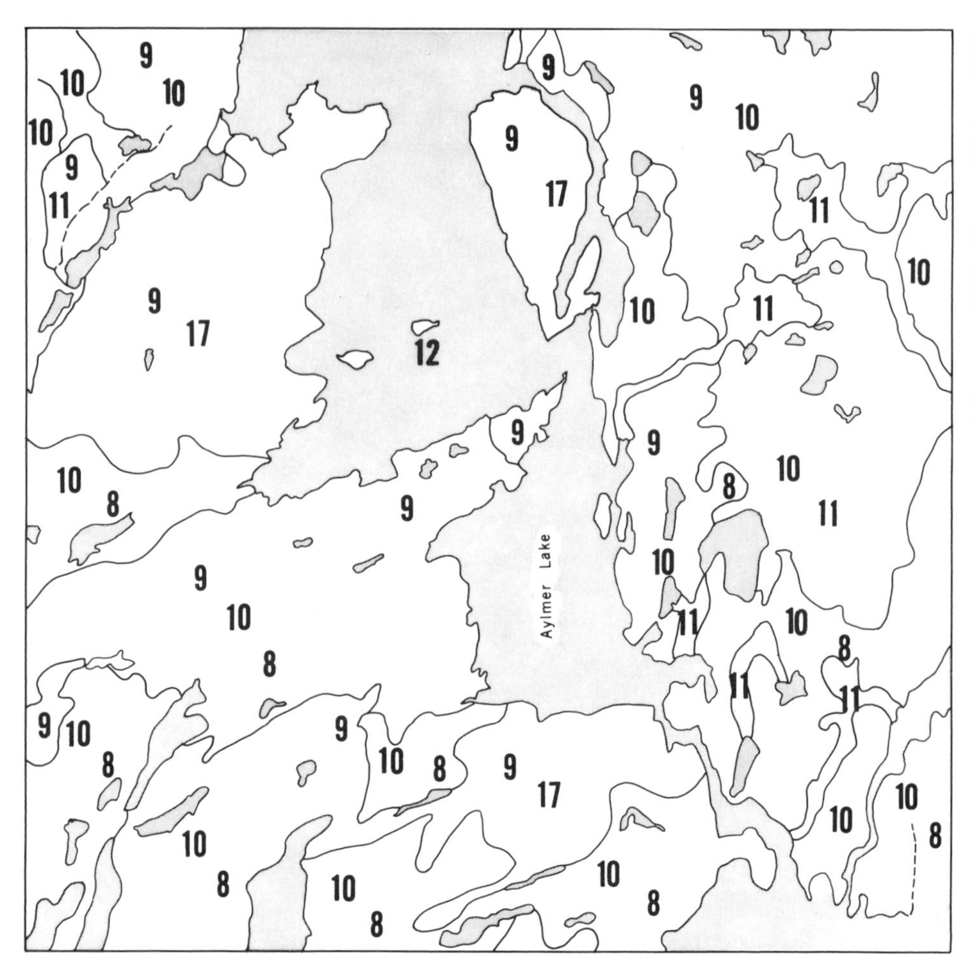

Figure 4.46. Key to airphoto of study area at the east side of the north arm of Aylmer Lake. For description of features indicated by the identifying numbers, see Fig. 4.2.

water. A traditional portage route east from Ptarmigan Lake to the Hanbury River follows along over relatively flat terrain (Fig. 4.40).

Clinton-Colden and Aylmer Lakes

Clinton-Colden (63°53′N, 106°57′W)
    Airphoto No. A 14370–103 (Figs. 4.41, 4.42)
Aylmer Lake (North end of north arm) (64°26′N, 108°20′W)
    Airphoto No. A 15704–168 (Figs. 4.43, 4.44)
Aylmer Lake (East side, north arm) (64°26′N, 108°20′W)
    Airphoto No. A 15494–183 (Figs. 4.45, 4.46)

Figure 4.47. View of a small clump of spruce in the area of the north arm of Clinton-Colden Lake, northernmost site at which spruce was found in the area (see Larsen, 1971).

The Lockart River joins the large lakes MacKay, Aylmer, Clinton-Colden, and Artillery into a chain, providing the traditional route for both winter and summer travel through the country north of the East Arm of Great Slave Lake. It was apparently first used as such by other than native hunters when George Back and his party traveled along it to the headwaters of what is now known as the Back River in their journey northward to the Arctic Ocean in 1834 (Back, 1836; King, 1836).

Descriptive notes accompanying a geological survey map of the Aylmer Lake area (Canadian Geological Survey Series Sheet 76C; see also Craig, 1964b, Wright, 1967) indicate that local relief may be as much as 250 feet above the lakes but commonly reaches no more than 50 or so feet. The general character of the surficial geology is summarized: "Glacial drift covers about 80 percent of the land area. It consists mostly of till, and commonly gives an indication of the bedrock .... Drumlins and drumlinoid features are common .... Well-scoured outcrops, eskers, and scattered sand and gravel deposits interspersed with numerous small rounded lakes mark the Pleistocene drainage routes."

Clumps of dwarfed black spruce are to be found at the eastern end of Clinton-Colden and the western end of Aylmer Lakes but are apparently absent, or at least insignificant,

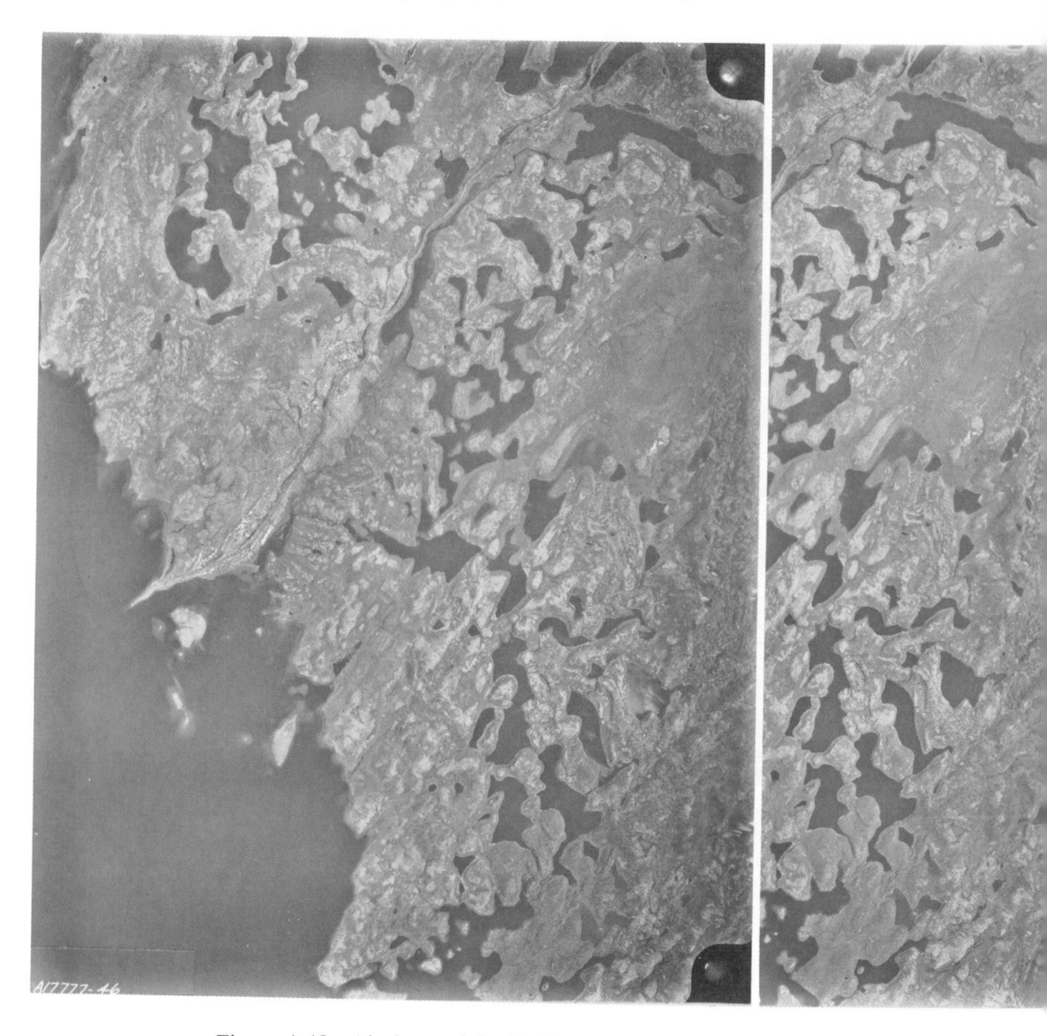

Figure 4.48. Airphoto of the Kasba Lake study area (A 17777-46).

in the area between. A low-level reconnaissance flight to perhaps 100 miles north of Clinton-Colden revealed the area to be uniformly barren of trees, with rolling rocky hills covered with a sparse vegetation, much of it representative of rock field communities. From the west end of Aylmer Lake, the forest border trends roughly northwesterly, and Cody and Chillcott (1955; see above) describe black spruce as "fairly common in sheltered valleys" around Matthews Lake (which lies just north of the northwestern arm of MacKay Lake).

Indians who have trapped in the barrens (personal commun.) point out that for them it is essential to survival that they know the location of these small clumps of spruce. Only then can they dig into the snow and obtain firewood for tent stoves. When temperatures are far below zero the Indians traveling in these areas must obtain at least a few

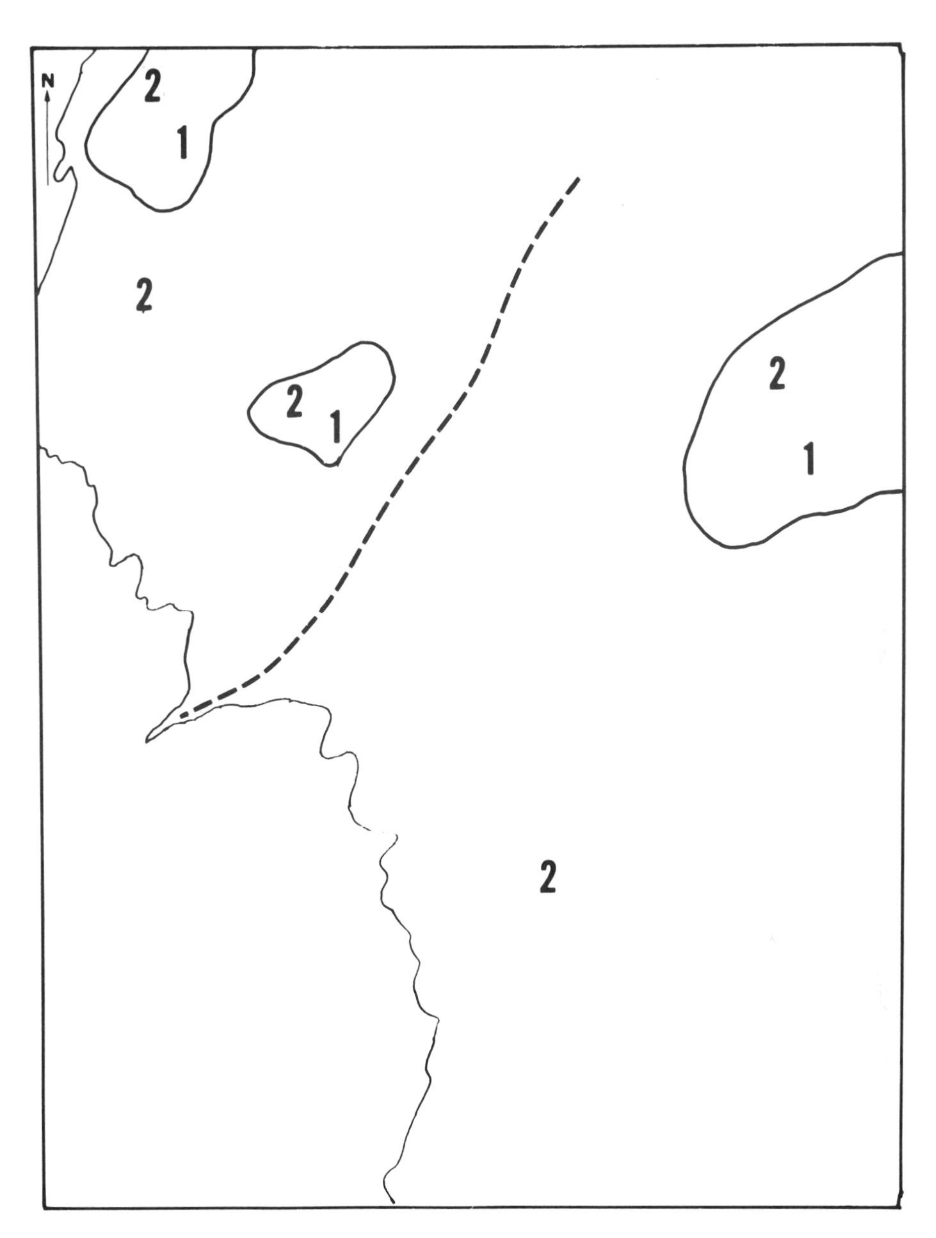

Figure 4.49. Key to airphoto of the Kasba Lake study area.

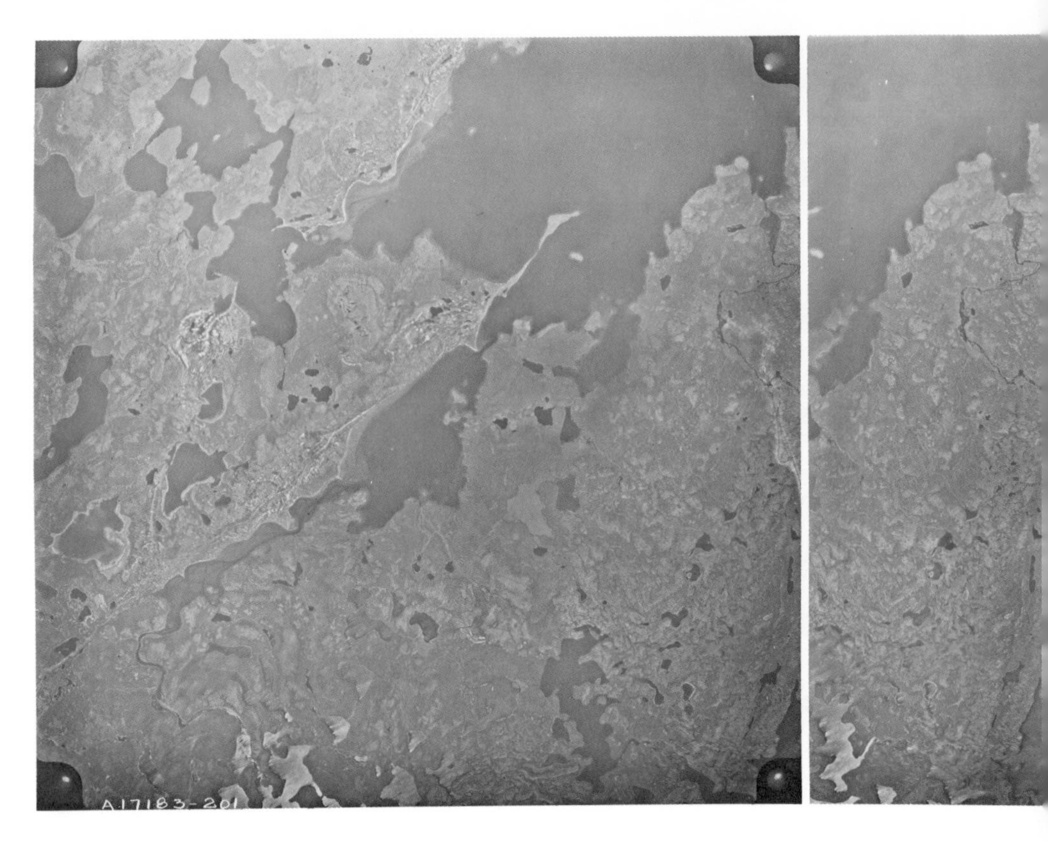

Figure 4.50. Airphoto of study area at the south end of Ennadai Lake (A 17183-201).

small pieces of wood which can be burned intermittently throughout the night. The spruce is rarely visible above the surface of the snow, however, and prior knowledge of the location of these small clumps must be acquired either from other trappers or during summer travel.

The vegetation throughout the area is rather uniformly dependent upon topographic position. Thus, rock field communities occupy the upper slopes and summits, tussock muskeg communities are found on intermediate and lower slopes, and meadows and muskegs are found on low-lying flat areas. The terrain is varied and often rugged in aspect when observed from the ground, but the landforms are repetitive over the entire area of photographic coverage and the total number of different features is relatively low. There is, however, a close juxtaposition of forms, and it is impractical to map each individually for the purposes of photointerpretation.

Spruce groves are in evidence at the south end of Artillery Lake (left in Photo A 14398-126). The north end of the lake is occupied by tundra (A 14371-36), and although spruce is present in very small groves these are most often too small to be visible on the photographs. Darker patches indicating a denser type of vegetation than that found over most of the area is usually an aggregation of willows and dwarfed birch with

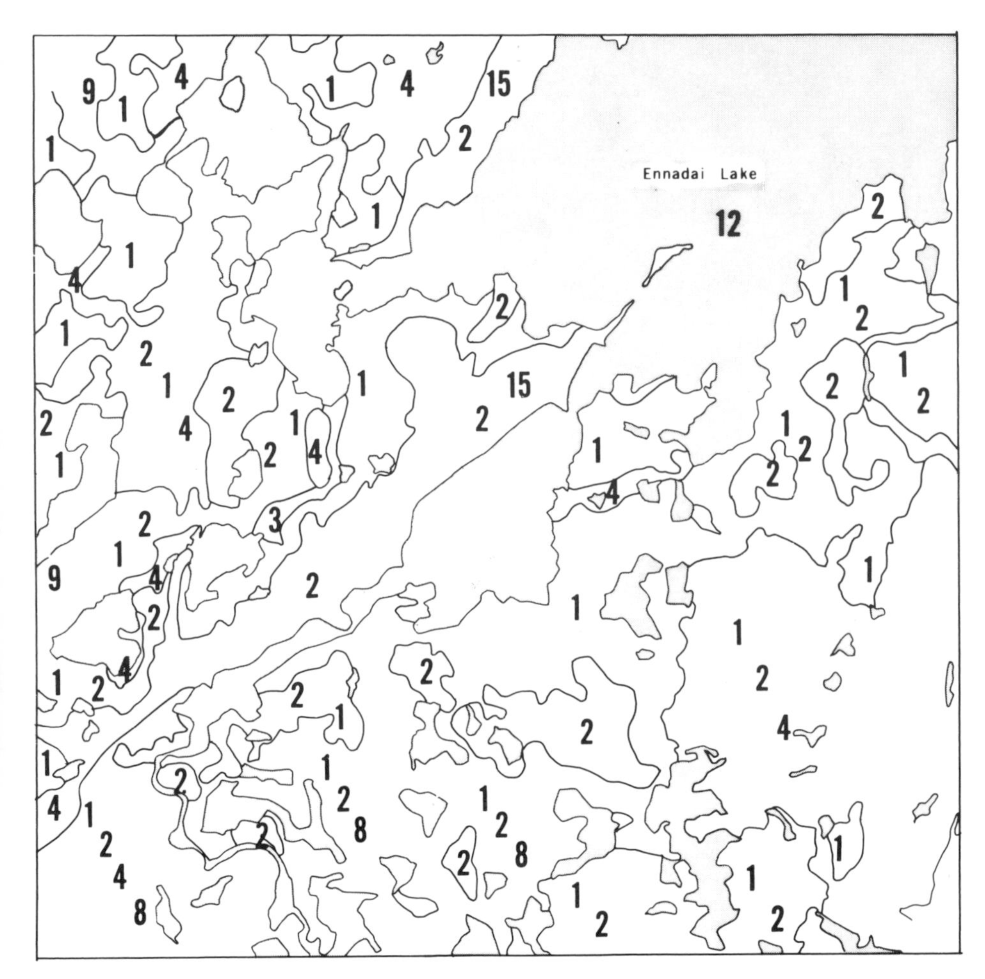

Figure 4.51. Key to airphoto of study area at the south end of Ennadai Lake. For description of features indicated by the identifying numbers, see Fig. 4.2.

an herb, moss, and lichen ground cover. Fort Reliance is indicated by the number 15 on Photo A 14357-45. Spruce forest is prevalent throughout the area around Fort Reliance wherever glacial till or weathering products of exposed rocks have provided a sufficient depth of soil for rooting.

The north end of Aylmer Lake (A 15704–168) is drift-covered and barren of trees. At the west end of the lake (A 15494–183), scoured bedrock, often with a thin covering of till in places, is more in evidence. Features of the Clinton-Colden area (A 14370–103) are similar to those of the Aylmer Lake area. The numbers 15 refer to small stands of extremely dwarfed black spruce in the drainage lines (see Larsen, 1971). Small clumps were found near the northeast end of Clinton-Colden (Fig. 4.47).

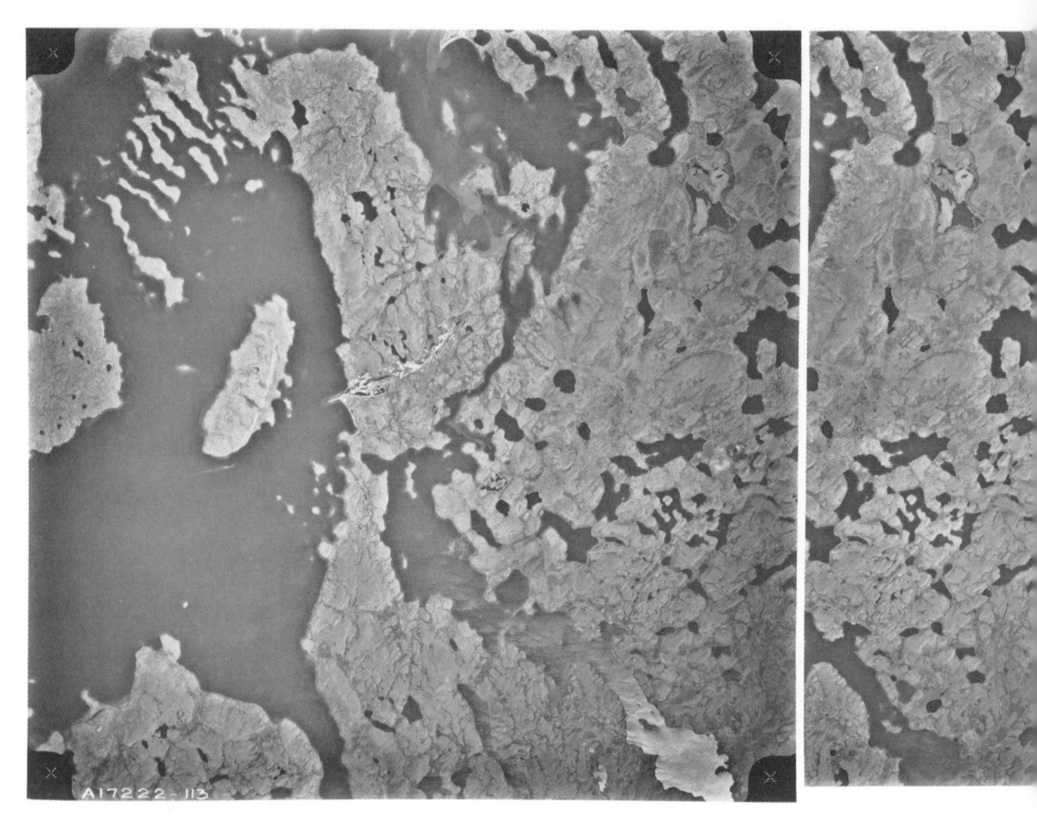

Figure 4.52. Airphoto of study area at the north end of Ennadai Lake (A 17222-113).

The Kasba Lake Area (60°13′N; 101°53′W)
 Airphoto No. A. 17777–46 (Figs. 4.48, 4.49)

The area is covered largely with spruce forest, either open lichen-woodland or a more closed upland forest. White spruce and black spruce are both present, the former in greater abundance on the eskers while the latter is the dominant species on the lower areas and presents a closed-canopy forest over much of the area. Familiar features of past glaciation are apparent and glacial till is present over most of the surface with outcropping bedrock not uncommon. The steep slopes indicate bedrock near or breaking the surface of the glacial till.

The Ennadai Lake Area

South End Ennadai Lake (60°43′N, 101°45′W)
 Airphoto No. A 17183–201 (Figs. 4.50, 4.51)
North End Ennadai Lake (61°10′N, 100°55′W)
 Airphoto N. A 17222–113 (Figs. 4.52, 4.53)
Dimma Lake 61°33′N, 100°38′W)
 Airphoto No. A 14852–39 (Figs. 4.54, 4.55)

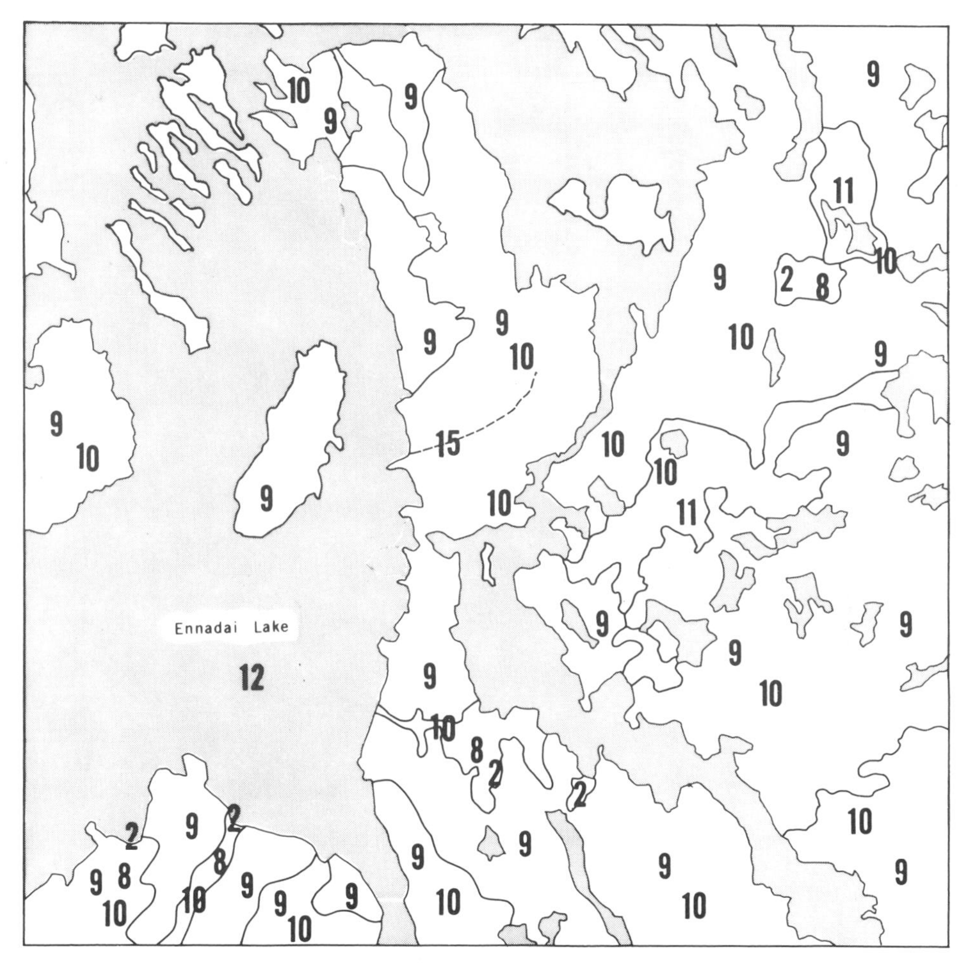

Figure 4.53. Key to airphoto of study area at the north end of Ennadai Lake. For description of features indicated by identifying numbers, see Fig. 4.2.

The physical features and vegetation of the Ennadai Lake area have been treated in detail in a previous monograph and will not be repeated in detail here (Larsen, 1965). The south end of Ennadai Lake is similar in surficial geology to the Kasba Lake area, and the vegetation of the Kasba Lake area and the south end of Ennadai Lake are similar. The land is virtually covered with spruce forest of varying density, from open lichen woodland to dense closed-canopy black spruce lowland forest. The north end of Ennadai Lake, however, is preponderantly tundra with only widely scattered small patches of dwarfed spruce. The remarkable abruptness of the forest/tundra transition in the Ennadai Lake area led to the detailed study of the forest border noted above (Larsen, 1965), and the similarity in surficial geology between the north and south ends of the Ennadai Lake area makes it apparent that environmental features other than substrate or soils must account for the transition from forest to tundra along the 45-mile

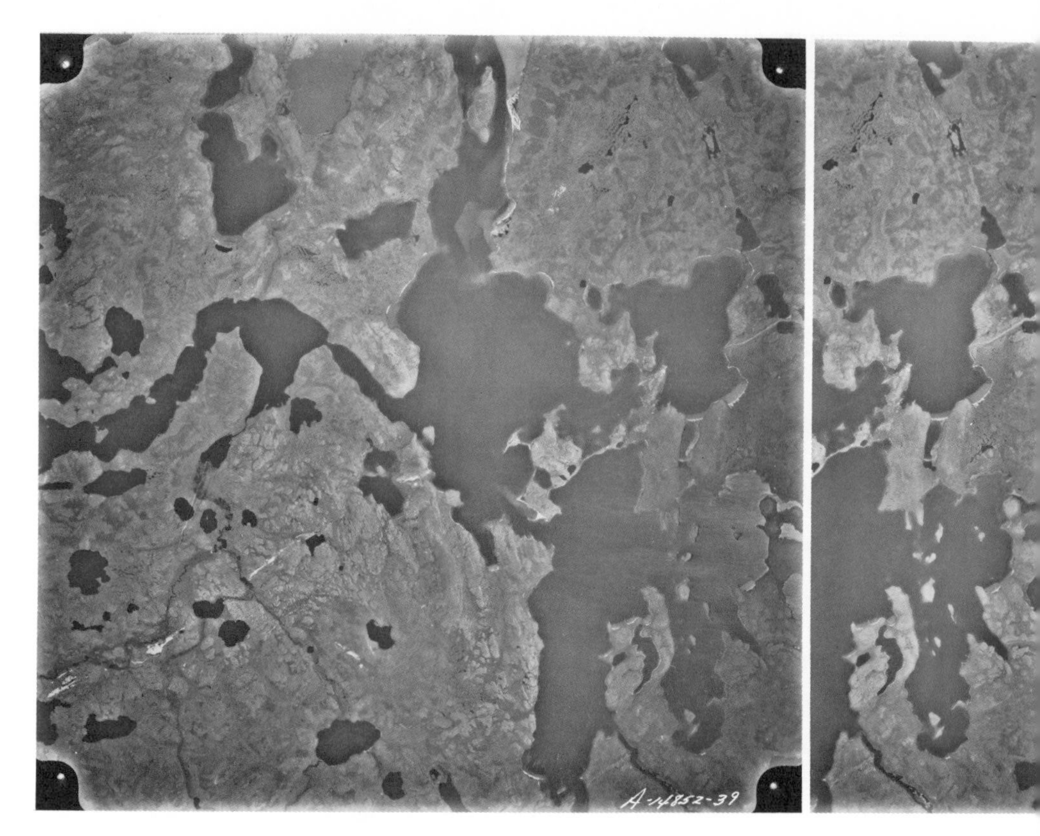

Figure 4.54. Airphoto of Dimma Lake (Kazan River) study area (A 14852-39).

extent of the lake. Climatic factors, it is believed, account for this transition, and these are discussed in other publications (Larsen, 1965, 1971a,b, 1971a,b,c, 1973, 1974, 1980).

Geology of the region is thoroughly discussed in Lee et al., (1957), Lee (1959), Lord (1953), and aspects of the botany of the general region are described by Baldwin (1951), Argus (1968), Maini (1966), Larsen (1965, 1971a,b,c, 1980), and Gubbe 1976). Representative vegetational sites in the region are illustrated in Figs. 4.56–4.60.

The transition from arctic tundra at the north end of Ennadai Lake to spruce forest at the south end is, thus, a most striking vegetational feature. At the south end of the lake, as at Kasba Lake, black spruce forest covers virtually the entire landscape, with the exception of the highest hills and low areas of bog and muskeg. Some 20 miles north, the proportion of spruce to tundra vegetation over the entire landscape is about 50:50, and at the north end of the lake, spruce is found only in favorable sites and is usually dwarfed or even decumbent. The summits of the slopes are usually covered with a vegetational community designed as rock field, the intermediate slopes are tussock muskeg, and the low level areas are occupied by low meadow communities. The communities at Dimma Lake, some 40 miles north of Ennadai Lake along the Kazan River,

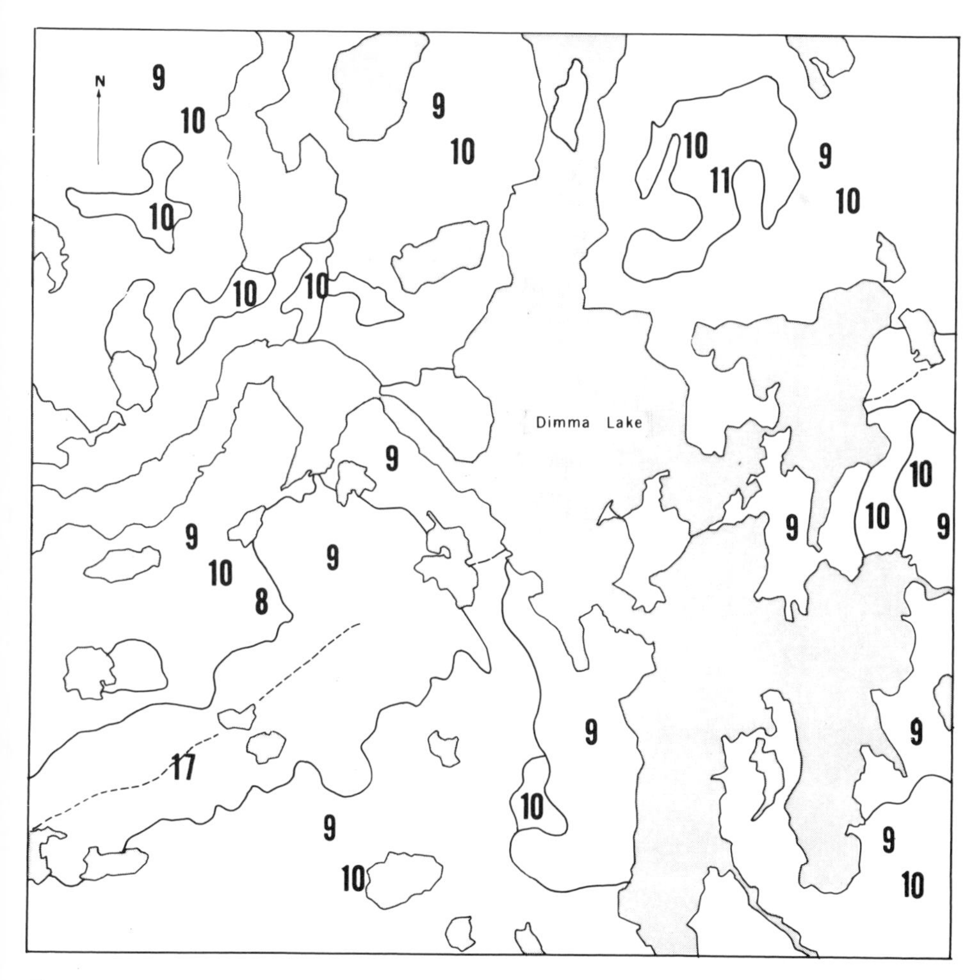

Figure 4.55. Key to airphoto of Dimma Lake (Kazan River) study area. For description of features indicated by identifying numbers, see Fig. 4.2.

are essentially the same as those at the north end of Ennadai Lake. Examples of these communities are shown in Figs. 4.56–4.60.

Dubawnt Lake Area

(64°01′N; 99°25′W)

Airphoto No. A 15059-20 (Figs. 4.61, 4.62)

Dubawnt Lake lies far into the interior of the barren lands. It constitutes the northern edge of the range of spruce in the region, with small clumps of very dwarfed trees scattered at wide intervals over the landscape. The terrain around Dubawnt and for many miles south is predominantly tundra, picturesque with rolling hills and a vast multitude of small lakes and streams running with clear cold water.

Figure 4.56. Tree-line on hilltop in Ennadai Lake study area. This view is approximately 10 miles north of the south end of the lake. The view, toward the northwest, shows a portion of the lake in the background.

In his description of the region, Tyrrell, the first European to descend the Dubawnt, noted that spruce are to be found along the river before it nears Dubawnt Lake, and he saw both black and white spruce in scattered clumps as far north as the small lakes (62°09′N) just south of Dubawnt. On Carey Lake, he writes, " . . . our camp is a hundred yards from the lake, near the edge of a bog, with a scattered grove of larch and small black spruce just behind us" (Tyrrell, 1896).

At Nicholson Lake, the last lake encountered before entering Dubawnt, Tyrrell found the shores to be "almost everywhere sloping and grassy, although at its north end are several small groves of spruce and larch and a few dead trunks are standing on the western shore." About 10 miles north of Nicholson Lake, the Dubawnt River flows through a valley that in places is some 200 feet deep and, here, " . . . on the steep hill-sides were some small groves of white spruce, the last that we were to see that summer, while the little patches of snow here and there in every direction would have kept us reminded that we had reached a sub-arctic climate, if the almost constant cold rain and wind had not made us thoroughly alive to the fact. On the hillsides, arctic hares were seen for the first time." The date was August 5. Always the careful observer, Tyrrell found the water temperature in the river to be 47°F.

Figure 4.57. A view from the shoreline in a small bay near the north end of Ennadai Lake; emergents along shoreline in the foreground, a small grove of spruce in the background.

Although this was the last of the white spruce to be seen by Tyrrell and his party, one grove of black spruce remained to be encountered. "On the north bank of the river," he writes, "half way between the above lake and Dubawnt Lake, is the last grove of black spruce on the river, where the trees are so stunted that they are not as high as one's head."

Tonal contrast on the Dubawnt area aerial photographs is minimal, making identification of contrasting vegetational communities somewhat difficult. Most of the area is covered with a vegetational community intermediate between rock field and tussock muskeg, with the former more pronounced on the sharper hill summits and the latter prevalent over larger flat areas. Darker areas are often tussock muskeg with a prominent component of low shrubs, and the lighter areas are tussock muskeg without a large frequency of shrubs or, on higher terrain, rock fields.

It should be noted that white spruce is present along the southern shoreline of Dubawnt Lake in small clumps of dwarfed individuals and black spruce is to be found in scattered clumps in protected sites inland. As pointed out in earlier monographs, there is every appearance that these are relict clones from a time some few centuries ago when the entire region was forested. If the climate were to ameliorate sufficiently to allow spruce to do so, it is apparent that a viable seed source is present for foresting

Figure 4.58. View of a rock field along a broad expanse of upland in the vicinity of the north end of Ennadai Lake. This view is a few miles (across the lake) from the site depicted in Fig. 4.57, hence very nearly the same latitude.

the entire area with both black and white spruce, but obviously this has not happened within recent time.

The area apparently has been heavily scoured by glacial ice and is covered over most of the area with at least a thin veneer of glacial till. A photograph north of Dubawnt Lake along the Dubawnt River and showing a portion of Wharton Lake (Photo A 14767-16, 62°45′N; 101°30′W, not shown) contains a large well-defined esker and a distinct pattern of ribbed recessional moraines. Maximum relief to the esker summit from the lake surface is 220 feet; esker sides approach 30% slope.

No spruce are to be found at the north end of Dubawnt Lake, and the area from here northward to the arctic coast is tundra barren of trees. To the east, however, at a few points in the Yathkyed Lake area, small groves of dwarfed spruce are to be found. These must be the northernmost extension of spruce in this area as a consequence of the fact that Baker Lake Eskimos were, as reported by Rasmussen, accustomed to traveling south to Yathkyed Lake to obtain supplies of wood for sleds and other tools in prehistoric time.

Figure 4.59. View of a rock field (foreground) and low meadow (background: near distant) with rocky uplands towards the distant horizon. This site is in the Dimma Lake study area.

Northern Quebec–Labrador

In the northern parts of Quebec as well as Labrador, the climatic influences, oceanic on the one side, continental modified by Hudson Bay and Hudson Strait on the other, result in a distinct combination of atmospheric conditions, and these evidently are a readily apparent factor involved in the proportionally greater extent of the forest-tundra and lichen woodland in this region compared to the west. As Barry (1967) points out, in the west the northern tree-line is quite abrupt but in northern Quebec and Labrador " . . . the forest boundary is replaced by an extensive forest-tundra ecotone and the treeline for individual species extends almost to Ungava Bay . . . " This contrasting width of the forest-tundra ecotone between west and east is clearly seen in Rowe's (1972) delineation of the vegetational zones in Canada, and the relationship of the northern forest boundary to regional climate in North America is discussed in Bryson (1965), Larsen (1965, 1972a, 1974, 1980), Barry (1967), Krebs and Barry 1970), Hare and Ritchie (1972), Hare and Barry (1974), Larsen and Barry (1974), Rencz and Auclair (1978), and others.

Figure 4.60. View of rocky ridge in foreground and low meadow to the right in the Dimma Lake (Kazan River) area.

Both black and white spruce are present in the lichen woodland, with the latter often predominant, but neither can scarcely be considered dominant species in the community since the individual trees are so widely spaced. The dominant in terms of ground cover is *Cladonia alpestris* (*Cladina stellaris*), but shrubs are also abundant, with *Ledum groenlandicum*, *Betula glandulosa*, *Vaccinium uliginosum*, *V. angustifolium*, and *Cornus canadensis* all represented.

Collections and vegetational sampling and descriptions of various areas in northern Quebec and Labrador are now available, and these are summarized (with abundant references) in publications by Makinen and Kallio (1980), Morisset and Payette (1980, 1983), and Morisset et al., (1983). There are a somewhat bewildering number of community types described in the literature, but white and black spruce are accorded dominant positions in all studies of the forests of the northern areas. Black spruce is the dominant in lowland forests throughout Labrador and northern Quebec, becoming increasingly frequent northward, with the pattern over the landscape broken by open boglands and low ridges. The distribution of black and white spruce is believed to be at least in part the result of a preference by white spruce for alluvium and slightly basic or neutral soils, with black spruce able to persist on virtually all types of substrate. The

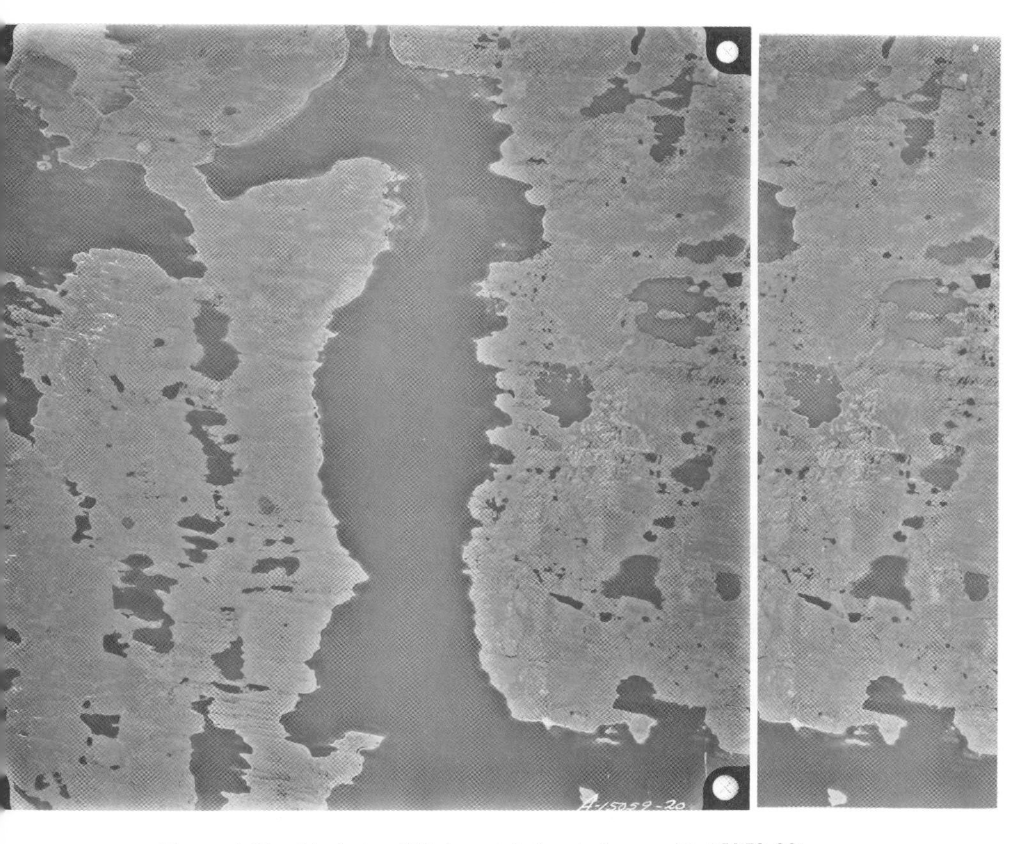

Figure 4.61. Airphoto of Dubawnt Lake study area (A 15059-20).

upper levels of the highest hills in the Schefferville area are bare of trees, the latter generally failing to grow above about 760 meters over sea level, and this altitudinal tundra then appears at successively lower levels toward the northern parts of the peninsula (see Fig. 4.63). Both Gerin and Irony Mountains near Schefferville possess a rich association of species representing both arctic and boreal affinities.

In a study of the distribution of boreal and arctic-alpine plant species in northern Quebec, it was shown that the area around Schefferville in north central Quebec, within the taiga or boreal forest as mapped by Payette (1983), is floristically the richest of the areas studied (see Table 4.1; Morisset et al., 1983). This is the consequence of possessing both the (relatively) rich boreal flora as well as a large group of arctic-alpine species. The latter are found mostly on the higher elevations in the area, Gerin Mountain and Irony Mountain in particular, and this list of 72 arctic-alpine species, long as it may be, is not as extensive as that of other collectors in the area. Hustich, for example, lists a half dozen or so more (Hustich, 1951), and, combining his list with that of Viereck (1957), lists 151 species collected on Gerin and Irony Mountains (Hustich, 1962), at least a large portion of which could be considered arctic-alpine

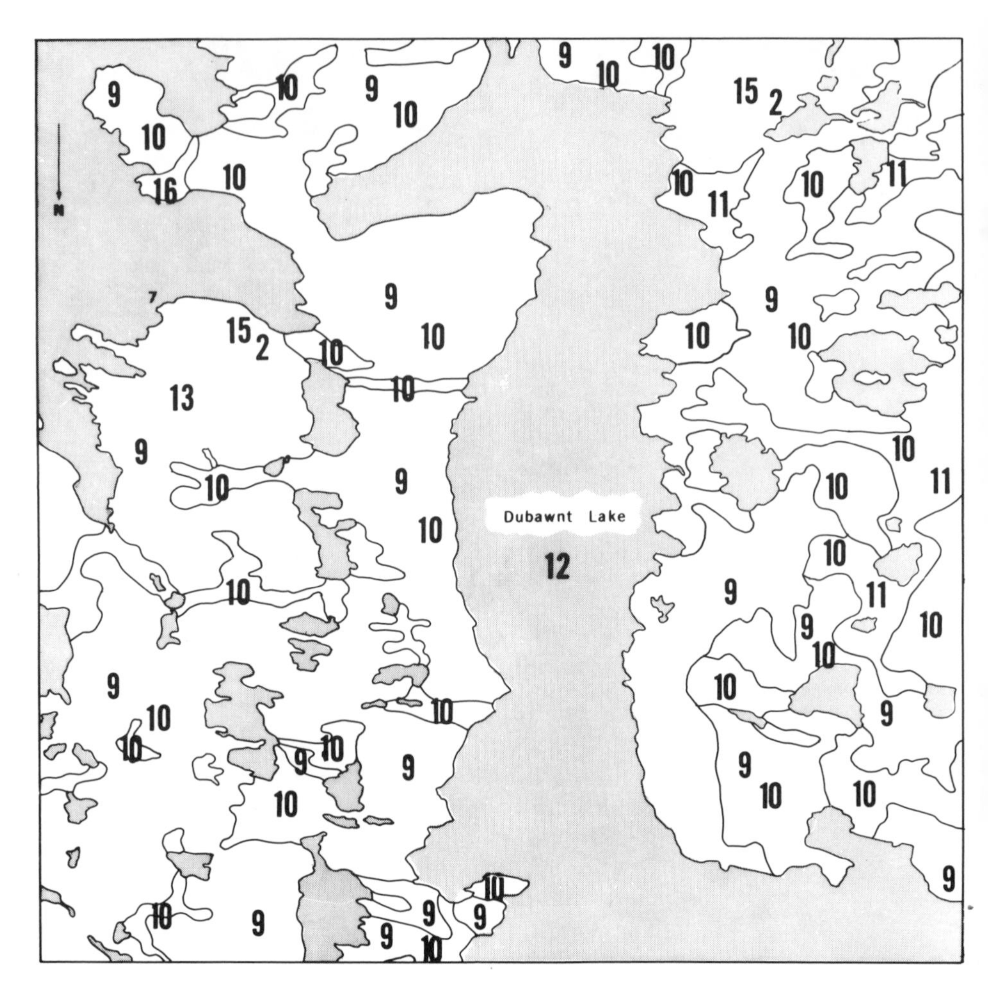

Figure 4.62. Key to airphoto of the Dubawnt Lake study area. For description of features indicated by the identifying numbers, see Fig. 4.2.

tundra species, 94 of which were also found on or near the summit of a mountain in Finland, Mt. Ounastunturi.

It is, thus, evident that arctic-alpine plant species are found south of the northern forest border and the forest-tundra ecotone in northern Quebec, wherever elevations are sufficient to create habitat conditions sufficiently comparable to those of the arctic to exclude other species and allow the arctic-alpine species to persist. Morisset et al., (1983) also indicate that the boreal taxa become less numerous along the gradient from the southern to the northern part of the subarctic forest-tundra and then northward away from the tree-line. As for the arctic species:

> Both the numbers of species and the relative importance of phytogeographical categories are quite variable between local floras. The main patterns emerging from the analysis are:

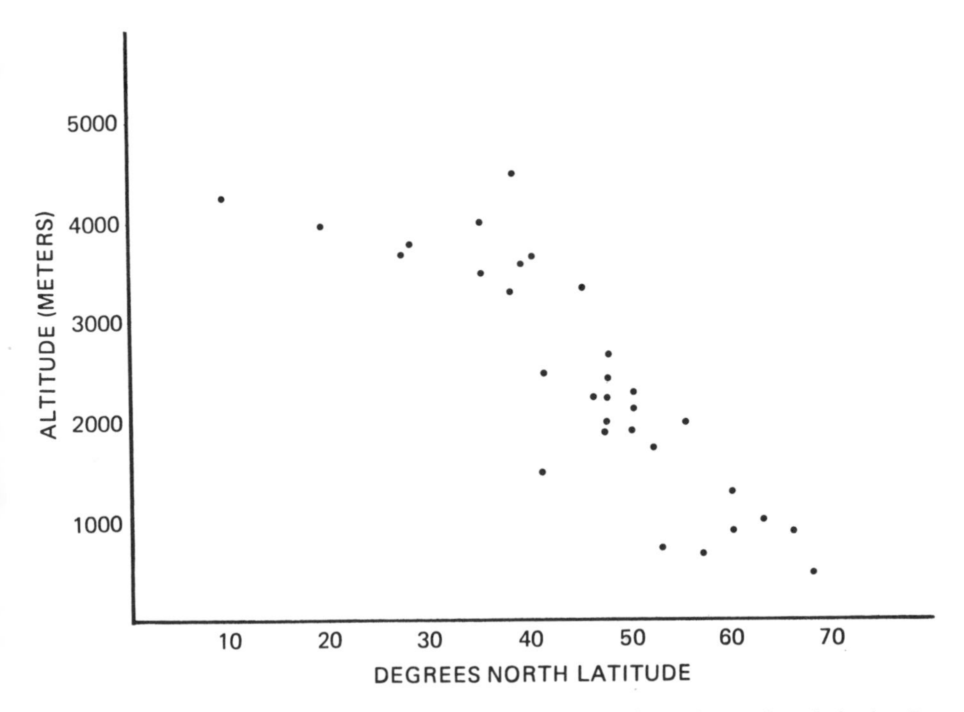

Figure 4.63. Altitude of tree-line on various mountain peaks located at various latitudes. Dots at far right, located at 1000 m or less, are for mountains in the Yukon of Western Canada and for Mts. Gerin and Irony in north central Quebec. Data are for mountains in North America, Asia, and Europe, and are taken from Wardle (1974), Hämet-Ahti (1978), Hustich (1962), and Arno (1984).

> (1) A clear-cut difference between subarctic and arctic floras in the relative importance of arctic and boreal species . . . ; (2) the gradual disappearance of boreal taxa from the southern to the northern part of the subarctic forest tundra . . . and northwards away from the tree-line; (3) the occurrence of a rather uniform number of arctic taxa in both the arctic tundra and the subarctic forest tundra; (4) the relatively high similarity of the arctic component throughout the region; and (5) the much lower similarities between arctic localities in the boreal components of the vascular flora.

In general, these observations correspond to those of Ritchie (1960a) at Caribou Lake in northern Manitoba and of Larsen (1972c, 1973, 1974, 1980) in central Keewatin and eastern Mackenzie, in which the flora in the vicinity of tree-line and northward was a depauperate one, with both few arctic and few boreal species. Had there been higher elevations in these areas, as on Gerin Mountain and Irony Mountain in the Schefferville area, it is likely that a greater number of arctic species would be found in these more western areas.

Despite the fact that exploration and settlement in North American moved from east to west, botanical exploration and accurate designation of the position of the northern forest border was not, in Quebec and Labrador, greatly in advance of that in the west.

Table 4.1. Arctic and Alpine Species Numbers

Distribution of Arctic-Alpine and Boreal Plant Species in Northern Quebec-Labrador (From Morisset et al., 1983).

Approximate Location	Arctic-Alpine	Boreal
Lat. Long.		
RM 61°30′ 73°00′ (Tundra)	68	4
MC 58°00′ 64°00 (Montane tundra)	67	27
BL 57°30′ 76°00′ (Forest-tundra)	51	92
ML 57°00′ 74°00′ (Forest-tundra)	48	125
SC 55°00′ 65°00′ (Boreal forest)	72*	240

*The species were collected at high elevations on the slopes and summits of Gerin Mountain and Irony Mountain.

Location names are as follows: RM = Raglan Mine/Asbestos Hill; MC = Merewether Crater; BL = Bush Lake; MC = Minto Lake; SC = Schefferville.

It is true that some writers described the plant life of "New France" at an early date (Boucher, 1664; Lescarbot, 1612; see Rousseau, 1966), but scientific observation and serious collection of specimens for established herbaria occurred much later. The latter might be said to have begun with the collections of surgeon-naturalist and explorer Sir John Richardson, who accompanied Sir John Franklin on his expeditions in the years 1819–22 and 1825–27 and led a search for the missing Franklin expedition in the years 1848–49. His collection encompassed 474 species of flowering plants and ferns, along with lower cryptograms, from the Northwest Territories (Porsild,1943). During 1819 an expedition under W.E. Perry, another British explorer, collected plants in the Canadian Arctic, and together the collections of Richardson and Perry formed the basis for W.J. Hooker's "*Flora Boreali-Americana*," which appeared in the years 1829–1840.

The coastal regions of Labrador and Ungava had missions and Hudson's Bay Company posts in the early years of the nineteenth century. As early as 1811, Moravian missionaries explored Ungava Bay, in 1828 Hendrey traversed the Ungava Peninsula, and in 1830 the Company's post at Ft. Chimo was established by Finlayson. The posts on the east coast of Labrador, such as Hebron, Okak, Nain, and Hopedale had all been established at an early date.

Early impressions of the land are always of interest, and a great many are included in Elton's study of the animal population cycles in Labrador-Ungava, based on early records of trapping returns and of events near the missions (Elton, 1942). For example, Finlayson, who built Fort Chimo, wrote in 1833 of the surrounding country:

> The surrounding country is the most sterile and mountainous imaginable, here and there intersected with ravines, swamps, and lakes, with now and then a sandy plain, enlivened with no verdure except patches of dwarfish pines, larch and willows in the ravines and swamps .... The Interior, as far as we have seen, is equally mountainous, rugged, and sterile ... but variegated with patches of pine and larch which grow to a good size. That

> on which the Outpost was established grows pine superior to the York* wood and in quality inferior to none in the country . . . . The Indians never leave the coast, where deer are most numerous, above four or five days journey, and that is when they go to look for birchrind for their canoes.

From Finlayson's account, it is apparent that spruce (his "pines") and larch were relatively abundant around Ft. Chimo and even more so a few miles to the south, growing to good size.

Further serious botanical exploration, however, awaited the pioneering expeditions of the Canadian Geological Survey and Topographical Survey, notably those of A.P. Low and Robert Bell and their associates in eastern Canada in the 1880s and 1890s, and of J.B. and J.W. Tyrrell in Keewatin and Mackenzie, and also of Macoun in this same period (Bell, 1879, 1881 (maps of tree-lines for various species); Low, 1888, 1895, 1898, 1899; Macoun, 1895; J.B. Tyrrell, 1896, 1897; J.W. Tyrrell, 1898, 1902). All made significant collections as well as general observations of the regions through which they passed (see Porsild, 1943), and since that time the number of individuals making collections and observations, and more recently geographical and physiographic studies, has increased rapidly in number (see Polunin, 1940, 1947, 1948; Porsild, 1955; Simmons, 1913; Raup, 1947; Hare, 1959; Porsild and Cody, 1980; Cooke and Holland, 1978).

The explorations of Low are of particular significance for he was the first of the scientific travelers in Labrador-Ungava to depart the coast for the unsurveyed interior, following the Indian river routes from the coast, across the inland "barrens" to Ungava Bay. As described by Elton (1942): "Such journeys put everything except light Indian travel in the shade. Within two years these two* men went nearly three thousand miles in canoes, five hundred with dog teams, and a thousand on foot all through wild and trackless country. The results are set down in a full, scholarly way, and from the broadest point of view." It is this work, Elton points out, that: "It is to the geological expeditions of A.P. Low and his companions that we turn for first-class knowledge of the trees and plants and some of the animals. Here is a noble pattern drawn from years of travel and clear insight of the land. We bless the Canadian Government Department of that time, that printed fully what he saw, ravishing the sacred pigeon-holes to do it."

Historically, the refinement of the definition of tree-line, forest border, and forest-tundra, as well as more accurate delineation of these vegetational zones on maps, is well illustrated by the evolution, in Labrador and northern Quebec (Ungava), of ever more precise and detailed delineation, from the maps of Bell (1881), through those of Halliday (1937), Hustich (1949), Hare (1959), Rowe (1959, 1972), Rousseau (1948, 1952, 1966, 1968, 1974), to finally the work of Payette in a series of papers, some with colleagues, between 1974 and 1983 (see refs.), as well as those of colleagues with whom he was joint author (Gagnon and Payette, 1981; Gilbert and Payette, 1982; Godmaire and Payette, 1981).

It is only through intensive effort that a precise delineation of vegetational zonation can be achieved, and it is to be hoped that similar efforts will be conducted in the region

---

*York Factory, northern Manitoba on Hudson Bay.

†Accompanying Low was D.I.V. Eaton, who did survey work.

west of Hudson Bay to the Mackenzie River. A start has been made in northern Manitoba (Ritchie, 1962), but the area yet to be covered is so vast in comparison to Labrador-Ungava and Manitoba that a clear concept of the position of tree-line across the entire continent will probably not be achieved in the near future. It is to be hoped, however, that it will eventually be possible. It is certain that the differences in the structure of the forest-tundra ecotone and the northern forest border along its extent from east to west will be most revealing of the environmental and ecological differences that must exist. For example: the transition from boreal forest through forest-tundra will be found to be more abrupt in at least some western areas as compared to the forest-tundra ecotone in Labrador-Ungava. The causes of this difference will be of great ecological and phytogeographical interest and significance. The questions to be asked are numerous. Are there, for example, barriers to migration in the interior, as apparently exist along the east coast of Labrador? Wheeler (1935) wrote as follows:

> It would seem that the tree line (or at least the northern limit of the ecotone between forest and tundra) comes a few miles north of the mouth of the George River in the rolling lowlands west of the Torngates. . . . Hebron Bay only a few miles further north, running equally far inland, and with deep valleys entering its head, should be an equally favorable locality. This sharp transition suggests the possibility that the limit of trees on the Labrador coast is not due so much to climatic conditions as to the absence of migration routes for the trees between the wooded valley of the George River and the valleys draining into the Atlantic on the northern section of the coast.

This view is supported by Hustich (1939, 1949) and more recently by observations of Elliott and Short (1979) and Payette (1983). Do similar barriers exist elsewhere? This and many other questions remain to be asked, and answered, by a new generation of phytogeographers, with modern techniques for plant community analysis, and improved instruments for environmental measurement; one looks forward to seeing what they will learn.

5. Forest Border Community Structure

In this volume, there has so far been little discussion of the boreal forest as a plant community. It is true also of recent symposia and compendia mentioned in the Preface that most boreal forest research has been concentrated upon matters related to silvicultural characteristics of trees, the nature of the soils, nutrient cycles, and to the productivity of the forest, and less to the aggregation of species making up the understory, relationships to one another, and to the climate and other environmental factors. It was my intent in previous reports to draw attention to the boreal forest ecosystem as a distinctive plant community, varying in composition from place to place in response to environmental variables, particularly the regional climatic regimes. The ecosystem was viewed in the light of statistical community composition and dynamics, and it is the purpose of the discussion undertaken now to add something more to what has been written (Larsen, see refs.) on boreal community structure, particularly at the northern edge of the taiga, the northern forest border, the northern edge of the forest-tundra, or the lichen woodland—whatever may be the terminology preferred.

It is generally accepted that there are causal and reciprocal forces at work among the natural entities, vegetation, soils, and climate. These causal links and reciprocal interrelationships are, indeed, the foundations upon which the community structure is based. They are, as a consequence, the topics of study in the various branches of ecology, from the simplest depiction of the range of the plant and animal species, as in phytogeography, to the physiological and biochemical responses involved in adaptation to the range of environmental conditions tolerated by the individual species.

The relationships among soils, climate, and vegetation are not, however, necessarily easy to discern and describe, nor are they usually linear in the same sense that a unit change, if such can be imagined, in soil results in a unit change in vegetation, such as one might think of in terms of one species taking the place of another in a given assortment making up a community. So the interrelationships of cause, response, reciprocal action, feedback, limiting factors, as well as population controls such as competition, disease, predation, and so on, are rarely clearly apparent, with the result that ecosystem dynamics are not easily translated into conceptual models, even with present-day techniques of model design and rapid computer processing of mathematical formulations.

It is, nevertheless, conceptually and aesthetically satisfying to paraphrase such mathematical ecologists as Margalef, whose elegant equations express in symbols those relationships which one intuitively senses must exist and which traditionally are expressed in words. The result, one must admit, is more often than not a statement of apparent but abstract relationships. One quickly falls back on the old dictum that it is too much to expect of theoreticians that they also be practical.

The inability to correlate the three entities—soils, climate, and vegetation—in any precise delineation of cause and effect was expressed long ago by Jenny, for example, in his *Factors of Soil Formation: A System of Quantitative Pedology*, published in 1941:

> The ultimate goal of functional analysis is the formulation of quantitative laws that permit mathematical treatment. As yet, no correlation between soil properties and conditioning factors has been found under field conditions which satisfies the requirements of generality and rigidity of natural laws.

The same can still be said of ecology in general, and of plant ecology in particular. In Jenny's scheme, properties of soil become dependent variables and can be expressed as functions of soil-forming factors, that is,

$$s = f(cl, o, r, p, t \ldots)$$

in which cl = climate, o = organisms, r = topography, p = parent material, and t = time, with the dots (. . .) indicating other factors that may have to be included.

Simple transformation of the equation can be made to designate the vegetation, that is,

$$v = f(s, cl, o, r, p, t \ldots)$$

in which v = vegetation, o = other organisms, and the remainder stay the same as given.

All are, with exception of time and parent material, perhaps of macroclimate, subject to feedback effects from the other variables, and in diagrammatic form each affects the others in a network of cause and effect, with waves of influence and response cycling through the system. Time has a position of interest in the scheme. It is uninfluenced by other factors in the sense that it progresses at a uniform rate, as measured by clock and calendar, but the rate at which events occur—the time interval between events—is affected by a number of factors such as temperature, moisture, energy flux (incoming long and short wave solar radiation and outgoing long wave radiation), nature of parent material (i.e., ionic composition), organisms, topography, and so on, and is, thus, in this sense at least, dependent to some degree upon the other factors, not as time is measured by the clock but as the rate at which certain of the events occur. Thus, if one were to measure time by the rate at which nitrogen ions are absorbed by a certain specific molecular complex in a root hair, time would vary from place to place, from plant to plant, from soil to soil, and so on, indeed meeting the definition that time is relative.

There are also such variables as the opportunities for migration of species and colonization of land newly opened as a result of deglaciation, or emergence from other disturbances that otherwise disrupted or prevented ecesis of plant species capable of pioneering such areas, the accidents of geotectonics, for example, or of fire, flood, drought, or other disturbance of the environment.

The consequence is that adequate formulations from the elegant equations devised to describe an ecosystem are difficult to achieve. For practical purposes, we would like the equations to be useful predictive devices, or at least accurately descriptive, possessing specific quantitative relationships expressed by some means capable of conceptual manipulation. With each factor—vegetation, soils, organisms, and other factors—measured, and probably measurable, only by different means and—at least as yet—described in different terms and units (species, chemical composition, color), how can these be so arranged, so organized in an equation, that the design makes sense?

There has been much thought given to this subject, as evidenced by the near-approximations implicit in the work of individuals pursuing the general field known as ecological land classification, and the principles involved in this effort have been well delineated by Rowe and colleagues (Rowe and Sheard, 1981; Rowe, 1984; Bradley et al., 1981), and the latter reference provides an example, a study of the Lockhart River Map Area (east of Great Slave Lake, where boreal forest grades into tundra northward), as well as by others (Hills, 1960; Bunce et al., 1975; Kessel, 1979; Rubec, 1979).

In what is termed the "landscape approach" the landscape units that are similar in ecological respects are identified and mapped, and in Rowe's words: "Areal differentiation is sought in the known interrelationships of climate, landform, biota, and soils at all scales in the landscape." Recognition of the pattern is based on prior knowledge, both of the species present in the vegetational communities and of their known physiological and ecological characteristics. The interactions and controlling influences of the various aspects of the ecosystem(s) under consideration are, however, difficult to measure and describe in quantitative terms even when they are recognized, which may often be a source of difficulty and error in itself.

Some plant species are wide-ranging and tolerant of marked variation in conditions, others are to some degree intermediate, and at the other extreme are those that are very demanding, with narrow limits of tolerance to environmental conditions. The former are those most often selected for studies of ecotypic differentiation; the latter are the best indicators of environmental conditions although they obviously tend not to be widespread and abundant (Hamberg and Major, 1968). Inferences or conclusions from expansion or contraction of range limits should take these individual characteristics into account. To compound the difficulty, the set of factors to which a given species is responding is not always easy to discern, as can be illustrated, for example, in the decline of red spruce in the mountains of the northeastern United States (Hamburg and Cogbill, 1988).

Theoretical insights into plant community structure also have a direct bearing upon the nature of community response to environmental conditions, disturbance, and change, and while this aspect has not yet been fully explored there is a varied literature on the nature of plant communities that is pertinent (Austin, 1985; Pielou, 1974, 1986; Auerbach and Shmida, 1987; Diamond and Case, 1985; Grubb, 1977; Pickett and White, 1985) as well as discussions of the effect of climatic variation on the boreal forest regions (Singh and Powell, 1986; Kauppi and Posch, 1985; Ball, 1986).

Most noteworthy are those species that are wide-ranging and abundant within the northern edge of the forest, with ranges that extend north to the edge of the trees but not far beyond, and these might well be studied as indicators of climatic change. Some of the species in this group are *Ledum groenlandicum*, *Vaccinium myrtilloides*, *Salix bebbiana*, *Equisetum sylvaticum*. On the other hand are species of arctic and alpine North America that reach southward at least to the southern border of tundra where it meets the northern forest border ecotone, or which may extend for a distance southward into the forest in suitable habitats or, more infrequently, as disjunct rarities in distant sites. Among the plant species in this group one might include the following:

Hierochloe alpina
Carex stans
Carex supina var. spaniocarpa
Carex membranacea
Luzula confusa
Salix herbacea
Salix reticulata
Sagina nodosa
Silene acaulis
Saxifraga hirculus
Saxifraga tricuspidata
Dryas integrifolia
Ledum decumbens
Loiseleuria procumbens
Arctostaphylos alpina
Vaccinium uliginosum
Pedicularis lapponica
Saxifraga aizoides

Some of these species are found relatively infrequently even within their range of greatest abundance, but others are of sufficient frequency to be readily observed by even a casual inspection of the communities in a region.

Since the vegetational communities are the topic of major concern in these pages, description of the characteristics of the vegetational communities found in the areas under study is preliminary to any investigation of the "conditioning factors" defined by Jenny. In other words, the composition (species present) and the structure (the proportion with which each of the species occurs as individual plants) of the vegetational communities is the perspective from which the ecosystem is to be viewed here. The functional relationships must await the studies to be carried on by others, as well perhaps of new and more sophisticated sets of techniques for analysis and modeling of genetic, biochemical, and physiological factors at work within the system.

Community Structure

The change that occurs between the forest and the tundra in northern Canada is visually a striking one, one in which the environmental factors that constitute the major influences are now beginning to be elucidated, and it appears that the change is predominantly conditioned by climate. The environmental factors affecting this transition from forest to tundra have been aptly summarized by Bradley et al., (1982): "Treeline represents a striking break in the patterning of regional vegetation, the change from woodland to tundra on the upland interfluves takes place within a short distance and is expressed over a circumpolar range. It parallels no known discontinuities in lithology, surficial geology, or relief. Therefore, it appears to be climatically induced, and most ecologists accept it as such."

The existence of the forest border in the position at which it is found is generally accepted to be the consequences of an inability on the part of the tree species to reproduce under climatic conditions prevailing northward. Why this should be true is a question that logically arises, and efforts to answer the question in physiological terms have been, to date, not exceedingly numerous but they do reveal at least some of the mechanisms involved. There is little need to repeat the conclusions here, however, and it should suffice to note recent work of a few authors and make reference to the full bibliographies contained therein: Savile (1963, 1972), Clausen (1963), Elliott (1979a,b), Black and Bliss (1980), Larsen (1980), Carter and Prince (1981), Tranquillini (1979), and Rowe (1984). The work on species other than trees has been even more sporadic, but it should be noted that studies by Karlin and Bliss (1983) reveal that accelerated germination of the northern species *Ledum palustre* (*L. decumbens*) in comparison to that of the closely related but more southern *Ledum groenlandicum* may be at least one part of the reason why the former has a more northern distribution.

It has long been noted that there is a demonstrable correspondence between climatic parameters and the position of the northern forest border (Nordenskjold and Mecking, 1928; Hopkins, 1959; Bryson, 1966). Mapping of climatic isopleths (length of growing season, temperature, average wind direction) often reveals a correspondence with the range limits of the northern tree species. Moreover, numbers of shrub and herbaceous species also show changes in abundance along climatic gradients (Larsen, 1971), affirming what was shown as early as a half century ago that certain interesting groups of species have range limits coinciding with the continental tree-line or northern forest border (Raup, 1947; Porsild, 1951, 1958). For example, Raup noted that a small group of about 19 species exhibits the most extensive plant ranges in Canada, occurring widely through the boreal forest and reaching far out into the tundra, e.g.:

Alnus crispa
Pyrola secunda
Betula glandulosa
Vaccinium vitis-idaea
Rubus chamaemorus
Epilobium angustifolium
Ledum groenlandicum

but that the remainder of the widespread forest species are more rigorously limited by the northern tree-line, including

Geocaulon lividum
Equisetum sylvaticum
Rubus strigosus
Arctostaphylos uva-ursi
Rubus acaulis
Mertensia paniculata

while others have their southern limit in the vicinity of the tree-line (except that they range southward on alpine sites), for example:

Hierochloe alpina	*Salix reticulata*
Cassiope tetragona	*Pedicularis capitata*
Pedicularis sudetica	

It has also been amply demonstrated that in addition to responding to climatic influences, plant species respond also to other environmental factors, and the response of individual species to the total environment has been employed by forest ecologists to evaluate sites in terms of their potential for tree growth. Thus, as pointed out by Corns and Pluth (1984), the presence or absence of certain indicator species can be used to evaluate the potential of sites for supporting good growth of lodgepole pine and white spruce. Used in conjunction with conventional soil and site observations, individual plant species or species groups may reveal subtle environmental characteristics important to tree growth not readily discernible from physical measurements. Among the important indicator species were *Rosa acicularis*, *Cornus canadensis*, *Rubus pubescens*, and *Calamagrostis canadensis*, as well as *Cladonia* and *Peltigera* lichen species. It is thus apparent that plant species respond to environmental conditions in characteristics ways, and that the response can be employed as a measure of the prevailing environmental conditions. This observation is pertinent to the designation of the ecotonal region of the northern forest border by means of the plant species found growing there in abundance.

The northern forest/tundra ecotone is coincident with the pathway of cyclonic storms along the average summer position of the arctic front, a climatic zone separating arctic air masses prevailing over the tundra of northern arctic Canada from the air masses of southern origin prevailing over the subarctic vegetation of spruce forest and treed muskeg (Bryson, 1966), and the same holds true over Eurasia (Krebs and Barry, 1970) and is, thus, circumpolar in extent with the exception that the coincidence does not hold as closely over northern Quebec due evidently to the influence of Hudson Bay and the Atlantic Ocean (Barry, 1967).

It was subsequently demonstrated that not only does the northern edge of continuous forest coincide with this frontal zone but there are also marked changes across this zone in the quadrat frequency counts of species making up the ground or understory vegetation (Larsen, 1971). Climatic variation across the boundary between the two regions is sufficient to account, at least in large measure, for the equally marked vegetational differences, since geological and topographical characteristics are generally uniform throughout the area of Keewatin in which the study was initiated (Lee, 1959), as well as in the areas in which studies were subsequently carried on (Bostwick, 1967). The most detailed sampling across this ecotonal boundary was carried on in Keewatin, from the south end of Ennadai Lake, within the northern forest border, to Dubawnt Lake, well into tundra, and thence farther north to Pelly, Snow Bunting, and Curtis Lakes (Fig. 4.12).

Most apparent in the data obtained along this transect was a northward increase in total number of species in the sampled tundra communities (Table 5.1). The dwarfed black spruce communities sampled show a change in species composition and frequency along the traverse, but there appears to be no great change in species numbers (Table 5.2), and species that begin to appear more commonly in the associations toward

Table 5.1. Numbers of Species in the Various Communities*

Tussock muskeg communities (data is for individual stands sampled)																				
Study area	2	2	2	2	2	2	2	2	2	2	3	8	10	10	10	10	10			
Total species	16	14	13	11	9	10	10	10	7	10	18	23	27	27	20	31	29			
Total arctic species	2	1	0	0	0	0	0	0	0	0	4	11	17	15	14	20	21			
Rock field communities																				
Study area	2	2	2	2	2	2	4	5	6	7	9	10	10	11	11	11	12	12	13	13
Total species	7	9	9	9	9	11	8	11	15	20	14	20	21	14	13	14	12	13	13	15
Total arctic species	2	4	4	4	4	6	4	6	10	15	9	15	16	11	9	11	11	11	13	13
Low meadow communities																				
Study area	2	2	2	2	2	2	2	2	2	4	7	8	10	10	11	11				
Total species	18	11	10	14	11	10	13	14	12	12	15	13	17	18	18	20				
Total arctic species	2	1	1	3	1	1	2	2	2	3	4	2	3	9	7	5				
Black spruce communities																				
Study area	1	1	1	1	2	2	7	10												
Total species	22	10	11	18	23	12	15	24												
Total arctic species	3	3	2	4	7	5	5	10												

*Numbers of species are listed by study site; numbers in the Study Area row identify the sites indicated in Fig. 4.63. Progression is (left to right) south to north.

Table 5.2. Species Frequency in Black Spruce Stands Along South-North Traverse in Keewatin

	Area							
Species	So. Ennadai Lake	So. Ennadai Lake	So. Ennadai Lake	So. Ennadai Lake	Mid Ennadai Lake	Mid Ennadai Lake	61° 59′ North	So. Dubawnt Lake
Vaccinium vitis-idaea	100	100	100	100	100	100	95	45
Vaccinium uliginosum	100	95	90	75	95	90	65	80
Ledum groenlandicum	100	10	95	90	85	55	95	40
Betula glandulosa	50	65	65	60	55	70	85	40
Alnus crispa	15							
Petasites palmatus	15			15	15			
Salix glauca	30	15	15	5	10			20
Picea marina (seedling)*	45	35	20	40	10	20	35	25
Carex sp./spp.	65						30	15
Empetrum nigrum	10	50			80	90	65	65
Arctostaphylos alpina	30	5			5			
Salix myrtillifolia	5			20	25			
Rubus chamaemorus	60			55	15		70	95
Equisetum sylvaticum	35		25	95	75		45	45
Carex vaginata	10							
Equisetum scirpoides	10		5		15			5
Juncus castaneus	5				5			
Salix planifolia	5		30	15	25	5	5	30
Larix laricina (seedling)	5			5				
Ledum decumbens	15	80	10	20	15	20	65	20
Lycopodium annotinum	5			5	5			
Oxycoccos microcarpus	5			20	20			15
Loiseleuria procumbens		40				5		
Grass sp./spp.			15	70	50	65		75
Epilobium angustifolium				5		10		
Pyrola secunda				5				10
Rubus acaulis					5			
Tofieldia pusilla					5			5
Calamagrostis canadensis					10		70	60
Carex rariflora					5			
Carex bigelowii						5		
Equisetum arvense							10	20
Poa arctica							5	35
Salix arbusculoides							5	
Salix arctophila								10
Carex stans								15
Salix reticulata								5
Total Species	22	10	11	18	23	12	15	24

*Reproduction primarily by layering but not differentiated.

the northern end of the traverse are species more generally arctic in geographical affinity.* They are absent, or present only very rarely, in the plant communities of the southern end of the traverse and, hence, are missed in sampling. Frequency values of the arctic species can be seen to increase northward, in general, although frequency values of some species at first rise northward and then decline again. The ubiquitous species usually maintain high frequency values throughout the traverse, although the frequency values may decline in the northernmost study areas as the data from Pelly, Snow Bunting, and Curtis Lakes reveal. Comparisons among the different communities, (1) dwarfed black spruce, (2) tussock muskeg, (3) rock field, and (4) low *Carex* meadow, in terms of total species present as well as total arctic species present, can be made using the data in Tables 5.3–5.5.

The species categorized as ubiquitous on the basis of their widespread appearance in the transect data of both forest and tundra study areas include *Vaccinium vitis-idaea*, *V. uliginosum*, *Betula glandulosa*, *Empetrum nigrum*, and *Rubus chamaemorus*. Among the species categorized as arctic in affinities are *Arctostaphylos alpina*, *Ledum decumbens*, *Dryas integrifolia*, *Cassiope tetragona*, *Diapensia lapponica*, *Salix arctica*, *Oxytropis arctica*, and *Polygonum viviparum*.

Southward of this transect, in the northern edge of the lichen woodland termed the forest-tundra transition zone or northern forest/tundra ecotone, there also exists forest communities with a relatively depauperate shrub and herbaceous understory, albeit the latter is dominated by lichens, and here are found these same ubiquitous shrubs and herbs that perhaps have in common at least one major adaptive characteristic, the ability to withstand stress in a severe, variable, and unpredictable environment (Rowe, 1984).

That these same species maintain their dominant position in communities across the northern forest ecotone from the east slopes of the Mackenzie Mountains to Keewatin, and thence to northern Quebec as well, is shown in Tables 5.6–5.8. The species occurring with high frequencies in communities of the ecotonal region across the continent include *Betula glandulosa*, *Empetrum nigrum*, *Ledum decumbens* (*palustre*), *Ledum groenlandicum*, *Vaccinium uliginosum*, *Vaccinium vitis-idaea*, *Rubus chamaemorus*, and with somewhat lesser frequency *Geocaulon lividum*, *Vaccinium* (*Oxycoccus*) *microcarpus*, *Equisetum sylvaticum*, and *Arctostaphylos* (*Arctous*) *alpina* (or *rubra* not differentiated).

Since species occurring with greatest frequency just north of the forest border generally are ubiquitous species, and since arctic species occur in increasing numbers and with increasing frequency values northward, it is postulated that this represents a response to climatic conditions prevailing in the region. It is, of course, inferred that

---

*Selection of species for inclusion in one or more of the tables was as follows: To be included, a species had to be present (1) with a frequency value of at least 15 in at least one stand, (2) present in at least two stands with a frequency value of at least 5 (the minimum value, indicating it occurred in one quadrat of the 20-quadrat transect) in each, (3) present in at least two community types (i.e., rock field and black spruce, for example), or (4), rarely a species has been included because it was of unusual interest, i.e., the occurrence of *Equisetum variegatum* and a *Viola* sp. in one stand or another. The latter instance is so infrequent, however, as to not have any appreciable effect upon the listings of total species numbers. Hence, also, the diversity values, given in Chapter 6, have been computed using tabulations of all species found in the transects run through the sampled communities, whether listed in the tables presented here or not.

Table 5.3. Species Frequency in Tussock Muskeg Communities Along South-North Traverse in Keewatin

Species	Area																
	Ennadai Lake	Ennadai Lake	Ennadai Lake	Ennadai Lake	Ennadai Lake	Ennadai Lake	Ennadai Lake	Ennadai Lake	Ennadai Lake	Ennadai Lake	61° 22′	62° 00′	So. Dubawnt Lake	So. Dubawnt Lake	So. Dubawnt Lake	So. Dubawnt Lake	So. Dubawnt Lake
Ledum decumbens	90	85	80	85	100	100	100	95	100	90	100	100	100	20	100	85	95
Vaccinium vitis-idaea	55	75	65	85	100	100	95	95	100	75	95	100	100	20	100	90	100
Vaccinium uliginosum	55	70	80	70	100	45	20	90	45	70	40	40	100	75	95	90	95
Salix arbutifolia								35				5					
Carex sp./spp.	45	70	80	25	20	15	80	95	25	95	100	100	100	100	35	100	95
Betula glandulosa	60	65	85	40		5		55		80	95	100	100	95	100	100	95
Rubus chamaemorus	55	55	45	85	100	80	95	55	95	85	90	40	5	75		20	15
Andromeda polifolia	50	35	75	40	50	15	20	55	50	100	85	15	15	65		35	40
Empetrum nigrum	25	25			80	65	5	20	45		10	35	65	15	80	75	80
Eriophorum vaginatum	40	20	25	20	30	10	15	10		20	15	20	5	70		15	10
Carex rotundata (group)	40	30	20	25	20	15	80			10	25	10	5	40		20	25
Salix arctophila	60	55	50	5						10	50	20		100		85	100
Oxycoccos microcarpus	10	5	30	15							5						
Carex chordorrhiza	5	25	25				45				5			5			
Potentilla palustris	5		35								5	10					

Arctostaphylos alpina	10	5		10	50	20	20	35	35
Pedicularis labradorica	5		15	10	5		35	10	
Juncus castaneus			5						
Pinguicula villosa			10						
Carex vaginatum			10		20	45	5	35	50
Grass sp./spp.				5			90	70	85
Carex bigelowii				80	10				5
Salix glauca				15	35		10		
Salix arbusculoides				25	25		50	40	55
Pyrola secunda				10			20	10	
Polygonum viviparum				5	30	45	10	35	50
Pedicularis lapponica				5					
Calamagrostis canadensis				5	5		10	15	15
Salix reticulata					5	50		35	50
Poa arctica					5				40
Dryas integrifolia					15	25		30	60
Stellaria longipes					10			10	
Tofieldia pusilla					10	5		20	25
Rhododendron lapponicum					10	80		15	25
Carex capillaris					5	15	5		20

Table 5.3. *Continued.*

	Area																
Species	Ennadai Lake	Ennadai Lake	Ennadai Lake	Ennadai Lake	Ennadai Lake	Ennadai Lake	Ennadai Lake	Ennadai Lake	Ennadai Lake	Ennadai Lake	61° 22′	62° 00′	So. Dubawnt Lake	So. Dubawnt Lake	So. Dubawnt Lake	So. Dubawnt Lake	So. Dubawnt Lake
Calamagrostis purpurascens													5				
Carex stans													5	70		50	40
Pedicularis sudetica														20	5	5	
Carex rariflora														15		15	
Scirpus caespitosus														5			
Salix planifolia														10	20		5
Carex atrofusca														10			
Eriophorum angustifolium														10			
Saussurea angustifolia															5		40
Salix herbacea															5		
Arctagrostis latifolia																25	
Equisetum scirpoides																20	25
Juncus albescens																10	5
Carex brunnescens																5	
Pedicularis flammea																	30
Pedicularis lanata																	15
Salix calcicola																	5
Total Species	16	14	13	11	9	10	10	10	7	10	18	23	27	27	20	31	29

Table 5.4. Species Frequency in Rock Field Communities Along South-North Traverse in Keewatin

Species	Area																			
	Ennadai Lake	Ennadai Lake	Ennadai Lake	Ennadai Lake	Ennadai Lake	Ennadai Lake	61° 29′	61° 42′	61° 48′	61° 57′	62° 21′	So. Dubawnt Lake	So. Dubawnt Lake	Pelly Lake	Pelly Lake	Pelly Lake	Snow Bunting Lake	Snow Bunting Lake	Curtis Lake	Curtis Lake
Vaccinium vitis-idaea	95	85	100	90	100	90	95	100	100	85	100	85	90	95	100	100	35	75		
Vaccinium uliginosum	55	60	35	30	40	90	45	85	70	100	100	100	95		35					10
Ledum decumbens	50	65	50	40	70	100	50	100	100	65	85	90	100	100	100	65		5		10
Betula glandulosa	30	30	45	10	70	45		45	25	55	25	70	45							
Empetrum nigrum	50	35	65	25	85	70	65	100	75	80	65	90	100	15	60	15				
Loiseleuria procumbens	15		55	30	60	35	10	90	20		5		5							
Carex sp./spp.						15				60	10	45	40	95	90	50	15	90	60	95
Grass sp./spp.		35	45	5	10	25	20			20	30	60	5	80	70	75	100	60	35	
Arctostaphylos alpina	70		25	20		65	15	90	45	70	85	80	90			5				
Hierochloe alpina		35	45	5	10	15	20			20	30	60	5	80	75	70	80	60	60	95
Salix reticulata												50	10							
Luzula confusa		15			10			10	65	10		5	10	15	20	5	40		30	15
Salix herbacea		15						5	15	10			5		10		90	85	5	
Carex glacialis						5														
Salix glauca														35	5					

Table 5.4. *Continued.*

Species	Area																			
	Ennadai Lake	Ennadai Lake	Ennadai Lake	Ennadai Lake	Ennadai Lake	Ennadai Lake	61° 29′	61° 42′	61° 48′	61° 57′	62° 21′	So. Dubawnt Lake	So. Dubawnt Lake	Pelly Lake	Pelly Lake	Pelly Lake	Snow Bunting Lake	Snow Bunting Lake	Curtis Lake	Curtis Lake
Agrostis borealis										5			5							
Carex bigelowii								5				5		20	60	25	5			5
Festuca brachyphylla								5		10			10							
Rhododendron lapponicum									60	80	30	50	45			5				
Salix arbusculoides									10	45	15	30								
Rubus chamaemorus									25											
Stellaria longipes									10			10				10		5		
Silene acaulis									10		5	15	10	30		40			45	5
Carex capillaris									5	60		35								
Poa arctica											25			10	10			35		
Tofieldia pusilla											5		50							
Saxifraga tricuspidata											10	5								
Polygonum viviparum											5								20	25
Dryas integrifolia													45				30	15	45	35
Pedicularis lanata													10							

Carex nardina													20							
Diapensia lapponica													10							
Cassiope tetragona														35	10		30	50	100	90
Salix arctica														5	5		70	30	65	75
Armeria maritima														5	25	15				
Salix brachycarpa																30				
Luzula nivalis																	15	5	15	10
Arctagrostis latifolia																	35	5		
Eriophorum angustifolium																				35
Lycopodium selago																			15	10
Oxytropis arctica																			5	
Carex rotundata (group)																				5
Total Species	7	9	9	9	9	11	8	11	15	20	14	20	21	14	13	14	12	13	13	15

Table 5.5. Species Frequency in Low Carex Meadow Communities Along a South-North Transect in Keewatin

	Area															
Species	Ennadai Lake	Ennadai Lake	Ennadai Lake	Ennadai Lake	Ennadai Lake	Ennadai Lake	Ennadai Lake	Ennadai Lake	Ennadai Lake	61° 29′	61° 57′	62° 00′	So. Dubawnt Lake	So. Dubawnt Lake	Pelly Lake	Pelly Lake
Carex sp./spp.	100	100	100	100	100	100	95	100	100	100	100	100	100	100	100	100
Andromeda polifolia	60	70	65	65	10	60	55	70	85	70	90	90	65	45	65	35
Eriophorum spissum	5			25	25	70		15	10	10		25	60	20	25	35
Carex rotundata (group)	20	55	25	50	60	85	50	50	35	55	15	25	60	55	20	40
Scirpus caespitosus				20		40	5		5	60		5	5	35		
Vaccinium uliginosum	35	50	45	15		10	15	70	15	5	45	25		10	10	25
Eriophorum angustifolium		15				10			65		35		5			5
Ledum decumbens	25		50	30	25	5		75		10			55	10	75	40
Carex chordorrhiza	85	90	40	60	80	25		20	90	85	90	30	60	45		20
Oxycoccos microcarpus	5			5	5	5	90	15					10			
Potentilla palustris	15	10			10		75	5	15		80	85	20			
Carex rariflora	20	35					15		15		75	100	5	15	30	50
Rubus chamaemorus			15	5	20		20	35					40			5
Pinguicula villosa				10												
Luzula wahlenbergii							10								5	5
Salix arctophila	100	85					35	90	95	55	95	100	65	95		
Salix arbutifolia															85	65
Empetrum nigrum		5	10					15								5
Betula glandulosa	55	40	50	30	20			40	15	40	30	35	50	60	40	15
Vaccinium vitis-idaea	5		15	5	30			40		5	10	5	10		60	15

Carex stans				50							5		5		70	20
Ledum groenlandicum							5									
Pedicularis labradorica							5									
Carex vaginata										10				35	5	
Polygonum viviparum											10	20		50	5	25
Salix arbusculoides											5					
Calamagrostis canadensis											5					
Lycopodium selago													5			10
Carex atrofusca														35		
Carex capillaris														15		
Arctagrostis latifolia														5		
Dryas integrifolia														10		
Arctostaphylos alpina														5		
Salix herbacea															5	
Poa alpina															5	15
Cassiope tetragona															5	5
Total Species	13	11	10	14	11	10	13	14	12	12	15	13	17	18	18	20

Table 5.6. Species Frequency in Black Spruce Stands Within the Northern Edge of the Forest Border–West to East Traverse

	Area								
	Mackenzie Mts.	Colville Lake	Inuvik	"East Keller" Lake	Ft. Reliance	Kasba-Kazan	So. Ennadai Lake	"Rainy" Lake, Quebec	Schefferville, Quebec
Number of Stands	5	3	2	2	7	6	12	2	5
Anemone parviflora	22								
Arctostaphylos uva-ursi	12								
Carex bigelowii	6								
Cassiope tetragona	44								
Equisetum variegatum	6								
Eriophorum spissum	19								
Hedysarum alpinum	21								
Lupinus arcticus	26								
Polygonum viviparum	10								
Salix reticulata	12								
Saussurea angustifolia	6								
Pinguicula vulgaris	4			10					
Potentilla fruticosa	29		37	17					
Rhododendron lapponicum	35		12	23					
Dryas integrifolia	76		7						
Andromeda polifolia	33	12	50	37	34				
Carex scirpoidea	22		20	12	13				
Tofieldia pusilla	5		22	15	19				
Salix arctica	8			5				18	17
Carex capitata			12						
Carex lugens			42						
Dryas octopetala			27						
Pyrola grandiflora			30						
Salix richardsonii			32						
Juniperus communis				10					
Rosa acicularis				10					
Equisetum scirpoides	14	12	92	10	37	23			
Pedicularis labradorica	16		15	7	9	7			
Arctostaphylos alpina (rubra)	61	22	67	27	55	37	3		
Betula glandulosa	49	47	47	27	7	26	37	40	43
Empetrum nigrum	28	32	35	25	41	44	46	83	54
Ledum decumbens	17	47	15	75	8	17	26		
Ledum groenlandicum	32	67	77	80	71	80	73	30	81
Vaccinium uliginosum	66	68	90	45	51	74	50	100	51
Vaccinium vitis-idaea	33	93	77	80	77	94	85	3	19
Salix sp/spp*	20	25	32	27	69	23	26	30	33
Salix glauca		18			19		6		
Salix myrtillifolia				12	36	10	4		
Salix planifolia		7		10		13	10		11

Table 5.6. *Continued.*

	Area								
	Mackenzie Mts.	Colville Lake	Inuvik	"East Keller" Lake	Ft. Reliance	Kasba-Kazan	So. Ennadai Lake	"Rainy" Lake, Quebec	Schefferville, Quebec
Number of Stands	5	3	2	2	7	6	12	2	5
Carex sp/spp*	47		74	24	43	4		100	56
Carex aquatilis (stans)				5		4			
Carex vaginata					26			13	18
Picea mariana (seedl.*)		20	52	50	20	37	32	33	21
Vaccinium microcarpum (see Table 5.7)		7		47		31	8	55	27
Geocaulon lividum	8			30	24			33	22
Grass sp/spp	7			12	15	7	20	15	7
Rubus chamaemorus		75		32	9	39	35	80	64
Myrica gale				15					
Scirpus caespitosus				7	4			20	5
Habenaria obtusata					7				
Larix laricina (seedl.*)					12	8	7		
Loiseleuria procumbens					3	9	4		
Lycopodium annotinum							17		
Chamaedaphne calyculata				2			3	40	17
Cornus canadensis			2					18	17
Smilacina trifolia				2		6		40	31
Rubus acaulis				7	7			15	11
Equisetum sylvaticum						16	23	63	57
Petasites palmatus							4	43	15
Gaultheria hispidula								25	33
Kalmia polifolia								73	40
Carex microglochin								15	9
Carex paupercula								40	16
Carex trisperma								23	8
Coptis groenlandica								48	19
Linnaea borealis								23	22
Listera cordata								3	3
Lycopodium selago								3	14
Solidago multiradiata								8	22
Trientalis borealis								8	2
Viola sp.								8	
Mitella nuda									18
Salix vestita									22
Lichens	99	75	100	97	79	79	43	38	51
Mosses	100	100	100	100	86	83	85	100	98

*Term sp/spp indicates that the figure includes frequencies of species of this genera listed separately as well as species identified only to genus or, in some instances, in the case of sedges to family. The term seedl. infers reproduction either by seed or by layering and the two are not distinguished.

Table 5.7. Species Frequency in Black Spruce Stands North of the Forest Border (Isolated Clumps)—West to East Traverse

	Areas					
	Winter Lake	Clinton-Colden Lake	Artillery Lake–Pike's Port.	Dubawnt Lake	Kazan River	No. Ennadai Lake
Number of Stands	2	1	9	1	1	8
Species	Average Frequency Values					
Chamaedaphne calyculata	45					
Vaccinium microcarpum*	27		3			
Betula glandulosa	47	90	50	40	85	53
Empetrum nigrum	12	100	69	65	65	50
Grass sp/spp**	32	30	14	75	70	19
Ledum decumbens	80	100	34	20	65	11
Ledum groenlandicum	75	60	65	40	95	57
Vaccinium uliginosum	40	90	67	80	65	78
Vaccinium vitis-idaea	100	100	98	45	95	78
Rubus chamaemorus	52		32	95	70	25
Carex sp/spp**	17		43	75	30	34
Equisetum sylvaticum	42		20	45	45	33
Picea mariana (seedl.)**	65		30		35	15
Salix sp/spp**		30	7	20		
Loiseleuria procumbens			9			3
Poa arctica				35		
Carex aquatilis (stans)						9
Carex brunnescens						6
Carex saxatilis						6
Equisetum arvense						4
Calamagrostis canadensis						16
Potentilla palustris						4
Salix planifolia						11
Lichens	50	n.d.**	73	n.d.	70	n.d.
Mosses	100	n.d.	85	n.d.	100	n.d.

*This is my own composite for the myriad names given this species by the many authors who have dealt with it. Perhaps the best reference is Hultén (1968) and Cody and Porsild (1980) and I am tempted to list my own polemic on the subject (Larsen, 1982). Many authors, of course consider it a separate genus, *Oxycoccus*.
**See Table 5.6 for definition of terms; also, n.d. = no data.

Table 5.8. Species Frequency in Tussock Muskeg Stands Beyond the Northern Edge of the Forest Border–West–East Traverse

	Areas									
	Canoe Lake	Carcajou Lake	Reindeer Depot	Coppermine	Winter Lake	Artillery Lake	Aylmer Lake	Dubawnt Lake	Ennadai Lake	Yathkyed Lake
Number of Stands	8	10	2	9	1	4	4	7	7	1
Species	Average Frequency Values									
Anemone canadensis	13									
Carex consimilis	16									
Dryas octopetala	4									
Petasites frigidus	4									
Polygonum bistortoides	13									
Cassiope tetragona	13	6	38							
Polygonum bistorta	19	6								
Salix pulchra	7	19								
Loiseleuria procumbens	6						19			
Carex scirpoidea		7								
Carex bigelowii	34	21								
Carex aquatilis (stans)	5			3					11	
Pyrola grandiflora	5		15							
Astragalus alpinus		12								
Carex atrofusca		7								
Pedicularis lanata		8								
Kobresia simpliciuscula		3								
Hedysarum alpinum		5								
Rumex arcticus		2								
Saxifraga sp.		7								
Arctostaphylos alpina	44	27	57	35		6	56	34		
Pedicularis sp/spp*	36	25	22	23	10		7	4		
Tofieldia coccinea		10								
Lupinus arcticus		14	25	15						
Luzula confusa		2	8							
Potentilla fruticosa		2	3							
Salix arctica		4		24				10		
Salix glauca		12	12	55				5		
Salix reticulata		17		48				20		20
Dryas integrifolia		57		30				19		
Rhododendron lapponicum	43		8		5		19			
Pedicularis labradorica		21	25			10		7		
Polygonum viviparum		8		21				24		
Pyrola secunda		5		3				4		
Tofieldia pusilla		2						9		
Hierochloe alpina		10		4				3		
Saussurea angustifolia		23		32				6		
Betula glandulosa	96	64	85	53	40	55	60	81	72	85
Carex sp/spp*	73	40	45	46	100	73	27	68	86	55
Empetrum nigrum	56	16	52	25	75	40	49	46	22	
Eriophorum spissum	35	43	50		35		16	30	9	

Table 5.8 continued on next page.

Table 5.8. *Continued.*

	Areas									
	Canoe Lake	Carcajou Lake	Reindeer Depot	Coppermine	Winter Lake	Artillery Lake	Aylmer Lake	Dubawnt Lake	Ennadai Lake	Yathkyed Lake
Number of Stands	8	10	2	9	1	4	4	7	7	1
Species	Average Frequency Values									
Ledum decumbens	83	57	75	30	100	96	100	83	95	95
Rubus chamaemorus	29		17			74	59	45	69	55
Salix sp/spp*	7	52	12	85		20	10	48	83	74
Vaccinium uliginosum	69	41	10	55	70	50	25	79	55	25
Vaccinium vitis-idaea	98	65	100	28	90	95	95	81	79	75
Andromeda polifolia		7			100	90	14	38	51	25
Grass sp/spp*		27	32	4			61	35	4	40
Carex lugens			30	11						
Anemone parviflora				3						
Carex misandra				7						
Equisetum arvense				11						
Eriophorum angustifolium				11						
Kobresia hyperborea				7						
Pedicularis capitata				13						
Pedicularis lapponica				3						
Salix richardsonii				22						
Pedicularis sudetica				7				4		
Carex vaginata				6				22		
Calamagrostis sp/spp*				3			7			25
Poa arctica				13				6		
Salix planifolia				4			10			
Senecio atropurpurea				11						
Pinguicula villosa					25					
Rubus acaulis					90					
Carex rotundata					30	7		11		20
Vaccinium microcarpum (see Table 5.7)					20		10			
Carex rariflora						16				
Salix arbutifolia						20		13	14	35
Luzula parviflora							15			
Arctagrostis latifolia								4		
Carex capillaris								8		
Equisetum scirpoides								6		
Pedicularis flammea								4		
Potentilla palustris									4	
Carex saxatilis									27	6
Salix arbusculoides									24	4
Salix arctophila									45	15

*For definitions see Table 5.6.

climatic conditions become increasingly "Arctic" northward owing to increased frequency of occurrence of air masses characteristic of, and originating in, the northern regions. Absence of weather data for the areas makes it impossible to make a detailed comparison of the frequency of occurrence of air masses of various types, but it can be inferred that northward the influx of air of more southerly origin occurs at relatively more rare intervals, increasingly rare northward, and that it contributes to a much lesser extent than arctic air to the environmental conditions to which plant life is subjected. A comparison of the radiation characteristics of energy budgets of forest and tundra has revealed differences of sufficient magnitude to warrant further study of the possibility that energy relationships on a gross forest-atmosphere scale are involved (Larsen, 1973).

Spruce stands found north of the forest border very likely represent relicts surviving from a period when the border lay farther north than it does at the present time. Small clumps of black spruce, probably clones since reproduction appears to be primarily by layering, thus constitute a potential source of propagules for aforestation of the entire area (Clarke, 1940; Larsen, 1965; Elliot, 1979a,b) were the characteristics of the environment to ameliorate, permitting the spruce to again occupy most of the available land surface.

The evidence indicates (Nichols, 1967) that the forest retreated from a former position at or near Dubawnt Lake about 3500 years ago. It appears to have fluctuated north and south of its present position at the south end of Ennadai Lake, for example, since that time, having produced two fossil soil profiles, one overlying the other, which have been observed at the south end of the lake (Bryson et al, 1965). These periods of soil formation were separated by a period in which aeolian sand blew over the profiles from nearby eskers, indicating an interval during which the area was drier, and probably colder, than it is at the present time. Species with arctic affinities, as well as species found more abundantly in boreal regions to the south, can be found in rare sites in the area, indicating that their absence in communities cannot be attributed to insufficient time for migration (Larsen, 1973).

There does exist one important series of climatological observations that may bear significantly upon the nature of the frontal zone in the region under consideration. Ragotzkie (1962) and McFadden (1965) reported on observations of dates of freeze-up and break-up of lakes in central Canada, with maps showing (for various periods in spring and fall) the position of the zone in which some lakes are frozen and some are not (i.e., the zone between the line north of which all lakes are frozen and the line south of which all lakes are open and free of ice). It is of interest that this zone is notably wider in the Keewatin area than in areas farther west. The floristically depauperate zone is correspondingly wider in the Keewatin area than it is in the area, for example, around the eastern arm of Great Slave Lake and Artillery Lake. In the Keewatin area, jack pine is absent for a considerable distance (perhaps 100 miles, according to my observation from the air) south of the forest border at Ennadai Lake, while it is found in some abundance at one point along the portage between the east arm of Great Slave Lake and Artillery Lake. Moreover, northern species, notably *Rhododendron lapponicum* and *Dryas* species are found at Fort Reliance, within the spruce forest, but to the east along the Kazan River they are not noted until one travels many miles north of the forest border. The same is true of other species of arctic affinities. Here there exists a wide belt in which only the more ubiquitous species are found in sufficient abundance to appear regularly with high frequencies in transects. Obviously, it will be of interest and perhaps of some considerable ecological significance to explore these relationships further and attempt to determine the physiological characteristics that account for the distinctive response on the part of these plants to the climatic conditions that prevail.

Dominant Species

Comparing these far northern forest border communities with more temperate boreal forest communities reveals some interesting differences. In previous studies, quadrat frequency data were obtained on the understory shrub and herbaceous species found not only in forest stands dominated by black spruce and white spruce but also stands dominated by aspen (*Populus tremuloides*) and by jack pine (*Pinus banksiana*). The variation in these, too, has been shown to be correlated, at least to a degree, with climatic parameters (Larsen, 1974, 1980). It can readily be discerned from the averaged frequency distributions that the communities of both forest and tundra are each characterized by a few dominants, a larger number of species of intermediate importance, and an even larger number of rare species occurring often only once or twice in the data of a sampled stand.

Grouping the data into logarithmically based classes makes it possible to discern the relative importance of the dominant, intermediate, and rare species (Fig. 5.1a,b). The frequency Class 1 includes all of the species occurring only once in a transect, Class 2 includes all species with an average frequency of 10% and half of those with a frequency of 15%, Class 3 includes all species with an average frequency of 20 or 25% plus half of those with a frequency of 15 and 30%, Class 4 includes all species with frequencies of 35 to 60% plus half of those with frequencies of 30 and 65%, Class 5 includes all species with frequency values greater than 65% and half of those with a frequency of 60%, and these latter are thereby classed as dominants.

Marked differences are now discernible between forest and tundra communities as well as between communities in each category, as revealed in Fig. 5.1a,b. In forest communities, the number of species occurring in each quadrat is markedly highest in aspen forests and lowest in jack pine forests. All forest communities demonstrate a general decline from large numbers of infrequent species to a few frequent species. The black spruce communities, however, have nearly equal numbers of species in each class, taking the entire sample of black spruce communities together. There are, however, major differences between the black spruce communities in the main region of boreal forest and those in the north. At the northern limit of forest (northern forest-tundra ecotone) the structure of the black spruce communities more nearly resembles the tundra communities, the latter quite distinct from the southern forest communities in the proportion of dominant species.

The distribution of species in the far northern black spruce communities closely approaches the distribution characteristic of the tundra communities, in which a larger proportion of species attain relatively high frequencies and fewer occur rarely. The ratio of dominant to rare species clearly demonstrates a correlation with a north-south climatic gradient across the forest-tundra ecotone.

Species dominant in their respective communities often also display the highest degree of relationships to airmass frequencies. In Table 5.9 are listed, for several communities, species showing both dominance and high correlations with airmass frequencies. These species increase in frequency northward throughout the forest-tundra ecotone and decrease northward in tundra. It is evident from range maps showing the geographical distribution of these species that here we have the interesting spectacle of species correlating positively with the frequency of arctic air in forested regions (thus decreasing in frequency southward) and with airmasses of southern origin in the tundra (decreasing in frequency northward). These are the species that also attain their highest frequencies in the communities of the forest-tundra ecotone.

In the black spruce, white spruce, and rock field communities especially, the species with the highest correlations with airmass frequencies are consistently among the dominant species in the sampled stands. In the black spruce community, species showing the highest relationship to arctic airmass frequencies are those that attain dominance in the more northern areas. Dominant species in the rock field community are those with a high degree of correlation with airmasses of southern origin; these are species that attain dominance in the more southern tundra areas. In the tussock muskeg community, the relationship becomes less apparent. None of the species showing high correlation with an airmass type are among the first 10 dominants in the community. Dominance and airmass relationships in the low meadow community are even less discernible.

It is of interest that the species in the rock field community highly correlated with arctic airmass frequencies do not attain dominance in the forest-tundra ecotone. These species include *Salix glauca*, *S. reticulata*, *S. arctica*, *Luzula confusa*, *Arctostaphylos alpina*, *Cassiope tetragona*, *Dryas integrifolia*, *Polygonum viviparum*, *Carex atrofusca*, *Arctagrostis latifolia*, and *Eriophorum angustifolium*, all of which are species with strong arctic affinities and obviously do not attain abundance in the rock field communities near the forest border but only much farther north. There is in the central Canadian region where this study was conducted, a notable absence of arctic species in the low-latitude tundra communities. In place of these species are those wide-ranging species previously noted that attain dominance at the northern edge of the forest and in the southern edge of the tundra.

Competition and Structure

It is generally observed that competition is of less significance in communities of northern plants than elsewhere, but this is not to say that, north of the forest border, competition ceases to be a factor in the ecology of the plant communities. Rather, it continues to be of importance, but declines gradually toward the higher arctic latitudes where, as one can see by observation, plants do not totally cover areas available for occupancy but seem to be scattered randomly over sites affording suitable substrate and habitat conditions. Between individual plants is bare ground. For reasons of chance or conditions

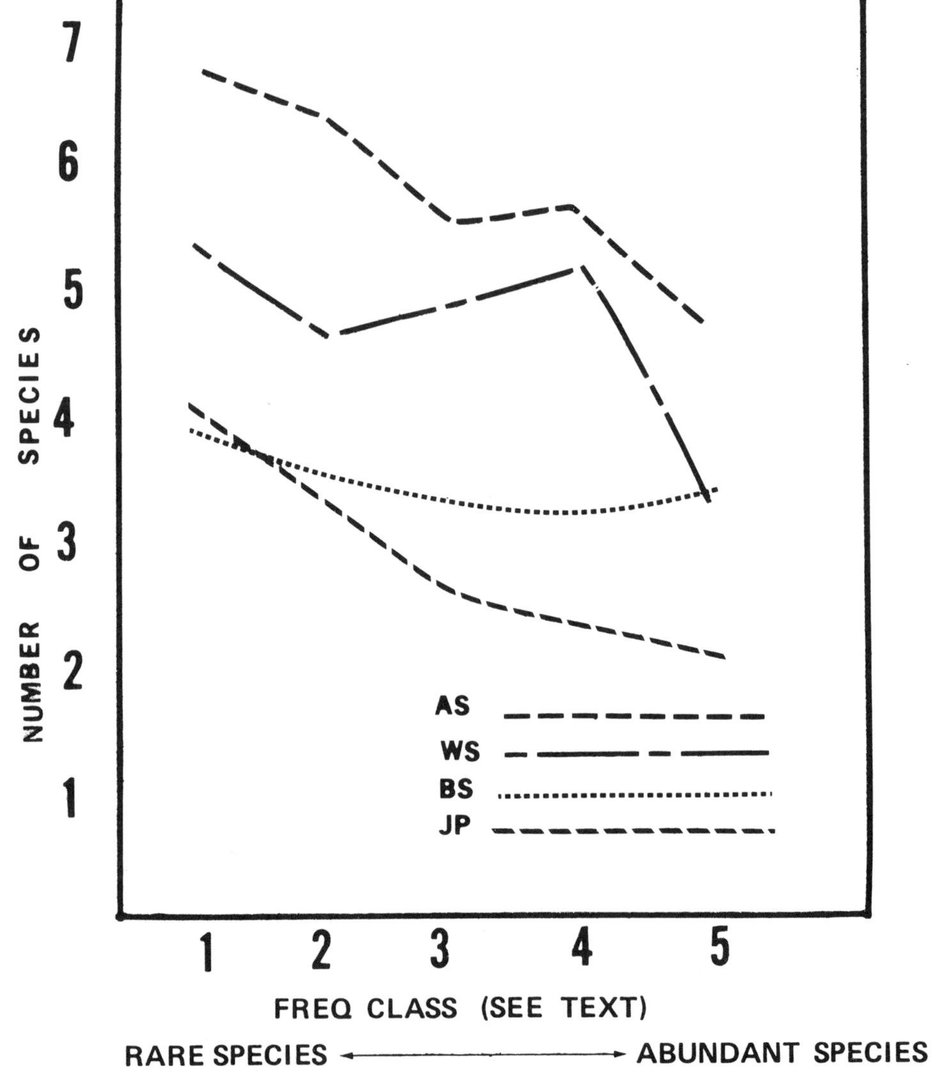

Figure 5.1A. Comparisons of numbers of species in the various forest and tundra communities. For descriptions of the forest communities see Larsen 1980, of the tundra communities, Larsen, 1965, 1971, 1972, 1973. The species frequency data are grouped into logarithmically based classes, making it possible to discern the relative importance of dominant, intermediate, and rare species in each community (see text). The communities are designated as follows: AS—aspen forest; WS—white spruce forest; BS—black spruce forest; JP—jack pine forest; TM—tussock muskeg tundra; LM—low meadow tundra; RF—rock field tundra. Noteworthy is the fact that tundra communities (and northern black spruce forest communities) possess relatively fewer rare species and more abundant species per average quadrat. Scale at left refers to the number of different species found per average quadrat (from Larsen, 1974).

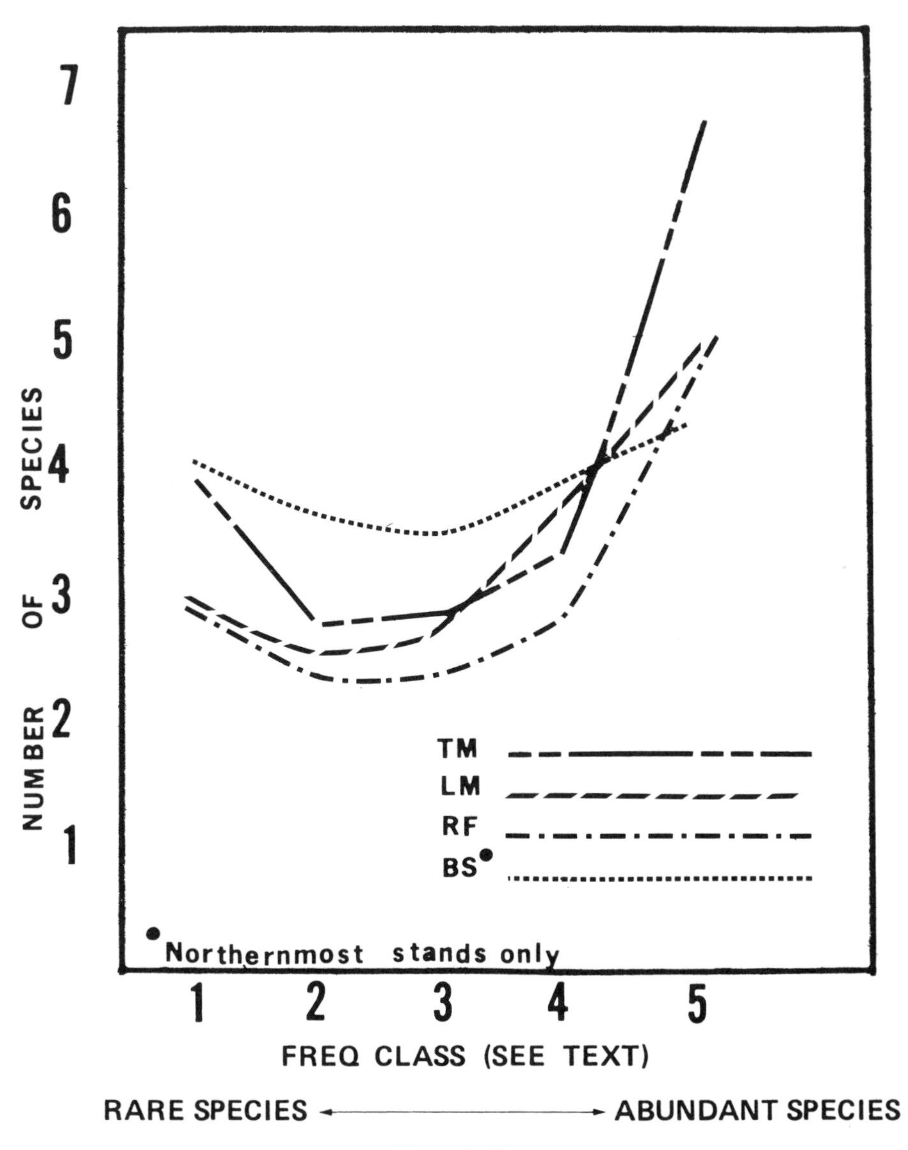

Figure 5.1B.

unseen and perhaps unsuspected, these areas have retained their status as land available for colonization.

There are, however, exceptions that would tend to invalidate any general rule that in the Arctic the principle of competitive exclusion has been waived altogether. At Pelly Lake and Snow Bunting Lake, for example, much of the terrain is occupied by a continuous mat mainly composed of *Alectoria* lichen species (*A. nitidula*, *A. ochroleuca*),

Table 5.9. Species in Various Communities Showing a High Incidence of Dominance and also High Correlation Between Species Frequency (in Community Transects) and Frequency of Airmass Types

Community	Number of Stands in Which Species Occurs	Number of Stands in Which Species Ranks 1st, 2nd, 3rd, 4th, in Frequency Value				Correlation with Frequency of Airmass Type
		1	2	3	4	
Black spruce						
Vaccinium vitis-idaea	104	63	22	10		0.65 (Arctic)
Empetrum nigrum	56	2	8	4	8	0.69 (Arctic)
Vaccinium uliginosum	48	9	8	10	7	0.56 (Arctic)
White spruce						
Vaccinium vitis-idaea	23	6	4	3	2	0.65 (Arctic)
Empetrum nigrum	18		4	3	3	0.76 (Arctic)
Betula glandulosa	12	4	1	1	1	0.67 (Arctic)
Ledum groenlandicum	25	1	3	1	1	0.51 (Arctic)
Cornus canadensis	23	5	1	3	1	0.48 (Pacific)
Rubus pubescens	15	3	1	2	3	0.67 (Southern)
Jack pine						
Vaccinium vitis-idaea	20	11	4	5		0.68 (Alaska-Yukon)
Maianthemum canadense	9	3			3	0.84 (Pacific)
Rock field						
Vaccinium vitis-idaea	65	32	13	6	2	0.86 (Pacific)
Ledum decumbens	45	12	8	9	9	0.68 (Pacific)
Vaccinium uliginosum	50	9	4	9	7	0.58 (Pacific)
Arctostaphylos alpina	44	6	7	8	2	0.56 (Pacific)
Empetrum nigrum	48	4	7	7	10	0.83 (Pacific)
Loiseleuria procumbens	32		1	6	2	0.60 (Pacific)

and it would seem as though such a carpet would stifle competition from other plants (competitive exclusion), an observation borne out by the nearly total absence of other plants on large areas covered by this dark fibrous lichen blanket (Larsen, 1972c). Such exceptions, however, serve to reinforce the rule holding more generally over most of the far northern regions that competition constitutes only a minor factor in the struggle for continued existence. The populations of plants are usually of low density, and it appears a major effort is devoted to wresting a livelihood from a difficult environment rather than from a neighbor.

In the low (latitudinally) Arctic and high subarctic, the available ground is largely occupied by plants which, individually, are in close proximity to other plants and, hence, could reasonably be considered to be in competition with them for existing nutrients and water. In the high Arctic, in contrast, plant densities are so low that root systems do not overlap, and root systems of the arctic species, unlike plants of temperate and semi-arid regions, generally occupy approximately the same areas as the aerial parts of the plants (Savile, 1960).

In the forest-tundra ecotone, however, high densities of plants appear almost everywhere, except on the summits of rocky or gravel hills. In the tussock muskeg communities especially the plants grow together in rather tight communities. This distinctive character of the tussock muskeg community is of considerable interest. In geographical range, the community is found with decreasing frequency over the landscape northward and southward of the forest-tundra ecotone, an observation first made by Tyrrell (1896) in his trips down the Dubawnt and Kazan in 1893 and 1894. In the course of the north-south transect in Keewatin, described on previous pages, the community was common at Ennadai Lake and Dubawnt Lake, but it was not found northward at Pelly, Snow Bunting, and Curtis Lakes, where at least moderately severe arctic conditions prevail. The community appears to be a relatively stable one, and although frost action may be a fairly common disturbance factor, recolonization of disturbed sites by species of the original tussock muskeg community is accomplished with what, for these northern areas, is probably a rapid rate.

The tussock muskeg community is, in physiographic terms, composed of hummocks, and around each hummock is a low wet area, the latter forming a network enclosing the hummocks which may be two or as high as three feet from base to summit. The low wet network usually is dominated by sedges, the hummocks by ericad shrubs and mosses. In the quadrats of the size employed in this study (which for statistical reasons were held uniform in shape and size throughout all sampling), both sedges and ericads occur with high frequency, which at first glance appears to be an improbable occurrence since sedges are abundant in low meadows and ericads are abundant on rock fields. Their co-dominance in the tussock muskeg community, however, is readily explained by the hummocky physiographic form of this community.

In summary, the decreasing frequency northward of atmospheric conditions related to the arctic frontal zone during summer months (and, hence, increasing frequency of arctic air masses) is associated with the increasing presence of arctic species northward in the regions studied, and it thus appears possible to define the extent of this climatic transition zone on the basis of the composition and structure of the plant communities of these areas.

Past and Present Climates of Tree-Line

Major factors in the limitation of plant life in the Arctic are severe climate and poor soils. The former contributes to the latter in restricting soil development and the formation of distinct soil horizons as well as the decomposition of detritus. Soil analyses show that subarctic woodland soils are notably deficient in nutrients (Moore, 1982).

Moreover, although the climate is strongly continental, with extreme temperatures ranging from −50°C to +35°C, the semi-arid conditions reduce the intensity of chemical weathering of the inorganic substrate. What nutrients are available often are locked in undecomposed organic material. Soil temperature measurements and moisture content indicate that black spruce forest ecosystems have the wettest, coldest forest floor (Van Cleve et al., 1983), resulting in the accumulation of materials which do not decompose and the appearance of shallow permafrost in the very thick organic soil surface layer (Viereck et al., 1983). What might be termed the mesic sites at the edge of the forest-tundra ecotone are occupied by tundra communities dominated by ericoid shrubs. Within this zone, changes in the relative amounts of major community types appear to be primarily controlled by differences in topography (e.g., see Ritchie, 1962).

The zonal categories of forest and tundra in northern Canada, especially on the Precambrian Shield, are aligned roughly in parallel with such climatic parameters as length of growing season and the mean daily temperature of the warmest month. Variation in vegetation locally is apparently controlled by historical factors (largely unexplored), disturbance by fire, and local topography (Ritchie, 1962). It was noted briefly in previous discussions that past fluctuations in climate have resulted in cycles of advance and retreat of the forest border, and hence of climatic conditions characteristic of this ecotonal zone, and much palynological work has been carried on to discern the general climatic conditions existing during various times of the Holocene (the time period since retreat of the Pleistocene glaciations) as reflected in the position of the northern tree-line. Moreover, the physiological and genetic basis for the rather narrow geographical expression of the limits of growth for the tree species, both in boreal regions and elsewhere in the world, have likewise been the subject of some considerable discussion (see Clausen, 1963; Tranquillini, 1979; Savile, 1963, 1972; Larsen, 1965, 1974, 1980; Elliott, 1979a,b; Black and Bliss, 1980).

On the basis of much palynological work carried out in central northern Canada, a rather complete climatic history of the region has been constructed, and the result has been summarized by Andrews and Nichols (1981) who reconstruct the July temperature history of the region as follows:

(1) temperatures were above average between 5500 and 4000 BP (before the present); (2) temperatures were below average (or average) between 4000 and 3000 BP; (3) temperatures were above average between 3000 and 2000; and (4) temperatures were below average between 2000 and the last few hundred years.

That the position of the northern boreal forest border is controlled by climate has also been corroborated by Carter and Prince (1981) who show that the structure of the border zone possesses similarities to a model in which the sites available for occupancy beyond the limits of the forest proper are of lower density than those within the forest. There are, thus, sites suitable for occupancy beyond the forest proper limit, but removal of the plants colonizing these sites by one factor or another occurs at a rate equal to the occupancy rate, unless conditions, usually climatic, change.

This structure also supports the concept that spatial isolation, as would occur in the occupied sites beyond the forest proper, makes possible hybridization and chromosomal reorganization in small marginal populations, which results in the appearance of ecotoypes, hybridization and introgression, and, one might imagine, occasionally even of forms that might be considered new species (Lewis, 1966; Eldredge, 1985). This also explains the presence of black and white spruce introgression apparently in evidence in at least the Ennadai Lake area (Larsen, 1965).

The hypothesis that the clones of northernmost trees in central Canada (Ennadai Lake, for example) are historical in the sense that they remain as relicts from a complete forest cover that existed at some previous time, is supported by the observations of Elliott (1979a) that (1) pollen viability is 0%, (2) seed viability at tree-line and in the forest-tundra ecotone is 0%, (3) there is a lack of a buried conifer seed store, and (4) an almost total lack of either sexually or asexually produced tree juveniles in the ecotone.

In contrast, however, it appears that the northern trees in Labrador-Ungava are in a healthier position, with abundant sexual regeneration occurring in and surrounding the northernmost trees, on, for example, the Labrador coast at Napaktok Bay (57°56′N, 62°35′W) and in an ecotonal stand above the Koroc River in northeastern Quebec (58°41′N, 65°25′W), as discovered by Elliott and Short (1979). The environmental conditions at tree-line in Labrador-Ungava, thus, are now apparently showing trends that are at variance from those existing in central Canada, although there has been an overall increase in the tundra component of the forest-tundra vegetation in the region during the past 3000 years (Lamb, 1985). Studies by Payette and colleagues in northern Quebec show that a climatic warming trend is registered in white spruce growth not only during the 1920–1965 period but also a few decades before. However, no major change in the position of tree-line has been observed in that study area for the period spanning the last 100 years, and it seems likely that the response to climatic warming has been an increasing density of preexisting tree populations (Payette and Gagnon, 1979; Gagnon, 1982; Payette and Filion, 1985).

Evidence of recent warming is also found in Alaska, where active expansion of tree-line by many young trees less than 40 years of age has been noted in the Tanana-Yukon uplands in the vicinity of Fairbanks and in several localities of the Alaska Range (Viereck, 1979; Cooper, 1986), the consequence apparently of above-average summer temperatures in the period since 1932 (Blasing and Fritts, 1973). In at least some areas of the eastern sector of the Alaska Range, however, there is no evidence of any recent change in tree-line (Denton and Karlen, 1977). As for the region in northwestern Canada encompassing the lower Mackenzie Valley, Black and Bliss (1978, 1980) show that minimum average monthly temperatures for June, July, and August of 11°, 14°, and 11°C are adequate for black spruce reproduction, and these suggest that the present-day forest line is out of equilibrium with the climate. These values were derived from both field and laboratory experiments, and tend to support the evidence that forest regeneration is not occurring in many areas as a consequence of the existing climatic conditions. Discussion of the autecology of the various species found growing at tree-line in North America by Ritchie (1984, p. 123) is most revealing of many currently puzzling matters and is part of his classic reconstruction of the postglacial vegetational history of northwestern Canada.

To summarize, the fact that dominant species are rather closely correlated in their quadrat frequency counts with the frequency with which various airmass types invade a given region makes it all the more apparent that climate must be considered a major controlling influence on species making up the plant communities in the region under consideration. It is perhaps of some contemporary significance that response to any changes in climate could be anticipated within a relatively short time, depending upon the life-cycles and the reproductive characteristics of the individual species. Moreover, these plant species might well respond to subtle climatic shifts that could, one imagines, indicate the start of major climatic changes but which might otherwise not be noticed for quite some time.

The correlations between plant frequency and prevalence of various airmass types also reveal that climate, and particularly the climate of the growing season, must be taken into account if—or when—efforts are made to implement in quantitative fashion the modeling of plant community dynamics as proposed in the discussion of Jenny's equation at the start of this chapter. As field ecologists are well aware, variability in vegetation is often so great as to virtually preclude high correlations between plant community parameters and environmental factors. This is a response to an exceedingly complex set of environmental variables, but in certain of the plant communities sampled during the course of the study described in previous pages, notably the rock field community and in two of the forest communities, one dominated by black spruce, the other by white spruce, correlations of frequency values of some of the more abundant species with climatic parameters were relatively high, giving some support to the possibility that the species/climate relationship is a predictable one within reasonable confidence limits. It is of significance, also, that modeling of growth characteristics of certain arctic and subarctic species has been undertaken by Ritchie (1984), based on earlier modeling proposals of Monteith (1981), for determining crop response to varied weather conditions.

As Ritchie continues: another type of simulation study "uses basic information on the autecology of the chief species involved in an area to generate hypothetical plant communities, expressed as estimates of species abundance .... The value of this approach is that hypotheses relating climatic control to biotic factors such as competition can be tested."

All of this encourages one to believe that plant ecology, throughout its history a field replete with impressionistic brush strokes, may yet become a science.

6. Northern Soils

Any comprehensive discussion of far northern boreal and arctic soils would require another volume equal in length to this one and would likely double the length of the list of references. Hence, the topic is one that must largely be excluded from detailed consideration here. For example, a recent compilation and review of significant papers on one aspect of soil ecology alone—northern forest soil microbiology—includes some 500 titles (Hendrickson and Robinson, 1982).

The northern soils are, however, surely a major aspect of the northern environment, and there are a few notable works that should serve as entry into what is now a vast accumulation of published material. These should be mentioned at least in outline if only to indicate that a literature on these soils exists, in comprehensive, and should be considered relevant to considerations of vegetational community structure. Soil properties are perhaps less influential in northern ecology than elsewhere, more a relevant branch of geomorphology than plant ecology (although this is another opinion that will evoke controversy) but the subject is far from barren—as any exploration of the literature will reveal.

Soil investigations in subarctic Canada have been carried on for some decades, albeit sporadically, and from the first it was apparent that podzolic processes were prominent in soil formation, although many of the soils had not progressed to full podzol (spodozol) development. By 1970, there had been found some 19 major soil regions across the northern parts of North America and Eurasia, of which (in the older terminology) podzol, gray wooded, brown wooded, and gley podzolic soils were important in forested regions and the arctic brown and tundra soils were found beyond the forests (Tedrow, 1970).

The accumulated nomenclature for the podzolization process has become very diverse, as discussed briefly below, but the essential process of podzolization is the decomposition of litter, releasing organic compounds that initiate removal of K, Na, Ca, Fe, Al, and other elements from the upper soil horizons. Readily soluble materials are leached by rainfall into lower levels, creating a light-colored layer beneath the upper organic material and a dark deeper horizon where the organic material and leached mineral compounds accumulate. Varying degrees of completion of the process result in soils known as podzolic, in which the horizons may not be visually as distinctive, or they may be detectable only by chemical analysis (Simonson, 1959; Retzer, 1965; Wilde, 1946, 1958; Moore 1974, 1976, 1978a,b, 1980, Nicholson and Moore, 1977). Once regarded as podzols, gray wooded soils are now recognized as a separate great soil group, developed from highly calcareous till under dense stands of mixed coniferous and deciduous trees (Nygard et al., 1952).

A number of symbolic notation systems for soil horizons are in general use, and for the general reader some confusion is inevitable. The upper layer of fallen leaves, needles, twigs, and other detritus as well as partly decayed humus, is known as the 0 or A_0 horizon (or L horizon in another system; L = litter), and these are further divided into the very uppermost layer, 00 or A_{00}, of recently fallen material, the 0 and A_0 of partly decomposed but still identifiable organic material, and the A_1 of virtually unidentifiable decomposed humus mixed with upper levels of mineral soil. The E or A_2 lies below the 0 or A_1 horizon, and is a zone from which most mineral elements excepting silica have been leached. The B_h or B_1 is a dark zone of accumulation of organic material leached from the A_2. The B_{fe} or B_2 is the zone of accumulation of iron and aluminum compounds, which characteristically concentrate just below the B_1, are typically brown in color, and may often harden into a relatively hard layer called hardpan or ortstein. The C layer is the unaltered parent material, derived from native rock, from which the soil has developed. The various zones are known by other symbols in other systems, some virtually interchangeable in the same publication, but the various zones in the podzol soils can be seen to correspond generally to the above description, although awkward discrepancies and differences will be discernible to anyone who looks into the matter closely enough (Soil Survey Staff, 1951, 1960; Canadian System of Soil Classification, 1978; Rieger, 1983).

An excellent recent discussion of the northern podzol and other soils, as well as the varied terminology in use in Canada, the United States, and the USSR is presented by Rieger (1983). It should be noted that the podzol soils are now termed spodosols in the US system, although the Canadian system continues to employ the term podzol, and the US system is further complicated by the addition of an entire new roster of names for the more traditional Canadian designations for the various podzol subdivisions.

The Podzolic Catena

Some insight into the role of the environment in soil genesis can be conveyed by a review of some early concepts of the formative relationships among the various northern forest soils as described by early soil scientists. Wilde and colleagues, for example, envisioned the existence of a schematic soil catena (topographic series) ranging from the upland regosols of crystalline rock outcrops, through sandy podzols, podzolized calcareous loams, melanized calcareous loams, wood peat, brown moss (*Hypnum*) peat, and finally to the lowland green (*Sphagnum*) moss peat soil with a very deep organic surface horizon (Wilde and Randall, 1951; Wilde et al., 1954; Wilde and Leaf, 1955).

Also diagrammed was a schematic soil catena for the interior of Alaska, with rock outcrops at one end, progressing through raw humus, podzolic, gley-podzolic, and finally to a forest-tundra soil, the latter two underlain by permafrost (Wilde and Krause, 1960; Krause et al., 1959).

The forest-tundra soil was seen as having a thick surface layer (5 in. or more) of raw humus; the A_1 had a very high content of incorporated organic matter. Parent material was primarily a gravelly sandy loam. Decomposition was known to progress slowly, although it was apparent that decomposition by soil microorganisms did occur, and later studies also revealed that numbers of bacteria, actinomycetes, and fungi are greatest in the upper zones, decline in the A_2, rise somewhat again in the B horizon, then decline to lowest levels in the C horizon of parent material (Boyd and Boyd, 1971).

Compared to regions southward, the Arctic has received little attention by soils scientists, but the characteristic properties of the soils are in general known and a basic understanding of the processes at work in arctic soils has accumulated. The soil patterns over the landscape are as complex as those elsewhere, with additional influences at work, such as permafrost and solifluction, not found in more southern regions (Rieger, 1974). There was an early inference that tundra soils are immature podzols, the result of cold temperatures, short summers, sparse annual rates of litter accumulation on the soil surface, slow rates of decomposition, incorporation of a high percentage of partly decomposed organic material in deep mineral horizons—in short, an arctic brown soil as defined by Tedrow and colleagues (Tedrow and Hill, 1955; Tedrow et al., 1958; Tedrow and Cantlon, 1958; Tedrow and Harries, 1960; Drew and Tedrow, 1957; Douglas and Tedrow, 1960; Hill and Tedrow, 1961).

Brown mineral upland soils with distinct profile features are found in the Arctic, particularly in the low-latitude regions not far north of the northern tree-line, where vegetational communities are characterized by relatively high densities of ericaceous shrubs. In these regions, field studies indicate that the forest-tundra ecotone does not mark a qualitative change in soil processes, only that rates at which the podzolization process occurs are influenced by low temperatures, lower rates of organic litter accumulation, low rates of decomposition, and so on (Tedrow and Harries, 1960; Tedrow, 1965, 1970). Farther north and well beyond the forest-tundra ecotone, soils are progressively less well-developed northward and weathering processes dominate (Hill and Tedrow, 1961; Douglas and Tedrow, 1960; Drew and Tedrow, 1962; Tedrow, 1966, 1968, 1977). Generalizations as to the intensity of podzolization processes north of the forest-tundra ecotone, or even within the ecotone, are not without controversy, however. There are those who maintain that podzol soils result from a combination of influences found only under forest conditions, with tundra environments so different as to exclude podzols from significance in tundra regions. Another even more extreme view is that toward the northern limit of the forest all podzol-formation processes disappear and are replaced by a tundra type of soil formation, characterized principally by weak interactions between organic and mineral constituents of the soils (Ponomareva, 1969).

Nutrient Availability

An early view was that forest soils developing under extreme cold of northern regions are influenced by parent material only to a limited extent, and that upper mineral layers of mature podzols are composed of nearly pure silica regardless of the nature of the parent material (Wilde, 1946). Evidence indicates that this view is, in general, true, and while exceptions most likely exist there are indications that calcareous parent material is, in time, leached of bases and eventually the upper horizons are transformed into typical podzol or podzolic soils. This is demonstrated by similarities in the pH of podzols in western North America, formed on Paleozoic sedimentary parent material, with those in eastern North America found on the crystalline rocks of the Canadian Shield, as shown below:

Number of Samples	Western (34)	Shield (7)
Horizon	Range of pH in Samples	
A_0, A_{00}, A_1	3.5–5.6	4.0–5.7
A_2	3.5–6.0	4.2–5.6
B_1	4.0–6.0	4.2–6.0
B_2	4.3–6.5	4.6–5.5
C	4.6–7.8	4.9–6.0

Nutrient Regimes

Lichen woodland is generally thought of as characteristic of the northern forest border, with widely spaced spruce trees and a ground cover of heavy lichen growth. Lichen growth is particularly thick and dense in the northern forests of Quebec, although it is present in varying degrees of ground cover across the northern hemisphere, encompassing some 4.4×10^6 km^2 in Canada alone (Rencz and Auclair, 1978, 1980).

Noteworthy is the fact that lichen woodland near the northern forest limit in Quebec possesses only a tenth the quantity of mineral elements in the organic soil horizons as does a northern boreal forest soil farther to the south. The mass of available P, K, Ca, and Mg in the mineral soil of the woodland averages some 29 times less than in the Alaskan taiga and muskeg (Auclair and Rencz, 1982; Van Cleve et al., 1981). There is, however, considerable variation in the levels of available or exchangeable nutrients in organic and mineral soils across the North American continent. On average, nevertheless, the level of available soil nitrogen in organic horizons of the lichen woodland is about a tenth that found in most forest soils southward. Soil temperature is considered to be the controlling variable in northern soil genesis, taking these areas as a whole, and there appears to be an additional factor of allelopathy by *Cladonia alpestris*, a dominant lichen in the lichen woodland, which inhibits decomposer fungi and reduces mycorrhizal activity (Brown and Mickola, 1974).

Among the more comprehensive studies of soil development in subarctic regions are those of Moore, carried out on soils at the forest border in northern Quebec. Accumulation of dead and decomposing organic matter on the soil surface is a distinguishing feature of the northern soils, and both the rate of production of organic matter and the rate of decomposition were considered by Moore to be major factors in pedogenesis in the northern regions. Litter and surface humus ranges are 7.0–14.0 kg m^{-2} in the low latitude shrubby lichen-heath tundra, less in spruce forests at the forest border (3.0–4.5 kg m^{-2})), and less also farther northward in arctic tundra (Moore, 1974).

Compared to steppe or prairie soils, litterfall in the podzolic zones is much poorer in base content. Moreover, the Ca, Mg, and K are rapidly removed from decomposing litter in northern forest soils, with P removed at somewhat slower rates but still far more rapidly than nitrogen, which is retained by the leaf litter from most species (*Betula glandulosa* is an exception) to a degree that limits its availability to growing plants severely (Moore, 1982). The Ca, Mg, K, and P are rapidly taken up by growing plants as they are released; this is made possible by the root systems which, in most northern species, are concentrated in the humus horizons of the soils (Strong and La Roi, 1983). That these elements are concentrated in the upper soil layer of both northern forest and tundra soils is shown by a number of analyses (e.g., in forest: Heilman, 1964; Moore, 1978a,b, 1980, 1981; Ponomareva, 1969; Payette and Filion, 1975; and in tundra: Tedrow, 1960, 1977; Tedrow and Cantlon, 1958; Tedrow and Hill, 1955; Tedrow et al., 1958; Douglas and Tedrow, 1959). Mineral elements are particularly concentrated in upper soil horizons following fire (e.g., Moore, 1981, 1982; Dubreuil and Moore, 1982), and decline as forest succession proceeds, particularly in the case of N, P, and K (Heilman, 1968).

Slow rates of decomposition and nutrient release from the organic detritus in upper soil horizons can be attributed in part to low temperatures and in part to the acidity characteristic of podzols. At pH levels below 5.0 there is restricted microbial activity resulting in nitrate release. The consequence, in general, is that very acid organic soils possess little nitrogen in a form available to plants. Below pH 4.5, other mineral nutrient elements also become of reduced availability: P, K, S, Ca, and micronutrients Fe, B, Cu, Zn, and Mo (Lucas and Davis, 1961). It is likely that there is no dearth of microorganisms capable of carrying out autotrophic nitrification, as studies have shown, but activity is limited by temperatures (Boyd and Boyd, 1971).

It is apparent that the northern plant species have certain physiological adaptations that make survival possible on organic soils deficient in nutrients (Small, 1972a,b), but it is also of significance that Auclair and Rencz (1982) found that Ca, Mg, and K were ample in *Picea mariana* tissues, evidently the result of efficient uptake and retention of these elements and effective recycling mechanisms. During excavation of root systems, they found that few *Picea* roots extended into mineral soil, indicating that nutrients were absorbed as they became available and before they were leached into the lower soil levels. This was not true in the case of N; acute deficiency was noted in *Picea* tissues, and available N in soil was low due to slow decomposition of nitrogenous compounds. The amount of P was marginally sufficient. Generally supporting evidence for these findings were reported from the opposite continental extreme, Alaska, where Dyrness and Grigal (1979) and Viereck et al, (1983) found that the black spruce ecosystem is situated near the end of the range of the cold, wet limits of tree growth, and the soils supporting black spruce are the wettest and coldest of the soils supporting tree growth in the taiga of interior Alaska, with the dominant forest floor species *Sphagnum* spp. and *Cladonia* spp., *Ledum groenlandicum*, *Vaccinium uliginosum*, and

V. vitis-idaea. In comparison, in Newfoundland, southward from both Quebec and Alaska, with a climate ameliorated over these northern areas, Damman (1971) showed that mineralization of organic matter was not excessively slow for any of the elements in a black spruce forest he studied. It is also important to note that a broad range of variation in substrate chemistry can exist in a relatively restricted area. Karlin and Bliss (1983) found, for example, that the pH and Ca content of water and substrate in six central Alberta peatlands ranged from, respectively, pH 3.5–8.2, Ca 2–120 mg L^{-1} and pH 3.3–7.8, Ca 4–138 m equiv. 100 g^{-1}. There were steep gradients in substrate chemistry, and these were followed by variations in plant community composition. Upper peat layers on hummocks, for example, were weakly minerotrophic to ombrotrophic and were occupied by shallow-rooted species such as *Ledum groenlandicum*, *Vaccinium vitis-idaea*, and *V. oxycoccus*. Plant species with deep root systems, *Rubus chamaemorus*, *Eriophorum vaginatum* spp. *spissum*, and *E. angustifolium*, were found in weakly to moderately minerotrophic sites. *Eriophorum viridi-carinatum* and *Scirpus caespitosus* were present only in the strongly minerotrophic peatlands.

In the arctic tundra, northward beyond the forest border, decomposition rates of soil organic matter are greatly influenced by moisture and temperature, the latter most noticeably, but ample moisture must be present for decay to take place. Thus, maximum rates of organic decomposition were found experimentally to occur at 19.5°C in arctic brown soil with 50% moisture content and in a silty, acid, mineral, upland tundra soils with 30% moisture content (percent of saturation). Organic matter in an arctic brown soil, dated by C_{14} techniques, was 2000 ± for humified material. The conclusion was that natural decomposition of a peat deposit 3 ft thick would require several thousand years (Douglas and Tedrow, 1959).

The consequence is that nitrogen nutrition is a limiting factor to plant production in tundra communities, shown by studies of primary production in a lowland wet sedge meadow and an upland birch-willow-heath community. Phosphorus supply did not necessarily limit production, but dilution of the soil solution in the wet meadow decreased availability. The low nutrient regime in the arctic tundra may partially explain, the author states, the high proportion of perennial plants, since these species can accumulate a nutrient pool over time in a deficient environment (Haag, 1974).

The low levels of decomposer activity in tundra soils is indicated by their structure, in which organic matter composition changes little vertically from upper to lower levels. There is a fairly high organic matter content in the mineral horizons of tundra soils, the result of transport by water percolating downward and from churning of the soil by frost action (Tedrow, 1965, 1977; Drew and Tedrow, 1962; Moore, 1978). Nitrifying bacteria are not common in tundra soils studied, and this accounts in large part for the low levels of available nitrogen (Gersper et al, 1980; Bunnell et al., 1980; Flanagan and Bunnell, 1980). Mycorrhiza are of importance in the uptake of nutrients by plants, and in the Arctic the roots are disrupted by freeze-thaw action and, in low areas, waterlogging of the soils.

It is of interest that studies of the coastal tundra at Point Barrow, Alaska, show that nitrogen fixation by blue-green algae is the major input mechanism for nitrogen in soils. The algae may be free-living, but in many instances are associated with mosses or occur as symbionts in lichens (Gersper et al., 1980). Elsewhere it has been shown that some nitrogen is fixed by nodulated plant species, including *Alnus* and other nonleguminous nitrogen fixers (Daly, 1966), as well as the ubiquitous lichen species *Stereocaulon paschale* and *Cladina stellaris* (*Cladonia alpestris*). This is shown by studies in northern Canadian subarctic woodlands by Crittenden (1983), which open the way for further investigations of soil nitrogen regimes in northern regions, as well as the extent to which rooting patterns of associated vascular plants may be affected by the presence of a *Stereocaulon* ground cover (see also Crittenden and Kershaw, 1978, 1979; Kershaw, 1977, 1985; Maikawa and Kershaw, 1976; Millbank, 1978; Lawrence et al., 1967; Lawrence and Hulbert, 1950; Boyd, 1958).

Permafrost

Permafrost has a major influence upon the soils of the forest border and the tundra northward, maintaining a permanently frozen substrate in all soils where it occurs, which is virtually everywhere in the tundra and beneath much of both tundra and forest in the forest-tundra ecotone (Day and Rice, 1964; Hopkins and Karlstrom, 1955; R.J.E. Brown, 1960, 1965, 1969, 1970a,b; J. Brown, 1967; Dimo, 1965; Tarnocai, 1984).

Permafrost impedes warming of the soil during the growing season and maintains the temperature of the rooting zone below the optimum; absorption of water and metabolic activity by roots are reduced. Tree species adapted to these conditions have lateral root systems, but frost activity in soil often causes movement of surface layers, and trees are dislodged from vertical and may even be felled (Benninghoff, 1952; Brown, 1970; Bird, 1974; Payette et al., 1976; Zoltai and Johnson, 1978; Price, 1972; Pettapiece, 1974, 1975, 1984; Nicholson, 1978a,b; Hansell et al., 1983).

In finer textured soils with permafrost and impeded drainage, there is a weakening of classical podzolization processes (Tedrow and Harries, 1960), but it has been pointed out that there is also an abrupt change in the nature of some of the soil-forming processes as well. The most obvious is cryoturbation, which must be considered a pedological process that results in the formation of hummocks, in solifluction, in formation of polygonal ground, and other consequences when the frost action is intense. Permafrost-affected soils should, some believe, be considered a separate group in any classification system (Pettapiece, 1975). The Canadian Soil Survey Committee (1978) classifies soils as within the cryosolic order when permafrost is within 1 m of the surface or within 2 m and affected by cryoturbation. When the latter occurs, turbic cryosols typically possess displaced soil horizons. Raw or decomposed organic material is displaced from the surface horizons and buried in the soil profile as streaks, clumps, or as humic material somewhat diffused throughout the lower mineral horizons. A weak B horizon may be found near the surface, but where frost action has been severe, any distinct zonation above the permafrost is lacking.

Geophysical characteristics of permafrost in Keewatin and tundra regions elsewhere have been studied fairly extensively, and it is apparent that the effects can be profound on not only the vegetation but on all activities related to human habitation and transportation (Shilts, 1974; Shilts and Dean, 1975; Van Eyk and Zoltai, 1975; Zoltai, 1972; Zoltai and Tarnocai, 1975).

Four summer's observation at Ennadai Lake afforded an opportunity to follow the development of a rather extensive area of tundra in which frost action was unusually pronounced (Larsen, 1965). An area of tussock muskeg with no initial evidence of frost action had become severely dissected by two years later, with the upper surface of vegetation and a 1 m layer of underlying peat lifted perhaps 8 ft above the level of the surrounding muskeg by a central core of clear ice. Hopkins and Sigafoos (1950) made note of similar features in Alaska and pointed out that such ice thrusting occurs where a relatively thin layer of peat overlies a sand substratum.

A record of permafrost depths in each of the major plant community types was maintained throughout a summer at Ennadai Lake. These data (Larsen, 1965, see Figs. 10–13) reveal rather uniform differences among the various communities, increasing in magnitude throughout the course of the summer and clearly differentiating communities by the latter part of July. Temperatures of the profiles from surface to the bottom of the active layer show that, in general, the level below the first few inches is distinctly colder than the surface layers, and sufficiently so to have a marked effect on physiological activity of roots. Comparisons of air at surface temperatures during daylight and night-time hours indicate that radiative cooling of the surface plays a major role in maintaining low soil temperatures, particularly on areas with minimal plant cover such as sand with sparse tufts of grasses.

In most, if not all, early papers and monographs dealing with patterned ground there appears to be an implicit assumption that polygonal and patterned ground phenomena are exclusively characteristic of tundra regions. During the course of sampling in the forest-tundra ecotone, it was my observation that some degree of patterning is a relatively frequent characteristic of soils in at least the northern portion of the boreal forest and the forest tundra ecotone. Moreover, Tedrow's extensive work in arctic soils indicates that podzolization processes are at work in soils of regions north of the forest-tundra ecotone, but these are often not clearly apparent because of the minimal development of the A_2 horizon, characteristically light-colored in podzol soils. In places, however, it was found that a thin light-colored A_2 was present in soils beneath tundra vegetation of dwarf birch and typical heath communities. (For photographs of well-developed podzols in tundra and patterned ground within forested regions, see Larsen, 1972.) J.C.F. Tedrow and Kenneth G. Taylor both indicate that they have also observed patterning under forest in northern Canada (personal communications). Frost mounds have also been observed near Fort Norman, Northwest Territories, within the forest zone (Van Everdingen, 1978).

Ecotonal Spruce

While upland tundra areas undoubtedly are subject to the greatest extremes of environmental conditions, exposed as they are to both winds and insulation as desiccating influences, it is nevertheless true that certain upland sites permit the continued tenacious growth of extensive clumps of black spruce in virtually all areas of the northern forest-tundra ecotone from east to west in North America. The problem of determining the limiting factors for growth of black spruce on such sites is one of the most interesting to be found in the forest-tundra ecotone, and at least some progress in elucidating the limitations on growth were made during studies at Ennadai Lake (Larsen, 1965). The special habitat enabling spruce to survive on upland sites at the north end of the lake (principally tundra over the large proportion of the landscape) evidently is best provided by a coarse rock substratum through which meltwater will drain during the early days of spring, when roots are still solidly encased in an ice lens but needles have undoubtedly begun transpiration. In such areas, conduction of heat by the boles of the trees during daylight hours is evidently capable of warming the tissues carrying on translocation somewhat above 0°C, thus providing at least a minimal water supply to upper portions of the trees. That the ice lens persists beneath the trees throughout the summer at Ennadai Lake is revealed in Table 8 in Larsen (1965), and similar perennially frozen mounds were observed under white spruce in Alaska by Viereck (1965) and Rapp, (1970).

In a study of the soils and vegetation in east-central Keewatin, Zoltai and Johnson (1978) found that seedlings of *Picea* and *Larix* were found only rarely, but when found they were almost invariably located on frost boils generally devoid of other vegetation, in effect a scarified seed bed. In all cases, the rare seedlings (perhaps established also by layering) were established close to a clump of the dwarfed trees, and it was apparent that no long-distance dispersal of seed was involved.

Zoltai and Johnson also point out that traffic over the areas studied would result in breaking up of the surface mat, making it susceptible to wind and water erosion. Stripping of the insulating vegetational cover would deepen the active layer, increase the moisture content of the soil, and susceptibility to subsidence and erosion. Under current climatic conditions, the clumps of trees in the area appear unable to regenerate when they have been removed by some disturbance. The stands, Zoltai and Johnson point out, are presently stable but in an easily disrupted balance with the environment, corroborating the evidence obtained in the Ennadai Lake area (Larsen, 1965; Elliott, 1979a,b) and in the far West of Canada (Black and Bliss, 1980). The effect of a single passage of a tracked vehicle in the Ennadai Lake area is shown in Fig. 6.1; the photograph was taken some six years after the tundra surface had undergone this disturbance.

Podzol Paleosols

The impact of climatic change and the controls exerted upon tree growth in northern regions by climate are described in detail and referenced in recent studies (Peterson et al., 1983; Jozsa et al., 1984). Edaphic evidence of climatic change is found in fossil podzol soil profiles in sites favorable for preservation north of the present-day forest border in Keewatin and elsewhere. The evidence was discerned initially at Ennadai Lake (Bryson et al., 1965) and is discussed in Chapter 1 (see page 2). Similar buried podzol soils have also been found at Artillery Lake (Larsen, 1972a) and elsewhere in Keewatin and Mackenzie (Sorenson et al., 1971; Sorenson and Knox, 1973; Sorenson, 1977).

During the course of sampling 40 sites in tundra and the forest-tundra ecotone, well-developed podzols were discovered in a number of places north of the present forest border, identifying areas northward of forest today that were once occupied by forest.

The evidence indicates that the forest border shifted on a number of occasions during postglacial (Holocene) history, lying 150 km north of the present forest border at maxima, at least 50 km to the south of the present border at minima. Significant periods of climatic deterioration and southward displacement of forest (and invasion by tundra), occurred between 3400 and 2900 BP, 2200 and 1800 BP, and 1100 and 800 BP. These periods were separated by milder conditions at about 2600 BP, 1600 BP, and after 800 BP. The conclusion was that the Arctic-subarctic furnishes a range of well-preserved evidence of the direction, character, and magnitude of climatic changes in the past (Sorenson and Knox, 1973). Changes in the atmospheric circulation patterns have been implicated in the displacement of the mean frontal zone crossing Canada during summer from northwest to southeast (see Chapter 9 on climate), causing latitudinal shifting of the vegetational boundaries (Sorenson, 1977).

Forest and Tundra Soil Genesis

An extensive effort was made by Sorenson (1973) to obtain sufficient numbers of samples of both forest-tundra and tundra soils, all taken on approximately similar well-drained upland sites in north central Canada, to determine whether or not there were pedogenic similarities between the tundra and forest-tundra soils. Later he used this knowledge to some advantage in obtaining evidence of past climatic change—for if forest soils were, indeed, markedly different from tundra soils, then the former should be found, still preserved one way or another, in tundra regions, and it could be concluded that forest once occupied what is now tundra and, hence, climate was once more favorable for forests (see above; also Sorenson et al., 1971; Sorenson and Knox, 1973; Sorenson, 1973, 1977). It is possible, also, to date the organic surface material of the soil or, better, charcoal from wood of a burned forest that once grew on the soil, using C_{14} dating techniques, and this was done.

Sorenson obtained soil samples from the various horizons in soils located at 40 sites in north central Canada, half in the forest-tundra, half in the low latitude arctic tundra, some of which overlapped into the extreme edge of tree-line. The sampling was centered principally in eastern Mackenzie, north and southeast of Great Slave Lake, and in Keewatin, around and between the Dubawnt and Kazan River drainages. Multiple regression analysis revealed 11 of 19 soil variables associated in varying degrees with air mass measures of climate, hence roughly a north-south axis based on the frequency of air masses of Pacific origin and of arctic origin during summer months.

Essentially, Sorenson found that podzols develop under forest in regions dominated by Pacific air masses in summer. Arctic brown tundra soils develop in areas dominated by cold, dry, windy air masses. The latter soils differ appreciably from forest podzols in genetic respects. For example, they are subjected to continual solifluction, cryoturbation, and deposition of wind-blown loess, all of which play a role in the incorporation of organic material throughout the soil horizons to a greater degree than is the case with the podzols, in which horizons with incorporated organic matter are distinct from those lacking organic material.

All measures of organic matter in the soils showed a general increase northward, and showed also greater variability from one site to another northward in the amount of

Figure 6.1. Photograph showing the long-term influence of a vehicle track on the vegetation and soils of moist tundra. Local reports indicated that the vehicle, a small tractor, had passed once over the site in this photograph about six years earlier. Area is in the low-latitude arctic tundra just north of the forest border.

incorporated organic matter. The pH of the soils, however, appeared not to vary greatly over the region studied, with some differences in response to the nature of the parent material. The shallowest soil profiles—the depth of the solum—were found in forests at the northern edge of the forest-tundra ecotone, on average deepening both into more closed forest southward and into tundra northward. The most intense leaching of materials from the A horizon occurred southward in the forested sites. Hence, intensity of podzolization increased with greater frequency of warm and moist air masses. Organic content of soils increased northward across the forest in consequence, presumably of decreased decomposition potential. Colder conditions inhibit the activity of organisms responsible for decay. Most forest mineral A horizons are leached of organic matter, with typical podzolic accumulations in the upper B horizon (Sorenson, 1973).

As with the forest and forest-tundra soils, the tundra arctic brown soils showed a good deal of group homogeneity. The thickness of the solum of individual profiles generally increased northward to the limit of the sampling area, although this presumably does not continue northward into regions where the soils begin to grade into the northern arctic regosols; Sorenson's study did not extend beyond the Arctic Circle in central Canada.

Total organic matter in the tundra soils was high, with the material generally dispersed throughout the soil rather than, as is the case with podzols, in horizons where it accumulates. There is also evidence that wind-blown transport of fine soil particles is important; where air currents favor deposition of such materials, accumulations of loess increase the rate of incorporation of organic matter into mineral components of tundra soil. Tundra soils appear to be rich or poor in organic matter depending upon site exposure, local winds, and precipitation regime. Also, soils near the forest-tundra ecotone have more organic variability than those of the more distant tundra. The distribution of iron (which is translocated in podzolization processes) was more uniform in tundra than in forest soils, indicating that they are only weakly podzolized, but nevertheless confirming Tedrow's (1977) observation that these influences are to at least a degree apparent in the arctic soils (Sorenson, 1973).

Because of the sharp changes in soil characteristics from forest, through the forest-tundra ecotone, to the tundra, Sorenson further concludes that the forest and tundra soils are genetically divergent. The difference essentially results from a transition from dominance by chemical processes in podzolic soil genesis to dominance by mechanical processes in genesis of the arctic brown soils. The latter can be regarded as possessing horizons, but these differ markedly in character from the horizons of the forest podzols. The combined forces of mechanical weathering, erosion, aeolian deposition, and frost action dominate the genesis of arctic brown soils, with the two latter influences most important. The importance of aeolian deposition increases sharply across the ecotonal zone from forest to tundra.

Regional Soils: Outlines

Most soils of northern Canada have formed on glacial deposits of late Pleistocene age, overlying a variety of ancient crystalline rocks making up the Canadian Shield in the East, sediments of Silurian age in the Hudson Bay lowlands (sandstone, shale, conglomerates), and sedimentary Devonian and Cretaceous rocks of the western Canadian sedimentary basin between the Canadian Shield and the Cordillera. The Canadian Rocky Mountains consist of Mesozoic intrusives together with Paleozoic and Mesozoic sediments (Prest, 1979; Retzer, 1965; Bostock, 1970; Tedrow, 1977). In forested uplands of eastern Canada, well-drained soils have podzolic features as a result of coarse soil texture, rapid percolation of water, high acidity, and low evaporation rates.

On the eastern side of Hudson Bay, nearly all of which is underlain by ancient crystalline rocks of the Canadian Shield, deep well-drained valleys in the vicinity of Great Whale River have developed a solum to a depth of about 2 ft, some with well-developed podzol profiles, but more often the soil has a brunisolic nature, with only minor horizon formation, and the vegetation is dominated by lichen woodland. The pH values usually range between 4.5 and 5.5, with the greater acidity near the surface. Surficial deposits of any size are confined almost entirely to the valleys; highlands are treeless but there are open forests in the valleys. Gleysolic soils are found where the water table is high and where permafrost is usually present at a depth of above 2 ft (Payette and Gauthier, 1972; Tedrow, 1977).

Intensity of podzol development has been related to the nature of the parent material in northern Quebec. Podzol features were observed under a wide range of vegetation, tree-line shrub, lichen-heath tundra, spruce-moss forests, and spruce-lichen woodlands, but development was greatest on quartzite-derived material and least on gneiss-derived deposits with a high content of ferro-magnesian materials (Moore, 1974; Nicholson and Moore, 1977; Belisle and Moore, 1981). The reason was held to be as follows: on quartzite, sesquioxides of iron and aluminum are mobilized by organic compounds and transported to lower levels with development of a podzol; on gneiss, release of iron and aluminum occurs, but only a small portion of the sesquioxides as such are translocated, and podzol features are weak or nonexistent. Thus, most soils analyzed by Moore fell into either the Orthic Humo-Ferric Podzol or degraded Dystric Brunosol subgroups of the Canadian classification scheme (Cryorthod and Cryochrept great soil groups, respectively, in the U.S. scheme), or podzol and brown-wooded, respectively, to use the older more general descriptive terms.

All but the extreme southwestern corner of Keewatin is in the tundra zone, underlain by continuous permafrost, with a variety of glacial landforms, with glacial drift dominating the landscape—eskers, kames, deltas, drumlins, morainic ridges, with large areas of well-drained sandy materials. Soils that resemble the Alaskan brown soil are found on well-drained sites, some with sola 8–10 in. thick. There are also very poorly drained soils, imperfectly drained meadow and upland tundra soils, and some thin organic soils on rock. Frost action is evident, and mineral and organic material is intermixed in the soils of many areas as a result of frost action and the formation of patterned ground (Lee, 1959; Craig, 1960, 1961, 1964a,b; Wright, 1967; Tarnocai et al., 1976; Tarnocai, 1977, 1984; Thomas, 1977; Cunningham and Shilts, 1974; Tedrow, 1977; Zoltai and Johnson, 1978).

In the high-latitude Arctic, at Hazen Lake on northern Ellesmere Island, for example, morphological and chemical characteristics of the soils indicate that there is no significant development of genetic horizons, thus soils are either gleysols on wet areas, or regosols on uplands. Low organic content of the surface soil is the result presumably of sparse vegetative cover and deficient soil moisture. Only where hummocks or perpetually wet soils were present, with a cover of *Eriophorum*, *Carex*, *Salix*, *Juncus*, *Dryas* and other species present, does the surface soil contain as much as 5% organic matter, although all of the soils had an organic matter content higher than 0.35% in the deepest layer. The highest organic matter content at many sites occurs in the deepest layers sampled, a characteristic of many tundra soils. Soils in this area are similar to those described in other parts of the northern Canadian tundra (Day and Rice, 1964; Rieger, 1974, 1981; Retzer, 1974; Rapp, 1970).

Brown wooded soils of the upper Mackenzie River area, west of Great Slave Lake, are described as coarse textured, well or imperfectly drained, with organic surface horizons, but lacking a distinct mineral-organic surface horizon. Another common soil, a brunisolic gray wooded soil, possesses profiles incorporating features of podzols in upper horizons and gray wooded soils in lower horizons. In both gray and brown wooded soils in this region, the parent material is calcareous. White spruce and balsam poplar are common associates in the forests of the area (Day, 1968; Pettapiece, 1975).

Soils northward in the uplands around Inuvik are of neutral pH, ranging from 6.9 to 7.4, beneath a continuous open spruce forest vegetation. Here the parent material is predominantly dolomite, limestone, and shale, which make up the Campbell-Dolomite

Uplands, major landforms of which are bedrock ridges and plateaus, stable slopes, shorelines, and depressions (Ritchie, 1977, 1984).

In the far west, along the Alaskan-Yukon border, in the interior and on the Arctic Slope of Alaska, and in the Firth River area, soils range from alpine regosols to upland calcareous tundra, meadow, and bog. Soils with podzolic features are present in the valleys and in the Alaskan interior, and arctic brown soils in various stages of development are widespread in well-drained tundra. There is a definite accumulation of deep organic material on the surface of forest soils (Drew and Shanks, 1965; Tedrow, 1970, 1977; Viereck et al., 1983).

Possibilities

Some interesting questions have been created by recent advances in knowledge of northern forest and tundra soils and their role in boreal and arctic ecology. One concerns the role of the mosses, particularly the *Sphagnum* species, of interest because they are found in the latter stages of forest successional development, both in wet and upland communities, and often constitute dominant ground cover species for extended periods of time, appearing in equilibrium with environment. They modify the conditions so that long-term persistence as members of the community is assured, they compete successfully with other species, and grow rapidly under conditions that appear detrimental to other species (Foster, 1984a,b). They also create an insulating ground cover that most likely raises the level at which the permafrost table is found in summer, decreasing the depth of the soil active layer, and, if not subjected to the influence of fire or other major disturbance, may eventually create conditions inimical to the continued growth and reproduction of forest trees. This, in turn, would result in eventual opening up of the forest canopy, warming and desiccation of the moss ground cover, deepening of the active layer, and reopening the habitat for reoccupancy by forest seedlings and ultimately trees. This, at this point, is a conjectural concept for cycling of the forest cover in certain northern boreal regions. Forests in these regions probably are subjected to fire too frequently for such cycling to occur naturally (at least very frequently), but it may occur in some particularly humid areas or those with high rainfall, and the possibility does create interesting possibilities as to the definition of mature climax forest (if such can be said to exist) in some boreal regions. In absence of fire, continued over long periods, there is evidence that a mixed black and white spruce forest might become a monospecific black spruce forest. It is, however, unlikely that fire would be absent for the long period of time needed for this change to develop (Strang and Johnson, 1981), and an even longer time would be required for any other hypothetical cycling to occur. There is also evidence that below-ground niche separation exists in boreal forest communities, based on root morphology and coexisting with the obvious need for shallow roots in soils with shallow active layers. Along with this observation, there is also some evidence that competitive abilities of different species are to a degree controlled by soil type, and whether these are climatically or genetically controlled is still uncertain (Strong and La Roi, 1983). Finally, in regard to podzolization processes in arctic soils, there still exists considerable uncertainty as to the genesis of the light-colored A_2 in some upland tundra soils (Larsen, 1972c; Tedrow, 1977) and whether the thin vegetational cover over soils in such areas provide sufficient chelating compounds to translocate iron and aluminum oxides to lower B horizons, or whether some other mechanisms are involved. All of these are interesting questions concerning the nature of the forest-tundra ecotonal environment that remain to be answered.

7. Faunal Community Relationships

The treeless interior of Keewatin remains one of the least known and most fascinating of the more remote regions of North America. Vast expanses of wild country stretching from Hudson Bay to eastern Mackenzie have borne the imprint of modern technology lightly and only within relatively recent time. Moreover, throughout much of the area, the only apparent modern influence has been a diminution of prehistoric caribou numbers usually attributed to the introduction of the rifle. The observations reported here were made during the course of field studies, conducted at various times during the summers of 1959–61, 1963, 1970–71, during portions of which it was possible to maintain a record of birds and mammals seen in the area as well as of the observations made on population density and behavior.

Ennadai Lake is situated in the southwestern corner of Keewatin (61°N, 101°W),, some 300 miles from the nearest outposts, Churchill to the southeast and Lynn Lake to the south. The village of Brochet, itself isolated, is 200 miles to the southwest. At the present time, all travel to Ennadai is by aircraft, primarily to supply the Aeradio Station of the Canadian Department of Transport at the northern end of the lake, operated by a staff of five persons. At one time, however, a small trading post existed at the south end of the lake, and even in recent decades a trapper reached Ennadai by dogteam from Brochet during the course of a winter. Trade at the post was conducted with a small band of Eskimos resident along the Kazan until the 1950s, but a tragic history of disease and starvation made it necessary for the Canadian government to evacuate these people to more accessible areas. During recent time, too, there was a trading post at Padlei, to the southeast, now abandoned, and there is a permanent, and fairly thriving settlement to the northeast at Baker Lake, some 400 miles distant.

While the earliest explorations of the region date to 1769 with Hearne's classic travels, culminating in his journey from Prince of Wales' Fort on Hudson Bay to what is now Coppermine, it was not until just prior to the turn of the 20th century that scientific investigation began, initially with geological explorations down the major rivers by Tyrrell (1897), whose detailed journals furnish a vivid account not only of geology but of the natural history of the areas, and which in many respects remain unsurpassed today. More recently, a number of excellent accounts of geological and biological surveys have appeared. Harper (1953, 1955, 1956) reported in detail on the birds and mammals of the Nueltin Lake area, Porsild (1950), utilizing Harper's botantical collection, listed the common plant species, Clarke (1940) gave an account of the region west and south of Dubawnt Lake, and Manning (1948) recorded birds and mammals seen over a large area of Keewatin and northern Manitoba. Lee (1959) summarized the surficial geology of the area, and Ritchie (1959 et seq.) provided a thorough account of the vegetation of areas in northern Manitoba. The aboriginal inhabitants of the Ennadai Lake area were evidently Eskimo at the northern end of the lake and reportedly Chipewyan Indians at the south end. The latter were probably only seasonal residents on hunting expeditions during the caribou migrations. At the present time, there is much evidence of past Eskimo settlement around the north end of the lake and extending along the Kazan River to Baker Lake.

Ennadai Lake itself is long and narrow, extending about 45 miles in a SW-NE direction and probably not exceeding 10 miles in width at any point. The area around Ennadai Lake is of considerable interest from a bioclimatological point of view because black spruce grows over almost all types of upland terrain at the southern end of the lake, while at the northern end spruce is to be found only on a very small proportion of the total land area. The terrain and vegetation at the north end of the lake are typically low latitude arctic tundra, except for scattered clumps of black spruce and dwarf birch (*Betula glandulosa*) in favorable ravines. There are countless small lakes and a rolling topography, possessing all of the common features of former glaciation. Permafrost is found continuously throughout the area, at depths ranging from about one foot in tussock muskeg to three or more feet in sand (in mid-August). The vegetation of the upland rock fields is composed typically of low arctic ericaceous species and lichens, grading into sedge meadows and tussock muskeg on areas of lesser slope and poorer drainage. A detailed account of the climatic features and plant communities of the area is given in an earlier publication (Larsen, 1965), and descriptions of other similar areas farther west have also been published (Larsen, 1972a,b,c, 1973).

General Phenology

In at least most years, all snow on uplands excepting that underlying spruce groves or patches, has disappeared by the last days of May. Many late snow patches, however, remain until two or three weeks later, and the last vestige of snow will persist until mid-July. In late May, the ice on Ennadai Lake was 67 in. thick in the spot measured and the lake is not clear of ice until the first days of July. Tundra ponds, however, begin to open as soon as the land is clear of snow.

The breeding season for the birds begins immediately following the melting of the snow over the greatest proportion of the land surface. The most spectacular displays are those of male ptarmigan, with perhaps a dozen or more to be seen at one time, each calling from a prominent point of the landscape, each of the birds still colored white excepting for the brown head and neck. Lapland longspurs are also in great abundance, performing mating flights at all times during the 24-hour period. At no other time of the summer will so many individuals of so many different species be visible over the landscape at one time.

According to the staff members of the Aeradio Station, caribou migration through the area begins in mid-May, and my observations note a marked decline in numbers by the first days of June, with no more than three seen in any single day until the last is seen around mid-June. By the third week in June, the behavior of the bird population had changed markedly. For a day in late June, my notes read as follows:

> The past two days have been windless and with temperature maxima of 75° and 77°F. The mosquitos have come out in some force. The entire landscape is now quiet, where a week ago it was filled with activity. Now it is as though nothing lived about except a few soaring jaegers and gulls and an occasional short-eared owl. A walk over the tundra will inevitably roust a hen sandpiper or ptarmigan and send her fluttering off in broken-wing decoy display, from which she makes a complete recovery once the danger is perceived as past. At this time the *Kalmia polifolia* is in first flower and the dwarf birch (*Betula glandulosa*) is about half leafed out.

By mid-July the population structure has begun to change noticeably. Harris sparrow and longspur young were visible in numbers, and semipalmated sandpiper adults feigned injury to decoy intruders away from young birds that appeared larger and considerably more vigorous than the parents. The first vacant Harris sparrow nest was observed July 5.

The longest periods of consistently good weather evidently occur during June, even though the lake proper is still mostly covered with ice. Ponds and smaller inlets, however, are ice-free, and ducks are seen in abundance. With the beginning of breakup, the weather becomes much more variable, and although warm days are often experienced, cool days with high winds, low clouds, and even rather frequent fog make lake travel throughout most of July difficult. By the beginning of August, the peak of summer definitely has passed. My notes of August 4 read:

> The numbers of birds have now declined greatly. Two afternoon hours spent in the field for the specific purpose of observing the numbers of birds revealed only a few juvenile longspurs and some arctic terns. Although an unusually high wind undoubtedly held birds in cover, these numbers would not be greatly exceeded under better conditions. Fall is in the air; the clouds are stratus and the wind holds constantly high and from the north. Lake travel is impossible. The *Arctostaphylos alpina* leaves have begun to turn color. *Vaccinium vitis-idaea* berries are crimson, and those of *Rubus chamaemorus* are now almost entirely ripe.

Variation from one year to another was most notable between 1961 and 1963, the latter year with spring at least a week earlier than two years before. The breeding season was well under way by the first of June, only a few male ptarmigan were still calling, and the caribou had all passed through. During a flight northward along the Kazan River, no caribou were seen 45–50 miles north of Ennadai, although they were then seen north of this point rather frequently but not in large numbers. It seems reasonable that the early migration was related to the snow depth, for the area had one of the lightest snowfalls on record during the previous winter.

Populations

Observations on changes in animal densities from year to year are subjective, although the changes were so marked they seem worthy of note. My journal of August 4, 1961 reads:

> There seems ample reason to believe that the bird populations, with the exception of ptarmigan, are down from last year—the result, perhaps, of higher juvenile mortality; the black fly and mosquito numbers are greatly declined over last year and constitute nowhere near the plague they were then. Thus, it might be reasoned that the food supply for the newly-hatched passerine birds has not been at last year's level. The ptarmigan, however, seem to be abundant and, additionally, the populations of *Dicrostonyx* and *Microtus* are also higher than last year.

In 1960 it was noted that the small birds fed heavily on the black flies. It was a common occurrence to have savanna sparrows and longspurs in our campsite, and from our actual counts we estimated that the young birds consumed, on average, 20 flies a minute while feeding. This all changed rapidly after August 15, however, initiated by a period of wet and windy weather. The flies virtually disappeared. The sparrows, redpolls, longspurs, and other species also vanished, presumably having begun the fall migration. Neither bird nor insect populations reached the levels in 1961 that they had in 1960.

Among earlier studies of the birds of the general region in which Ennadai Lake lies are those of Manning (1948), Mowat and Lawrie (1955), Harper (1953), and farther to the west, studies in the bend of the Thelon River by Norment (1985) and in the region of the Thelon Game Sanctuary by C.H.D. Clarke (1940) and Kuyt (1980). Numerous studies have been conducted in the Churchill, Manitoba, area, with that of McLaren and McLaren (1981) most attentive to description of the vegetation in which the birds were found, although only one of the study sites was actually in the tundra, having greater emphasis on forested regions to the south. Another study, by Cooke et al. (1975) records the birds observed in the tundra east of Churchill along the coast of Cape Churchill. In these studies, there is a noteworthy effort to describe the pertinent features of the habitat, but in these studies and others carried on in the general region (Jehl and Smith, 1970; Taverner and Sutton, 1934; Jehl and Hussell, 1966) the deficiencies noted by Erskine (1977) are not entirely absent and generally only two or three dominant species and physiognomic aspects of the vegetation are noted.

Erskine, in his study of the birds of the boreal forest, drew upon all major known works on avifauna in the conifer forests to relate the groupings and densities of birds to the major vegetation types, designated as conifer forests (spruce, fir, hemlock and

Table 7.1. Relative Frequency of Bird Species in Vegetational Communities. The value 100 assigned to the vegetational type in which the species is most abundant and other types assigned a percentage relative to this value.

Species	Total	RF*	TM	LCM	ESK	SC	BBC	SHO	W	IND
American pintail	6								100	
Old Squaw	3								100	
American Scoter	10								100	
American Merganser	8								100	60
Rough-Legged Hawk	3			50						100
Willow Ptarmigan	10	40	40	20		100				
Semipalmated Plover	13				50		12	100		
Least Sandpiper	9				28			100		
Semipalmated sandpip.	53	66	14	38		10		100	10	14
Long-Tailed Jaeger	7				16					100
Herring Gull	16			7						100
Arctic Tern	5							100	100	50
Horned Lark	12	100			50					
Raven	4									100
Gray-Cheek Thrush	4				100	100				
American Pipit	8	14			100					
Redpoll (?sp.)	18	20			20	100	30	10		
Savannah Sparrow	21	57	57			100	85			
Tree Sparrow	15	9	9			100	18			
Harris's Sparrow	37	30	8	4		100				
Lapland Longspur	179	100	27	11	40	14	6	3		
	441									

*Vegetation types are as follows: RF = rock field; TM = tussock muskeg; LCM = low Carex meadow; ESK = esker; SC = spruce clump; BBC = dwarf birch clump; SHO = shoreline; W = water; IND = indeterminate.

white or red pine, jack pine), bog forest (tamarack and/or black spruce), broadleafed forests (trembling aspen, balsam poplar, paper birch), and open habitats (early succession shrubs, grasslands), wetlands, open water, and riparian shrubs (shorelines). Erskine indicates that habitat descriptions employed in most bird censuses are largely inadequate, particularly when such descriptions are confined to numerical comparisons of the two or three major tree species. He indicates that the intrusion of a different habitat, making up as little as 10% or less of the total area, may introduce one or more major bird species and several minor ones which have nothing to do with the habitat represented by the general description.

The vegetational studies in the Ennadai Lake area furnished an opportunity, not fully realized at the time, to provide a detailed description of the habitat frequented by the various bird species in the area. In the Ennadai Lake area study (Larsen, 1965), a survey of the plant community relationships of the bird species inhabiting the areas was undertaken in 1961 during the period June 17 to July 14. While walking to and from botanical study sites—over distances often of at least several miles—a count was kept of the number of individuals of each species sighted on each of the major vegetational community types. These types are categorized as rock field, tussock muskeg, low *Carex* meadow, dwarf birch clumps, and shorelines. Categories were also established for birds

Table 7.2. Coefficients of Similarity Showing the Relationships Among Bird Communities*

	RF	TM	LCM	ESK	SC	BBC	SHO	W	IND
RF		52	26	29	34	31	19	2	3
TM			34	10	35	47	6	3	4
LCM				4	12	4	15	3	22
ESK					26	14	22	0	3
SC						36	4	2	2
BBC							9	0	0
W									23

*Values are coefficients of similarity among bird communities, i.e., relationships between communities on a 0–100% scale. Thus, the bird assemblage frequenting the rock field community (RF) is most similar (52%) to the assemblage frequenting the tussock muskeg (TM), and, in declining order of importance, to the bird communities frequenting the spruce clumps (SC, 34%), the dwarf birch clumps (BBC, 31%), the eskers (ESK, 29%), low Carex meadows (LCM, 26%), shorelines (SHO, 19%), water (W, 2%), and the indeterminate (IND, 3%), the latter usually high-flying individuals with no apparent relationship to any of the vegetational communities.

sighted on the water and those in flight at some height over the land surface. During the month, 460 individual birds were recorded in relationship to their surroundings. The most abundant, those observed more than three times, are listed in Table 7.1. The value 100 was assigned to the vegetational types in which the species was most often seen and the frequency of the other types were assigned a percentage relative to this value.

A coefficient of similarity was computed for each bird community with all others, and the result is presented in Table 7.2. This reveals on a percentage basis the interrelationships among the communities.

Only with the recent emphasis on habitat preferences of the bird species did the significance of this effort become apparent, and I have included the material here. It is, as far as I am aware, the only such survey carried on in the region. Certainly it was the earliest, since the vegetational community analysis utilizing quantitative methods was the first such study carried on in the Northwest Territories. The vegetational communities are described completely in Larsen, 1965, and similar studies of other tundra and tree-line areas are also pertinent (Larsen, 1967, 1971a,b, 1972a,b,c, 1973, 1974, 1980).

It should perhaps be noted that none of the more common bird species were restricted to a single vegetational community, and a number are shown to frequent several of the vegetational communities. However, the ordering of the bird communities by means of an ordination shows a strong correlation with habitat type, similar to that in the data gathered during Erskine's study of the bird communities in the main range of the boreal forest to the south.

There is, thus, a continuum within the habitat preferences of the bird species, just as it is also a characteristic of the vegetational communities, as can be inferred from the data in the vegetational study, a continuum of communities rather than discrete associations as envisioned by some plant ecologists, a matter of some controversy (Curtis, 1959; Whittaker, 1973, 1975; Cottam et al. 1973).

Thus, for example, the bird community inhabiting the high, rather barren rock fields is most closely interrelated with that on tussock muskeg and least related to that frequenting the water areas. The linear ordination, Table 7.3, places these bird communi-

Table 7.3. Linear Ordination of the Bird Communities. See text for description and discussion.

Ordination Number	0	1	3	7	11	14	15	25	29
Community	ESK	RF	SC	BBC	TM	LCM	SHO	IND	W

ties in relationship to one another (Bray and Curtis, 1957; Beals, 1960, 1985); those communities closest together or juxtaposed being most closely interrelated and those at different ends of the ordination being the least related. An ordination of the vegetational communities on a similar basis (Larsen, 1965) indicates the following relationships between the vegetational communities:

ESK–RF–TM–LCM

and demonstrates that the vegetational communities are arrayed along a continuum from the high esker slopes to the low *Carex* meadows, virtually coincident with the ordination of the bird communities.

Account of Species

Only a limited amount of collecting was done and, hence, no effort was made to identify the birds beyond the species level, and the same holds true for the animals observed. The listing of birds and animals observed is given in Appendix F along with brief remarks on some of the sightings. Additionally, an Eskimo phonetic pronunciation is included for some of the species, not according to any official anthropological scheme but merely as recorded by an unidentified Aeradio Station staff member at some time during the latter years of Ennadai occupation by the inland Eskimos. It may prove useful for comparative purposes to preserve what may now be a virtually lost dialect or local variations of the Eskimo language in the interior of Keewatin (see also Birket-Smith, 1941).

Weather Events

The Ennadai weather records and recollections of the Aeradio Station personnel provide some indication of the correlations that are probably significant to an understanding of the fluctuations in density of the bird and mammal populations during the years of observation.

The winters of 1959–60 and 1960–61 were marked by average or above-average snow depths in the Ennadai area, and these were both followed by summers in which bird and animal densities were at high levels, with a lemming high in either 1961 or 1962. The winter of 1962–63, however, was remarkable in two respects: (1) a notably light snowfall, and (2) a freezing rain on April 16 followed by extremely low temperatures. These two incidents would have an obvious influence on the populations of animals that are winter residents, and may well account for the low densities of small mammals in the summer of 1963. The low amounts of surface water may account for the observed reduction in densities of black flies and mosquitos, which constitute a main item in the diet of the juvenile birds.

Briefly: Some Implications

There is thus demonstrated a close parallel between the birds of the Ennadai Lake area and the vegetational communities, a relationship similar in principle to that found in southern Wisconsin (Bond, 1957), northern Wisconsin (Beals), and in boreal habitats of Canada (Carbyn, 1971; Erskine, 1977; other refs. therein). Years of increasing lemming populations, and evidently ptarmigan as well, were characterized by winters with normal or higher than normal snowfall and snow cover. The last summer's markedly diminished population had been preceded by a winter in which snowfall was much less than normal and with an unusually heavy freezing rain in early spring followed by a cold period.

It seems apparent that whatever other factors may at least partially influence the fluctuations in numbers in the northern bird and mammal species, weather events must be an important and perhaps dominant factor. In areas where extremes in weather conditions at all times of the year are a common occurrence, there exists few alternatives except abrupt losses in population density when a major period of severe weather is encountered. There is evidence also that the lemming density fluctuations are not synchronous between Ennadai and Baker Lakes, just as they are not synchronous between Baker and Aberdeen Lakes (Krebs, 1964). This supports the possibility that densities primarily reflect environmental events that can be local rather than regional in extent. The exceptions, of course, are those highly damaging effects on the environment of a high population, such as the eat-outs observed in other parts of the Arctic (Thompson, 1955), in which lemmings consume all available vegetation. Such influences on vegetation were, however, not evident in the Ennadai area.

There is now some doubt that the fluctuations observed in northern animal populations are cycles in the strict sense of the term, and it thus seems reasonable to look to environmental events for fundamental explanations. Cyclic periodicities averaging 10 years between peaks in some species (Hickey, 1954; Cole, 1954) may be an expression of the probability with which favorable conditions recur. The wide geographic synchrony observed at times may indicate that the particular climatic conditions have prevailed on a continental rather than a local scale. This observation corresponds to the inverse correlation demonstrated between winter snow conditions favoring heavy snow crusting, cold wet spring conditions, and ruffed grouse densities in northern Minnesota (Larsen and Lahey, 1958). The changes in population densities may well be the result of fluctuations in environmental conditions which appear in particularly severe combination at given average intervals. This view is confirmed, at least to some degree, by recent significant studies of birds in the region, such as those of Kuyt (1980) in the Thelon River region and of Norment (1985) in the Warden's Grove area and also along the Thelon River, as well as more general observations of the ecology of animal species in northern regions (Pruitt, 1960, 1966, 1970; Marchand, 1987; West and DeWolfe, 1974).

8. Essay on Diversity and Dominance

It is reasonable to state that the importance of a species in a given ecosystem is generally proportional to the amount of energy flowing through the species. A species through which a high proportion of energy flows will be by definition a dominant of the system. The terminology is widely accepted that ecosystems with greater numbers of species possess a greater "diversity" than ecosystems with fewer species. Diversity is also influenced by the degree of equality with which species are represented: those ecosystems which have every species represented by equal numbers of individuals possess greatest diversity.

Let us represent here the simplest possible case using two diversity indices commonly employed:

Index	Community Diversity Values	
	30 Species (10 individuals each)	10 Species (30 individuals each)
$D = -\Sigma\, p_i \log_2 p_i$	147	33
$D' = \dfrac{S - 1}{\mathrm{Log}_e\, N}$	13	1.6

Obviously, the complexity of a community composed of 30 species each represented by 10 individuals is greater than a community of 10 species each represented by 30 individuals. (S = number of species; N = number of individuals; p_i = proportion of the total number of individuals represented by each species. For instance, in the case of 10 species each represented by 30 individuals, each species contributes 0.10 of the total, so to obtain the index value the $\log_2$ of .10 is multiplied by .10 for each species and the results are summed, giving 33.22.) It might be noted that the negative before the summation sign in the first index is a convention employed to make the index a positive, prime value since logarithms of all proper fractions are negative numbers.

Usually D and D′ correlate well when large numbers of indices are computed, although the former should be used only when certain conditions in the data are met (Margalef, 1968). Both are merely mathematical devices to describe in relative terms the innate complexity of a given system. The actual numbers expressing the index value have little or no biological meaning but they have a certain crude utility in comparing community complexity or, as we have said, diversity. From observation, there appear to be certain rather important differences in the general characteristics of communities of high diversity and those of low diversity. There are hints that some fundamental underlying principles may be at work, but we know too much to make arbitrary rules and too little to state anything with certainty. There are these possibilities, however:

Communities with High Diversity	Communities with Low Diversity
Ecosystem more highly ordered and predictable	Less ordered: less predictable
Ecosystem more stable with time	Less stable
Energy channels more complex	Less complex
Greater number of trophic levels	Fewer
Reproductive dispersal slower	More rapid
Environment more stable	Less stable

A statistical disadvantage exists in the use of these equations. As larger numbers of species are represented in samples from which the index is computed, the curve tends to flatten after the first dozen or so most frequent species. These differences in D are greater between, say, samples containing 8 and 15 species, respectively, than between samples with 20 and 30 species. Sager and Hasler (1969) have shown in sampled phytoplankton communities that this effect can result in identical diversity indices (D) for communities containing a range of from 16 to 38 species. The D index also tends to plateau with sampled communities containing more than 12 to 15 species. These difficulties often make it impossible to estimate the diversity of a community with real accuracy.

It seems reasonable that a measure of the number of potential interrelationships between individuals would be a more useful index for the description of a community and for comparisons between communities. One such statistical device is described here and employed below in comparisons of the various community types sampled during the course of study of vegetation in boreal and arctic regions. A simple ratio gives the relationship of potential *inter*specific relationships compared to potential *intra*specific relationships. To illustrate with the simplest example, let us take a community of five species in which two of each are present in the average quadrat. We can then represent the number of potential interactions between the species in the matrix

	a	a	b	b	c	c	d	d	e	e
a	–	–	1	1	1	1	1	1	1	1
a	–	–	1	1	1	1	1	1	1	1
b			–	–	1	1	1	1	1	1
b			–	–	1	1	1	1	1	1
c					–	–	1	1	1	1
c					–	–	1	1	1	1
d							–	–	1	1
d	(Redundant)						–	–	1	1
e									–	–
e									–	–

in which it is evident that the total number of possible individual interactions are 60 (aa, aa, ab, ab, ac, ac, . . .), of which 20 are intraspecific and 40 are interspecific. We have, thus, an index value that will increase steadily as the number of both species and individuals are added to the sample, with obvious similarities to the diversity indices considered above but without the flattening effect to render it insensitive to rare species. Moreover, it is now possible to derive other useful measures such as a comparison between the densities of the potential interspecific (Q) and intraspecific (P) relationships:

$$Q = (40 + 20) = 60 \text{ (total interrelationships)}$$

$$Q/P = 40/20 = 2 \text{ (ratio of inter- intraspecific relationships)}$$

And, to use another example, in which the distribution of species differs:

	a	a	a	a	a	b	b	c	d	e
a	–	–	–	–	–	1	1	1	1	1
a	–	–	–	–	–	1	1	1	1	1
a	–	–	–	–	–	1	1	1	1	1
a	–	–	–	–	–	1	1	1	1	1
a	–	–	–	–	–	1	1	1	1	1
b						–	–	1	1	1
b						–	–	1	1	1
c								–	1	1
d	(Redundant)								–	1
e										–

in which

$$Q = 66$$

$$Q/P = 34/32 = 1.06$$

The interrelationships can be expressed as number per individual. Computation of interspecific Q can be carried out by determining the number of possible paired relationships. In practice, this constitutes simple multiplication of all density values in a column by all those below it and summing the result. The number of possible intraspecific interrelationships can be calculated using the combinatorial equation n!/r! (n - r)!.

Other applications of the ratio are conceivable. Using iterative computing techniques, rise or decline in densities of a given species could be predicted using coefficients assigned to species on the basis of species' competitive characteristics. Weather, nutrient regimes, disease infestations and so on could be used to predict changes in a community on the basis of coefficients and by employing feedback loops. Density levels of each species would be determined for each iteration, the latter representing a given time period. Nearest-neighbor data obtained from field observation or on the basis of random expectation might be incorporated into the model for establishing species–species relationships.

This discussion of the possibilities of Q is based on the apparent structure and relationships of the communities of the boreal forest and arctic tundra, characterized by relatively few species occurring in high densities. The technique may be of little utility in delineating relationships of temperate or tropical plant communities were large numbers of species are represented by few individuals.

Another interesting possibility for the technique, one that has not been explored in detail here, is the possibility that number values can be introduced into the matrix, weighting the inter- and intraspecific relationships. The weighting could be based on a value obtained from field measurement, such as distance to nearest neighbor (averaged for each between-species pair or between -individual pair in the case of intraspecific relationships) or a value representing some other aspect of associative or competitive interrelationship.

The matrix can be expressed as a set of differential equations which relate the inter- and intraspecific relationships of the species to the environment. To employ an expression of Margalef (1968), in such a matrix the term dN_i/dt represents the rate of change in the number of individuals, N_i, of any species, and dE/dt the rate of change in the intensity of environmental factors, E, a response proportional to the sum of products representing the inter- and intraspecific interactions:

	E	N_1	N_2	N_3
$dE/dt =$		$-aEN_1$	$-bEN_2$	
$dN_1/dt =$	$+eEN_1$	$-fN_1^2$	$-qN_1N_2$	$-hN_1N_3$
$dN_2/dt =$		$+iN_1N_2$	$-jN_2^2$	$-kN_2N_3$
$dN_3/dt =$		$+1N_1N_3$	$-N_2N_3$	$-N_3$

giving the matrix of possible cross-products:

E^2	EN_1	EN_2	EN_3
EN_1	N_1^2	N_1N_2	N_1N_3
EN_2	N_1N_2	N_2^2	N_2N_3
EN_3	N_1N_3	N_2N_3	N_3^2

and a matrix of coefficients:

$$\begin{matrix} 0 & -a & -b & 0 \\ e & -f & -q & -h \\ 0 & i & -j & -k \\ 0 & 1 & -m & -n \end{matrix}$$

expressing the strength of the interactions. The rates of change in the densities of species represented in the community depend upon the initial density values and the sign of the coefficients. Numerical methods could be employed to predict the steady state the system would approach after introduction of appropriate feedback controls, time lags, nonlinear effects, and higher order interactions. Externally induced changes in E and N_i would represent environmental changes, as well as changes in species densities, caused by external forces such as might result from grazing or other natural disturbance not otherwise accounted for. Often it appears that species interacting feebly do so with a great number of other species. Strong interactions are found often in systems having a smaller number of species, and these latter systems are often characterized by strong fluctuations.

Values of the coefficients representing strength of interactions between species may vary greatly with environmental circumstances, with the obvious consequence that the complexity of the system is, in nature, greatly enhanced over this simple model. To illustrate, an experiment performed by Ellenberg (Ellenberg, 1952, cited by Evans, 1963) shows that the response of a species to external conditions in the field may differ not only from that of single plants in the laboratory but from that of pure stands of the species in the field. The response of plants to an environmental factor is changed radically in the presence of competing individuals of other species. Physiological optima may reveal little of the performance of plants in the field in competition with other plants.

Diversity Relationships in Forest and Tundra

The relationships between these various indices and with the characteristics of the original community, particularly the numbers of species and individuals, reveal some of the properties of the indices as well as their differences and similarities. These relationships can be expressed in terms of correlation coefficients. In Table 8.1 some of the more interesting comparisons between indices and data are presented for the several community types studied.

Since density values are required for these computations, the relative frequency values of the original field data were transformed into density values employing the conversion table presented by Greig-Smith (Appendix; see also pp. 10–15, 78–79, Greig-Smith, 1964). Although justification for this procedure is based largely upon expediency—the field data are in terms of frequency and not density—it nevertheless has at least one pertinent theoretical advantage. In obtaining quantitative field data, the individual plant is, by field observation, an uncertain unit at best. This is especially the case in the arctic vegetation (and in prairie vegetation as well) where reproduction is often vegetative and there are no meaningful definitions of what constitutes an individual plant. Difficulties are compounded in the Arctic where a single plant may be

Table 8.1. Community Type

Correlated Values	JP	BS	WS	A	RF	LM	TM
				Correlation Coefficient			
N, Q	.96	.96	.91	.97	.97	.96	.99
N, P	.86	.84	.75	.94	.91	.82	.91
S, $-p_i \log_2 p_i$	.92	.91	.91	.92	.91	.92	.90
S, $S\text{-}1/\log_e n$	.85	.70	.78	.77	.84	.95	.98
S, Q/P	.71	.66	.72	.84	.66	.72	.71
Q/P, $-p_i \log_2 p_i$	.90	.84	.91	.96	.89	.90	.92
$S\text{-}1/\log_e n$, $-p_i \log_2 p_i$	.88	.68	.78	.90	.75	.84	.83

Values indicate correlation between diversity indices listed at left. See text for explanation of symbols N, S, P, Q, and so on. Letters signify community types as follows: JP = Jack Pine; BS = Black Spruce; WS = White Spruce; A = Aspen; RF = Rock Field; LM = Low Meadow; TM = Tussock Muskeg.

spread over a considerable area with branching portions covered by mosses, lichens, or peaty accumulations. Moreover, rhizomes often join a number of plants which otherwise appear distinct. Only intensive study of each shoot will reveal whether it represents a single plant or a portion of another plant some distance away.

The conversion from frequency to density is, moreover, based on the assumption that the population has a random distribution. From the data there is no way of ascertaining if the sampled populations were random or not. There is actually little if any data available on pattern (non-randomness) in arctic and subarctic vegetational communities, but in view of the limitations of the principle of competitive exclusion in the Arctic and subarctic, it is to be expected that communities here are characterized at least by a more nearly random distribution than in temperate or tropical zones. Moreover, the pattern of the vegetation, if it exists, would most likely be on a scale smaller than detectable with the one-square-meter quadrats used in the sampling. Even if the distribution is not random there is a definite relationship between frequency and density demonstrable by empirical means.

For these reasons it is apparent that density obtained from frequency is essentially an abstract value. In the context here, it is a useful and manageable value with pragmatic utility that can be interpreted as possessing biological meaning. It may actually be more constant and comparable than direct density measurements.

Relationships between indices, averaged for each community type, are shown in Fig. 8.1. Jack pine communities possess the lowest diversity with increasing values for the indices representing black spruce, white spruce, and aspen communities. Of tundra communities, rock field and low meadows possess lowest diversity values, both distinct in this respect from tussock muskeg. Values for tundra communities generally fall into the range occupied by forest communities of intermediate diversity.

The Q/P values demonstrate that, similarly, jack pine communities possess predominantly intraspecific interrelationships among individuals, with an increase in the proportion of interspecific relationships through black spruce, white spruce, and aspen communities. Rock field and low meadow values are similar, with a greater proportion of interspecific relationships in tussock muskeg by comparison. Relationships with

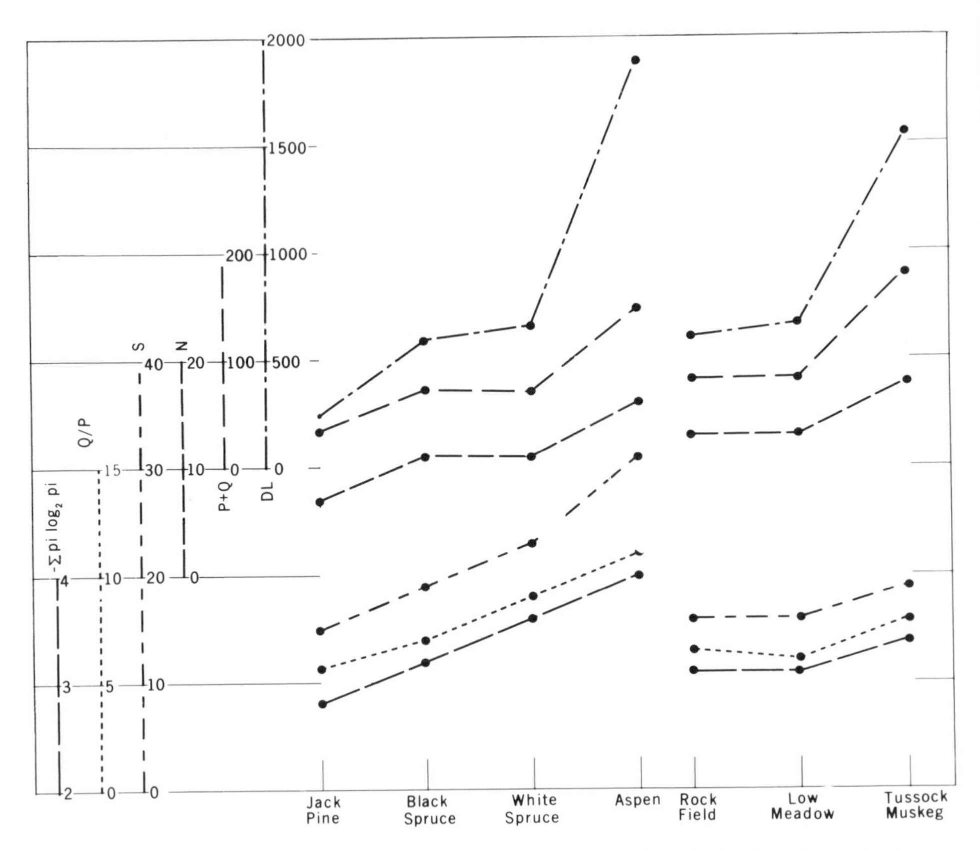

Figure 8.1. Average diversity values for the various communities obtained by six methods (see text). S refers to the number of species found in the average stand (in 20 1 m² quadrats). N refers to the number of individual plants in the average quadrat (from Larsen, 1974).

average number of species (S) in each stand sample (by community type) and number of individuals per average quadrat (N) are also shown.

These averages, however, provide no indication of the wide range of values shown by each of these indices for the stands in each community type. The figures relating individual stand indices with distance of the stand from the forest border (and hence the average position of the arctic front in summer) demonstrate clearly that in each community type there is a wide range of diversity and Q/P values (Fig. 8.2a–e). This wide individual variation in the diversity and Q/P ratio in stands is presumably the consequence of successional status (at least in the forest communities), disturbance, site conditions, and perhaps a variety of other factors.

Diversity, Succession, and Evolution

The interrelationships between species become increasingly complex as the number of species present in the community increases. These relationships are known to every

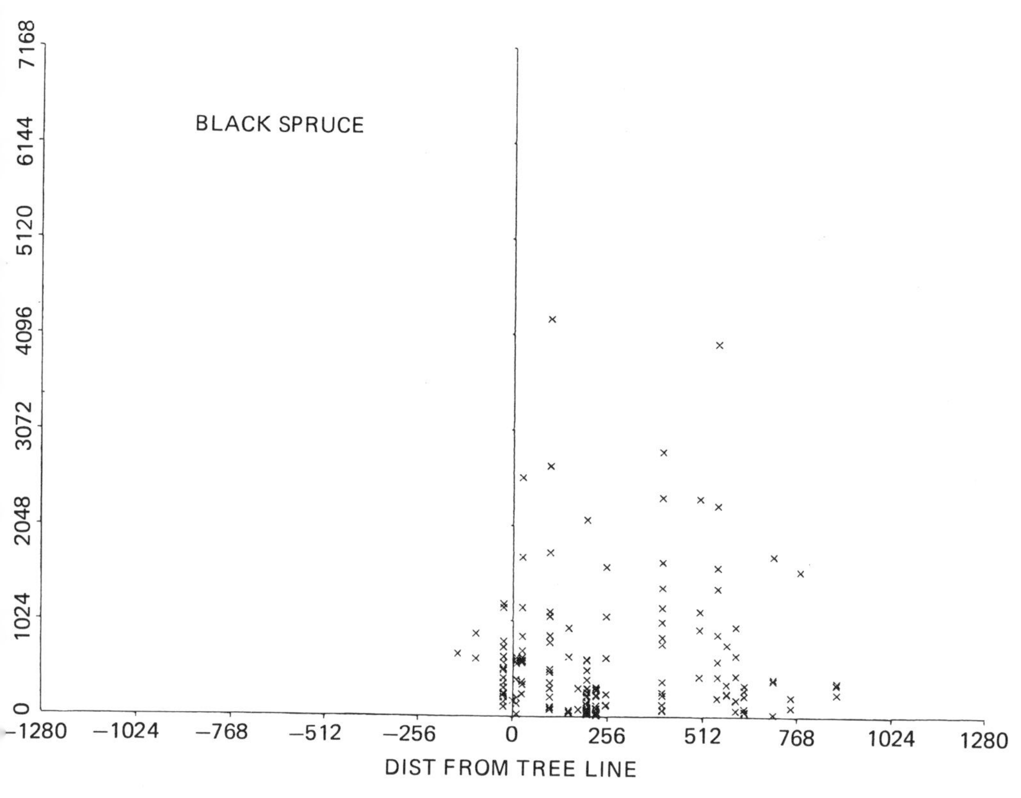

Figure 8.2.a-e. Diversity values for stands of each of the communities sampled, according to distance from the forest border. The negative values indicate stands north of the forest border, positive values are south of the forest border (in miles). Shown are diversity (D_L) values for stands of black spruce (a), white spruce (b), tussock muskeg (c), rock field (d), and low meadow (e) communities. Note change of scale for the D_L values. (*Continued on pp. 174–177.*)

ecologist familiar with succession. Pioneer vegetation is made up of fewer rapidly growing species. Communities in the more advanced stages of succession are characterized by a larger number of species joined in a complex integrated manner.

The genetic and evolutionary implications of succession and the occupancy of different habitat niches by each species have been discussed by McNaughton and Wolf (1970). The initial factor to consider is that there are two competitive interfaces in communities, the interface between individuals of the same species and the interface between individuals of different species. There are greater genetic differences between individuals of a dominant species when competition is largely between individuals of that species. Greater genetic uniformity is shown within species when the competition is predominantly interspecific and species are confined to narrow niches. In advanced successional communities, then, there is greater diversity in the community and greater genetic uniformity in individual species. In successional communities, there is lesser diversity, greater intraspecific competition, greater genetic variation within species ordinarily associated with early successional stages. Dominance and diversity are inversely related. Successional species are capable of occupying a wider habitat range.

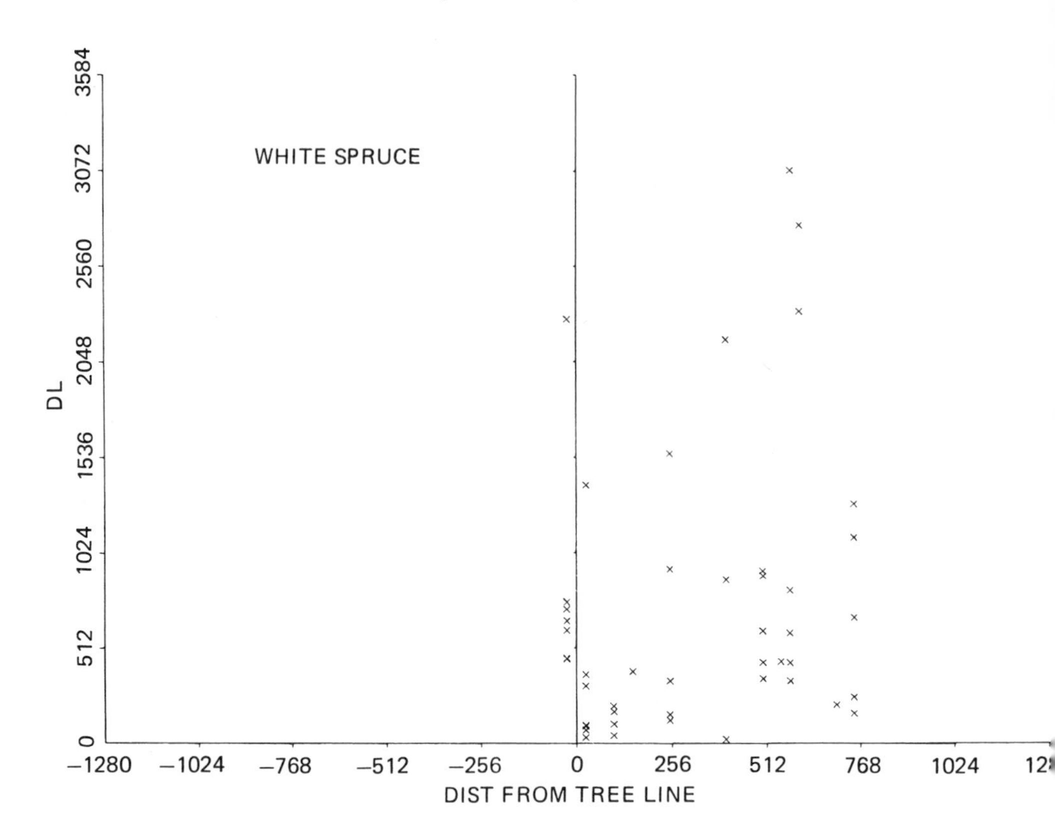

Figure 8.2b.

Individuals are able to occupy narrower portions of the environmental range than does the species as a whole.

We have, thus, seen that with an increase in diversity there will be a decrease in intraspecific competition. With low diversity there will be higher proportions of intraspecific competition. In the former case, evolutionary pressures will be to increase competitive ability of individuals of a species in relation to other species. In the latter case, pressure will be to increase the variability of individuals within a species. Rare species experience little intraspecific competition. Abundant species experience great intraspecific competition.

Under natural conditions, of course, competitive and environmental conditions are usually intermediate between extremes. Moreover, they change with time as succession or disturbance occurs, and species respond to increasing or decreasing diversity and competition of the various types, striking some kind of optimal or even fluctuating balance over long periods of time if they are to survive.

Perhaps in these relationships we can find tentative answers to at least a few of the simpler questions posed by the characteristics of northern forest and tundra communities.

First, let us look again at the depauperate zone in tundra. Here the diversity increases in general from Ennadai to the region northward along the Kazan River, in most

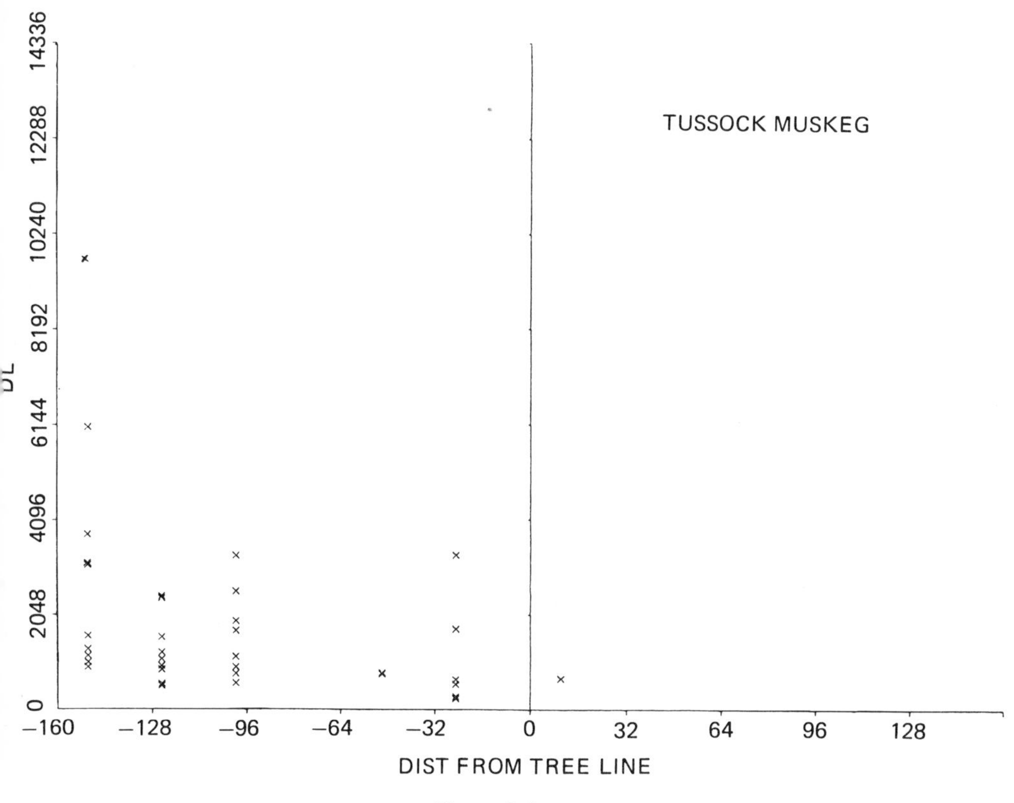

Figure 8.2c.

instances (but not all) then attaining a maximum at Dubawnt Lake. Farther north, in the apparently more severe climatic regime of Pelly, Snow Bunting, and Curtis Lakes (augmented by an altitudinal effect at the latter) diversity once again is usually shown to decline (Table 8.2).

It is thus apparent that there are significant differences between the communities of the ecotonal zone and those farther northward in the tundra. It seems reasonable to anticipate that these differences will be related eventually to the fluctuating and variable nature of the atmospheric conditions prevailing in this ecotonal and frontal zone.

A number of other ecological relationships are demonstrated by the data for the various community types. These are summarized below.

Black spruce. Diversity indices show a slight decline in average diversity for the spruce communities centered in the region about 250 miles south of the forest border. While the range of diversity values for understory communities is as great here as anywhere, there is nevertheless a noticeable decline in average diversity. This can be seen also in the values for P + Q and in the ratio P/Q. It is apparent that in this region there exist communities with a greater intraspecific competitive component than exist elsewhere. Communities with greater interspecific competitive intensity are centered around the forest border and in the region centered about 500 miles south from the

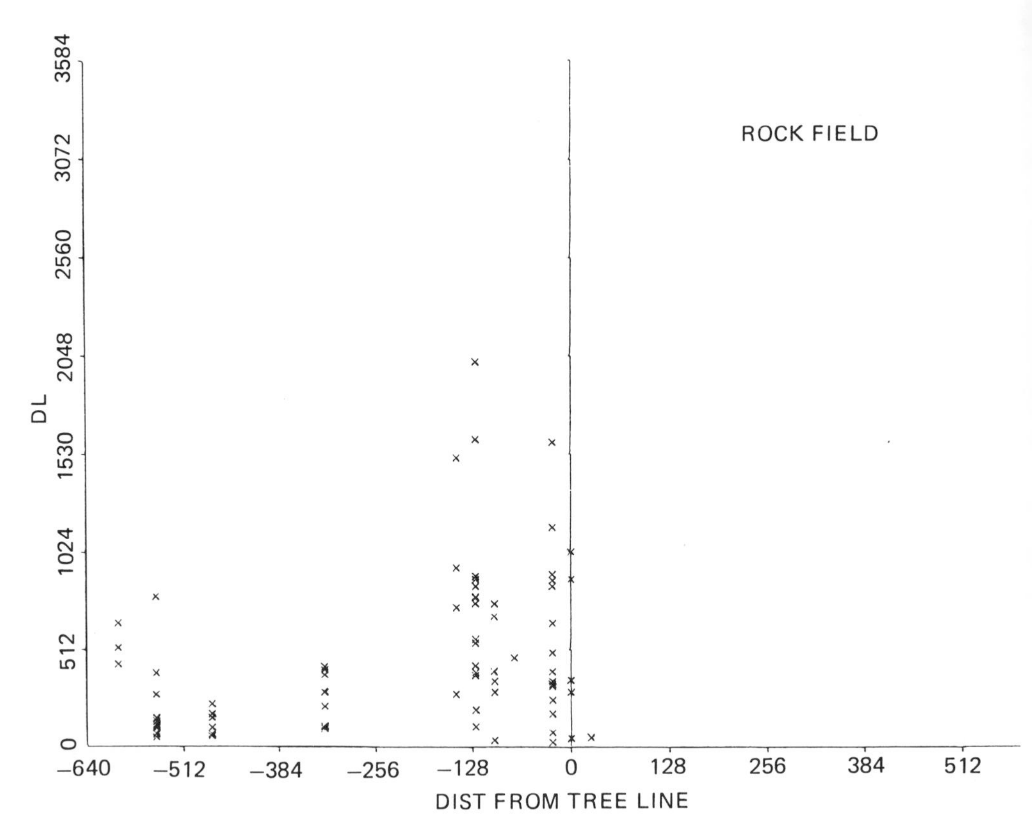

Figure 8.2d.

forest border. It is of interest that total basal area per acre in the black spruce stands is positively correlated (0.49) with distance from the forest border.

White spruce. White spruce stands show a somewhat increased diversity at distances of from 100 to 500 miles from the forest border in comparison to average diversity values for the understory community in stands near the forest border. There is, moreover, a greater increase in diversity range of stands southward. Stands approximately 100 miles south of tree-line show the lowest average diversity as well as range, corresponding to the black spruce minimum at 250 miles. These actual distances are probably not significant but it is apparent that in the region south of tree-line both black and white spruce understory community diversities attain a minimum in both average values and range of values. In total basal area per acre, the white spruce stands are positively correlated (0.60) with distance from the forest border.

Aspen. In no other forest type do diversity values have a relationship to trees per acre or total basal area per acre. In the aspen stands, however, rather striking relationships appear between overstory and understory characteristics. Correlation between understory community parameters and number of aspen trees per acre are as follows: Q, .59; N, .58; P, .64; P + Q, .59. Highest correlations of parameters with total basal area per

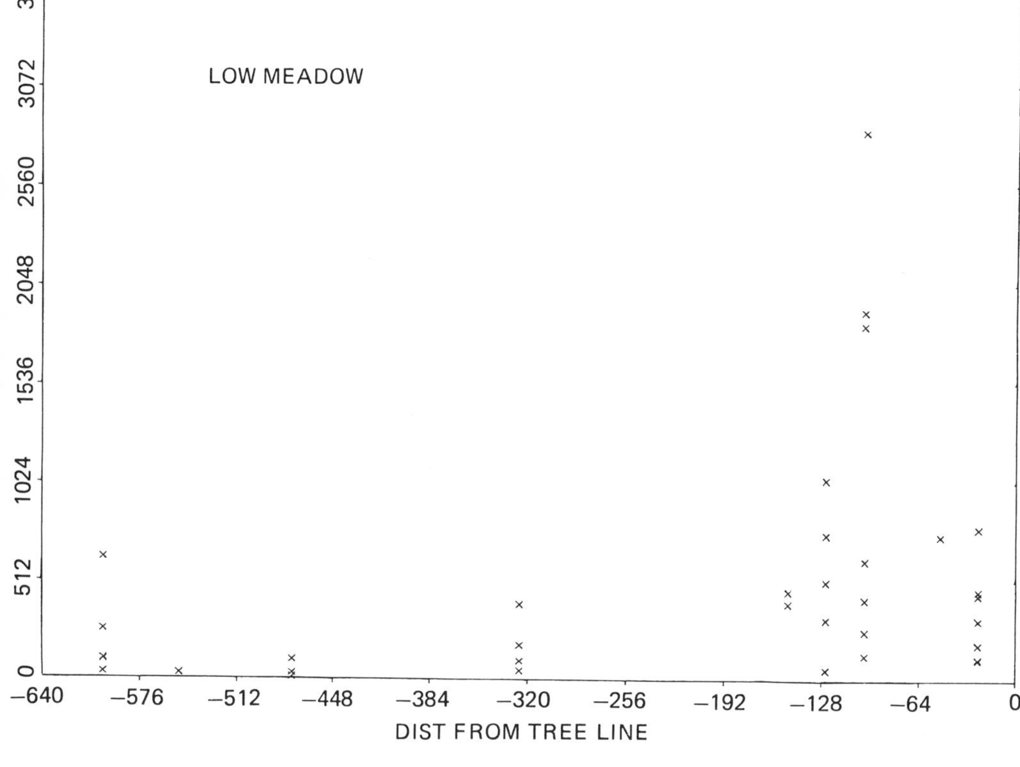

Figure 8.2e.

acre are as follows: D, -0.53; D', -0.54; P/Q, .52. This can be interpreted to indicate that the younger aspen stands have understory communities with greater diversity. As stands mature, basal area increases, number of trees per acre decrease, diversity declines, and intraspecific relationships show a proportional rise. It is apparent that in this case, community complexity does not increase with stand age. It should be noted, of course, that the range of aspen does not approach the northern forest border. This characteristic may be more prevalent within more temperate forest zones; it is included for purposes of comparison.

Jack pine. The jack pine stands demonstrate an increase in diversity of understory communities with distance from the forest border. There is a correlation of .66 between basal area per acre and distance from the forest border in the jack pine stands.

Tundra communities. Very little variation in average diversity or of range of diversity values is shown for any of the tundra communities when plotted according to distance from the forest border. Depauperate zone communities are included here and account for a somewhat greater clustering of lower values near the forest border. Elsewhere the differences in diversity between the various stands of a given community and between community types are minimal. Since diversity values in tundra communities are within

Table 8.2. Average Community Diversity By Study Area

Community*		Ennadai	Kazan	Dubawnt	Pelly	Snow B.	Curtis
	RF	2.384 (7)*	2.828 (4)	3.131 (3)	2.679 (9)	2.371 (7)	2.595 (13)
$-\Sigma p_i \log_2 p_i$	LCM	2.829 (4)	2.895 (4)	3.453 (2)	2.919 (4)	2.097 (3)	1.983 (1)
	TM	2.690 (3)	3.224 (2)	3.653 (7)	–	–	–
	RF	3.167 (7)	5.214 (4)	4.541 (3)	4.278 (9)	3.599 (7)	4.899 (13)
$s\text{-}1/\log_e n$	LCM	4.986 (4)	4.763 (4)	7.286 (2)	6.964 (4)	5.351 (3)	4.219 (1)
	TM	3.065 (3)	6.804 (2)	7.644 (7)	–	–	–
	RF	3.627 (7)	4.781 (4)	6.974 (3)	4.224 (9)	3.282 (7)	3.594 (13)
Q/P	LCM	4.663 (4)	5.128 (4)	6.960 (2)	4.227 (4)	2.022 (3)	1.696 (1)
	TM	4.727 (3)	6.369 (2)	9.363 (7)	–	–	–

*Number of stands averaged are indicated in parentheses. Communities indicated as follows: RF is rock field; LCM is low *Carex* meadow; TM is tussock muskeg.

the range of diversity values in forest understory communities, it appears that the common conception of a general decline in community diversities northward may not have much validity, at least within the geographical region (Low Arctic) here encompassed.

In general, the diversity relationships of the communities studied are neither simple nor readily apparent, with the exception of the few cases already noted. Competitive relationships are, similarly, only now beginning to be revealed and their significance appreciated. Research on community relationships in temperate zone plant communities is advancing with considerable speed, and it will be of interest to learn eventually whether the ecological principles discerned in these communities can be extended without modification to encompass boreal and arctic communities or whether, in these latter, the overwhelming forces of the physical environment take precedence over all other factors.

Dominance Matrix

These estimates of diversity have, of course, no intrinsic biological (ecological) significance. They are but a means of comparing communities of different sites, an index value, and the general significance of the latter in terms of ecological stability as well as other community characteristics are still matters for investigation.

Moreover, the estimates apply only to communities of proscribed habitats as indicated, not to possible total species numbers existing around the study site. Total abundance in the landscape may have an influence on numbers encountered in quadrat sampling, by ecesis of occasional individuals from nearby rare or unusual habitats.

In studies conducted along a proposed pipeline route through Keewatin, roughly north from about 125 km west of Hudson Bay at the Manitoba border (thus east of Ennadai Lake by about 200 km), Zoltai and Johnson (1979) found that a site just north of the Manitoba border possessed some 79 species, and one farther north contained 128, which they attributed to greater diversity of habitat in the latter area, exceedingly high levels of nutrients present in the soil as a consequence of a dolomitic substrate, and a base exchange capacity roughly three times that in soils of other study areas. Thus, it is obvious that these are factors that must be taken into consideration in making comparisons between plant communities of different areas. Higher species diversity is related to higher diversity of habitat and higher levels of soil nutrients, and thus these factors are to be evaluated when attributing community structure to certain other environmental influences such as climate. As for the total numbers of species (vascular plants) present in the region, Zoltai and Johnson indicate that they collected 217 in their study; Gubbe (1976) found a total of 230 species along the pipeline route from the Manitoba border to the Boothia Peninsula.

One aspect of community structure in the northern regions stands out as of particular interest. A small group of species form a matrix of dominants throughout forest, forest-tundra ecotone, and into southern ("low arctic") tundra. This matrix of species attains high frequency in most communities, and does so throughout a wide geographical range—actually circumpolar if one will accept substitution of closely related species in communities of North America and Eurasia. In North America these species include *Vaccinium uliginosum*, *V. vitis-idaea*, *Ledum groenlandicum*, *L. decumbens*, *Betula glandulosa*, *Empetrum nigrum*, as well as, but less abundantly, *Arctostaphylos alpina*, *Salix glauca*, *Rubus chamaemorus*, and *Andromeda polifolia* (see Tables 4.2–4.10).

Evidently these species possess a certain set of characteristics that permit adaptation to the prevailing conditions (see Appendix 1), and it is of interest here that they appear to have counterparts in community structure in certain other global vegetational formations, notably Mediterranean climate shrublands and many natural grasslands (Gubbe, in Diamond and Case, 1986). They contrast with such species-rich communities as montane and lowland rainforests, as well as certain African and Australian heathlands, in which there are no species that appear as constant community dominants.

In summary, there are certain distinctive characteristics of the far northern vegetational communities, not the least important of which is the relative simplicity of expression of biotic processes. They are of great scientific value for this alone, to be added to their other estimable traits. Moreover, as Heal and Vitousek (1986) point out, northern communities fall at an end of the continuum from tropical through temperate to taiga and tundra, and the clarity of their ecological relationships gives a degree of illumination to processes at work, not only here but on vegetational communities elsewhere.

9. Environment: Atmosphere

By 8000 years ago the Pleistocene continental glacial ice still covered a large area in north central Canada as well as northern Labrador and Ungava, but closed-crown spruce forest had become established over far northwestern Canada and the Mackenzie Delta (Ritchie, 1984) and was colonizing southern Quebec and the Maritime Provinces (Macpherson, 1985; Mott, 1985; Filion, 1984; Andrews, 1985). The contrast between the warmer air masses of the Alaska-Mackenzie region and colder air masses of the Hudson Bay region persists to the present day, the consequence of the western (ridge) and eastern (trough) pattern of wide amplitude in the upper atmospheric belt of westerly winds. The forest-tundra is found in the zone where frontal conditions prevail between northern masses of arctic air and more southern air masses of Pacific origin, a zone that fluctuates as a result of weather conditions but which is generally found not far north of a modal position that extends in summer from the Mackenzie Delta to the southern edge of Hudson Bay and thence in a somewhat more diffuse band across northern Quebec and Labrador. The closed-crown circumpolar boreal forest is found generally south of this climatic transition belt, in regions dominated by warm air masses in summer in North America, and warm continental air over Siberia. Tundra is found in regions northward dominated by arctic air masses.

The continental glaciation was rather obviously associated with temperatures lower than those of the present, since low temperatures were required to permit snow cover to persist through the summer season, but the characteristics of the glacial and postglacial atmospheric circulation patterns over North America and the nature of the shift between glacial and postglacial regimes remain largely matters of speculation and controversy (Larsen and Barry, 1974; Barry et al, 1981). It is generally accepted, however, that the retreat of the glacial margin at the close of the Pleistocene coincided with what is known as a Milankovitch thermal maximum. This was a period of maximum solar radiation in high latitudes of the Northern Hemisphere that occurred about 10,000 years ago, followed by disappearance of the glacial ice in large parts of North America. The earliest vegetation was in the west where fossil pollen evidence obtained by Ritchie (1983, 1984a,b) and others shows that *Picea mariana* and *Betula papyrifera* spread rapidly and formed woodlands in northwestern Canada between 9000 to 5000 years ago (Ritchie et al., 1983; Ritchie, 1988; Harington, 1985).

Hare (1976) questioned speculations concerning retreat of the ice during this time period, indicating that, given the estimated thickness of the glacial ice deposits and postulated climatic patterns, insufficient energy was available to carry out the required melting. Subsequent studies, however, show that mid-latitude westerlies over North America were stronger in the early Holocene as compared with late Holocene (the present time), although there apparently are no calculations to show that this would account at least in part for the discrepancy detected by Hare (Halfman and Johnson, 1984; Dean et al., 1984). Another study indicates that repeated thermal maxima occurred in the early Holocene and these, too, may have influenced the melting of the glacial ice (Davis, 1984).

Earlier movements of vegetational zones over the land mass that is now North America, occurring in late Eocene, Miocene, Pliocene, and Pleistocene, were evidently the consequence of continental drift. A slow shift in climate took place as portions of the continent pushed northward (Löve and Löve, 1974), and later the vegetational zones shifted again during repeated advance and retreat of the continental glacial ice during the Pleistocene. The conclusion to be drawn as to the response of the vegetation to these events is that the species demonstrated a sensitivity to temperature range and perhaps other aspects of the physical environment, and that many of the species had become cryophiles during their evolutionary history in the Eocene and Miocene. Some were able to survive in the northern areas that remained ice-free (Löve and Löve, 1974; Ives, 1974; Ritchie, 1988), others migrated south before the edge of the advancing ice and along the north-south Rocky and Appalachian mountain chains. At the present time, many or most have reached equilibrium with the climate. What then are the climatic conditions at the northern edge of the forest that account for the transition from closed conifer forest to arctic tundra?

Experimental physiological studies have been conducted primarily with temperate zone and tropical plant species, but the relatively sparse literature on the physiological responses of arctic and boreal species reveal that the same environmental factors are influential in controlling growth and reproduction of northern species: temperature, moisture supply, cloud cover, day length, wind, depth and duration of snow cover, number of frost free days from spring to fall, and so on. The extreme difficulty of analyzing and expressing the aggregation of meteorological factors affecting plant growth makes it understandable that meaningful correlations have not always been achieved, and it is easy to agree with an observation made long ago by Turesson (1930) that the species are far more sensitive to environmental factors than is generally supposed. Moreover, species respond in an individualistic manner to the factors, and this is borne out by the observation that the range boundaries of particular species often do not coincide with the range boundaries of the regional vegetational types. Why this should be so is not known in any detail. Climatic interpretations of the range of species do not yet include a good understanding of the physiological and genetic response of species to the meteorological factors, except in the most general terms.

The correlations between the range limits of individual species and climatic factors are, however, sufficiently provocative to have inspired studies of the climatic characteristics of areas occupied by distinctive plant communities and individual species. These were undertaken in anticipation that they would reveal the environmental requirements and factors limiting the range of the species, and it is in this descriptive phase that study of the biogeography of the northern forest-tundra ecotone largely rests at the present time. It is, moreover, apparent that at least some aspects of climate are influential in controlling the geographical range of northern boreal forest plant species as well as those of the ecotonal transition to northern arctic tundra (Larsen, 1965, 1974, 1980).

Supporting evidence exists in studies of alpine tree-line, one in which permanent climatic stations were established along an altitudinal transect from forest to alpine tundra. It showed that air and soil temperature controlled initiation of bud flush in *Picea engelmannii*, *Abies lasiocarpa*, and *Pinus flexilis* (Hansen-Bristow, 1986). Alpine climate at tree-line is characterized by a short growing season, low air temperatures, frozen soils, drought stress, high levels of solar radiation, irregular snow accumulation, and strong winds. These factors, most of which are also present in the arctic ecotone, produce stress that can be measured by means of variations in needle length, shoot elongation, and cuticular development. Similar comparisons across the northern tree-line would be most revealing of the limiting factors involved. A brief discussion of the climatic characteristics of the alpine and northern forest-tundra ectone (Larsen, 1980) mentions certain other distinctive characteristics present in each regime.

Ecotonal Macroclimate

Climatic data for the more heavily populated parts of Canada are now abundant and of reasonably long record. For the northern regions and the Arctic, however, the record is less complete both in terms of numbers of meteorological stations and length of the data record at each. Information now available is sufficient to furnish a reasonably accurate record of weather conditions at such remote places as the Arctic Archipelago and the interior of Keewatin and Mackenzie. Where once there existed only hazy descriptions by explorers and traders, there are now data sets of hourly records obtained at 20 or 30 northern outposts.

Perhaps the single most important observation to be drawn from recent research is that northern regions have variable weather conditions, similar in this respect to the rest of the Earth, and that northern weather from year to year is affected by a variety of influences, such as degree of land coverage by snow, of water by ice, cloudiness, and by climatic events in regions distant from Arctic and subarctic. This includes volcanism, which releases large amounts of particulates into the atmosphere, as well as industrial pollutants from sources quite distant from arctic and boreal regions. These are topics of great interest to the general climatologist, and numerical models incorporating satellite data and detailed instrument records have now been developed, leading to better understanding of the northern climates. The literature on these and related topics is today too extensive to describe in detail here, but certain aspects of the knowledge now available bear directly upon the ecological relationships of the transition from closed-crown forest and the transition quite abruptly to open lichen woodland and tundra northward.

Understanding of global climatic regimes has made it possible to discern not only long-term climatic changes but to explain them in terms of atmospheric dynamics (Kelly et al., 1982; Kukla et al., 1977; Yamamoto, 1980). Floristic responses to three climatic regimes of known characteristics in the southwestern United States have been studied by Neilson (1986), who showed that past floristic changes were related to global weather patterns known to be associated with north Pacific sea-surface temperature anomalies, the latter taken to infer the conditions during the late Pleistocene glacial maximum, early Holocene warming, and late Holocene cooling.

By means of the information thus obtained, Neilson was able to show that the present-day communities, established during the "Little Ice Age" of ca. 1550–1850, are only marginally adapted to the present climate, and that clusters of weather events in the different climatic regimes control onset or release of dormancy, success or failure of flowering, seed set, and germination. Each measure the degree to which the various species are in equilibrium with the prevailing climate, in much the same way in which trees of alpine and arctic forest-tundra ecotonal regions are controlled (Bliss, 1962; Billings and Mooney, 1968; Savile, 1972; Hansen-Bristow, 1986; Payette et al., 1985; Busby et al., 1978; Hamburg et al., 1988; Marchand, 1987; McKay et al., 1970; Dingman et al., 1980; Slaughter and Viereck, 1986; Woodward, 1987).

Primary Role of Sunlight

Solar radiation is of primal significance in both the biosphere and the atmosphere. In the former, it is the source of energy driving the chemical reactions essential to the green plant's photosynthetic system. It is likewise the driving force of the global atmospheric circulation and, hence, regional weather patterns. Solar radiation is not received uniformly over the surface of the Earth. Much of the energy received in tropical and temperate zones is transported via atmospheric and oceanic currents to polar regions and re-radiated to space as long wave (infrared) radiation, thus maintaining the overall global temperature regimes in equilibrium. Hence, in arctic regions, for much of the year the radiational balance is negative rather than continually positive as in equatorial regions. Total incoming radiation is termed "global radiation" and the balance between incoming and outgoing at any given point is "net radiation." In tropical regions the balance is always positive, in subarctic and arctic regions it is positive only during summer months, negative during winter. North of the Arctic Circle it is negative 24 hours a day during winter months when the sun does not rise above the horizon; it is also negative during most of the days when the sun is low over the horizon. At these times, the heat transported into the Arctic and subarctic comes by way of the atmospheric circulation—southerly winds—and oceanic currents, the former by far the more significant since the Arctic Ocean is, unlike the Antarctic, relatively inaccessible to oceanic currents originating in tropical regions.

At a critical point northward—the northern forest border—there is insufficient energy received during the short northern summer for tree species to support adequate photosynthesis for tall woody boles, and trees are stunted, decumbent, and ultimately absent northward, giving way to small herb and shrub species characteristic of arctic and alpine regions. This occurs even though, in far northern regions, there is incoming solar radiation 24 hours a day during summer months and in the subarctic the day length is somewhat less but still exceedingly long by temperate zone standards. The days may be long in these regions, but solar radiation is at fairly low levels during all but the hours of late morning, noon, and early afternoon.

Studies have revealed a close correlation between the net radiation received during the course of a year and the position of the northern forest border. Hare and Ritchie (1972) pointed out that the northern edge of the forest-tundra ecotone rests approximately along the isopleth representing an annual net radiation of 15 kilolangleys (Kly). Northward of this zone is tundra, southward is forest-tundra ecotone or lichen woodland extending to the zone with a net radiation of 20 Kly per year. From here southward is found closed-crown boreal conifer forest extending to the zone with net radiation of 35 Kly annually, south of which is found the conifer-hardwood and temperate zone deciduous forest communities.

These general observations of Hare and Ritchie have been confirmed by more recent data published by the Canadian Climate Program (1982) in which solar radiation data summaries for the years 1951–1980 have been made available. In the data summaries presented therein, the energy units are megajoules per square meter rather than langleys, making comparison of the data with Hare and Ritchie's possible only after conversion, but the same general conclusions apparently hold true. Although the weather recording stations at which the data were obtained are not positioned precisely equidistant on, above, or below (latitudinally) the forest tundra border, the data from these stations confirm that tundra exists above the 15 Kly yr^{-1} (628 MJ m^{-1}) zone, the transition ecotone exists between the 15 Kly yr^{-1} and the 20 Kly yr^{-1} and the 20 Kly yr^{-1} (628–837 MJ m^{-1}) isopleths, and the boreal forest is found between 20 Kly yr^{-1} and 35 Kly yr^{-1} (837 - 1465 MJ m^{-1}). This appears to be consistently true at least in areas where equilibrium has been reached between climate and vegetation, i.e., where postglacial boundary changes have come to an end, postglacial plant species' migrations have largely ceased, and stability has been attained.

Relationship to the General Circulation

The net radiation described above is the product not only of the latitudinal distribution of global radiation but also of other atmospheric factors such as cloudiness, precipitation, and prevailing winds. These are, in turn, the product of large-scale atmospheric events discussed in every textbook on climatology, more specifically the formation of air masses and the occurrence of frontal conditions as these move across the continent and abut one another. As Barry and Hare (1974) point out, a well-defined arctic front is usually distinguishable in the upper troposphere, and this is distinct from the major band of westerlies just southward. The arctic front and the arctic jet stream are major determinants of the climate of the subarctic and Arctic as well as southward, shifting northward in winter and south in summer. During the latter season the jet stream generates a distinct belt of maximum frontal frequency from northern Alaska to James Bay (Reed, 1960; Hare, 1968), coincidental, too, with streamline fields of surface winds and the northern forest border in Canada (Bryson, 1966). Although the correlation is less well defined in Quebec and Ungava (Barry, 1967), it is clearly apparent in Eurasia (Krebs and Barry, 1970). Net radiation in this frontal belt shows a broad minimum roughly parallel to the front and some 600–1000 km north of it (Hare, 1968; Barry and Hare, 1974). Precipitation ranges from 15–20 cm in the tundra to 25–50 in the forest-tundra and northern boreal forest, half of which falls in summer, although there is a degree of uncertainty in the annual precipitation data particularly for northern Canada since the measurement of snowfall is difficult in windy northern regions. The mean daily temperature is below 0°C from at least late September to late May or early June, and the frost-free period is in the neighborhood of 60 days at most recording stations in or near the frontal zone and forest-tundra ecotone. Strikingly apparent in the data from these stations is the great variability in conditions from year to year. For example, at Port Harrison (Inoucdjouac) on the west coast of the Ungava Peninsula (east coast of Hudson Bay), annual snowfall varied from 70 to 464 cm in 17 years of measurement.

The climatic record for a number of stations in parts of Canada at or near the forest border is now of sufficiently long duration to furnish a reasonably good understanding of the atmospheric conditions prevailing throughout the year, and the large-scale characteristics of the atmospheric circulation are also now well-established. It can be expected that ever more detailed analyses of correlations between climate and vegetation will be forthcoming. These will be important not only for an understanding of the ecological relationships between life and environment in the northern lands, but will also furnish at least a kind of minimal baseline comprehension of what can be expected if, and when, climatic changes, both anthropogenic and naturally induced, occur in the future. The northlands are becoming the site of ever more intensive exploitation and, increasingly, of mining and industrial activity, transportation, and habitation by a year-round resident population not only of native peoples but also of industrial workers, technicians, scientists, and government specialists. Advance indications of what a climatic change would mean in terms of human habitation of these boreal and arctic regions would possess great value (Ball, 1984, 1986; Simgh and Powell, 1986; Kauppi and Posch, 1985). Not only would a knowledge of incipient changes in temperature and precipitation be of great significance, but so would the probable changes in cloudiness, sea ice, permafrost, length of winter, and snow-free season, to name a few. Any or all of these would most likely be affected by climatic change, whether brought about by the increasing buildup of CO_2 in the global atmosphere as a consequence of industrial expansion, a period of exceptional volcanic activity, or some other cause initiating a rapid deterioration in climatic conditions such as existed in, for example, the so-called Little Ice Age, AD 1550–1850, which caused havoc in Europe and elsewhere. An increase in temperature globally would also result in great changes, with a general poleward shift of cyclonic activity and cloudiness (Barry, 1984; Crane and Barry, 1984; Barry et al., 1984).

The impact of the "Little Ice Age" and other post-Pleistocene climatic events on the northern forest border is shown in the general paleoclimatic history of central and northern Canada, and in greater detail in terms of spruce morphology and regeneration by Elliott (1979a,b), Payette et al. (1985), Filion (1984), and others. In these latter studies, older-growth vegetation in the region studied, northern Quebec, was associated with very cold windy winters, thin snow cover, and cool moist summers during the "Little Ice Age," and more recently with milder and snowier winters. A 600-year chronology reveals that marked differences are revealed between dead and living spruce trees which indicate responses to the climatic change that occurred during this period.

Other evidence indicates also that there have been changes in the seasonal pattern of climate, in which warmer summer temperatures were associated with past global glacial minima (Crowley et al., 1986; Barry, 1981; Mott, 1985; Andrews, 1985; Andrews and Barnett, 1979; Andrews et al., 1972). Climatic minima may have had an importance influence upon the vegetational history of advance and retreat of the mixed-conifer hardwood forest and of other events such as the advance and retreat of permafrost. Recent temperature trends in some northern regions are reflected in a gradual warming of permafrost, generally in the range of 2–4°C during the last few decades to a century in parts of northern Alaska (Lachenbruch and Marshall, 1986), indicating that a general warming trend, at least in Alaska, is evident, although data from other northern regions is conflicting. A 2°C annual cooling of mean air temperature since 1940 at Tuktoyaktuk, on the coast in the western Canadian Arctic, has resulted in a 40% reduction in the thickness of the soil active layer above the upper surface of the permafrost in summer (Harris, 1981, 1987), indicating that warming in some areas is accompanied with cooling in others.

Future Atmosphere

The multitude of forces at work influencing the global atmosphere, at present largely cloaked in uncertainty as to the degree to which each affects the climate, coupled with the complexity of the global atmospheric and oceanic circulation, all make it manifestly unwise to predict the future course of the global climate. This is most clearly apparent in the diversity of opinion, and the interpretations derived from existing data, among those who might be considered most capable of attaining some degree of unanimity concerning the global climatic future.

In general, however, there appears to be a consensus that warming of the northern hemisphere and probably of the globe as a whole occurred from roughly 1880–1890 to around 1940, after which there occurred either stability or a brief and minor cooling, which reversed around 1970 when there appeared a general increase in the variability of weather and a series of very warm years after 1980 (Jones et al., 1986; Gajewski, 1987; Kalnicky, 1974). It has been noted recently that a recurrent event—great variability of weather prior to a major climatic change—was long ago recognized by H. H. Lamb. Such an event, or series of events, appears to be a fairly good predictor of impending shifts in climatic patterns. Recent trends, as well as assumptions and inferences concerning these trends, have been described in a relatively abundant literature (see, e.g., Lamb 1966, 1977; Bolin et al., 1987; Jones et al., 1986; Hamburg and Cogbill, 1988; Brinkmann, 1976; Woodward, 1987).

The recent climatological literature is exceedingly voluminous and impossible to summarize here, but the major categories can be outlined according to the following natural causal influences involved in climatic change: variations in amount of energy received from the sun; variations in received energy due to orbital and elliptical positions of the Earth in relation to the sun; intensity of volcanic activity; changes in oceanic circulation patterns and effects upon the stabilizing influence of the large thermal inertia of the oceans; random, chaotic, or periodic fluctuations in climatic patterns due to inherent instabilities; variations in the interaction between oceans and atmosphere. There is also the influence on climates of the past—in geological time—of such events as continental drift and even possibly of extraterrestrial influences of one sort or another.

Imposed upon these physical factors, which may all contribute in some, probably variable, measure to the characteristics of the prevailing climate, are a legion of biological and man-made technological influences: the building up of carbon dioxide content of the atmosphere from fossil fuels; sulfates, water vapor, nitrates, and other gaseous emissions largely from industrial installations; dust and aerosol particles from exposed soil in agricultural lands, forest fires, industries, barren lateritic soils; methane from rice paddies, domestic animals, and landfills; chlorofluorocarbon compounds; and perhaps a multitude of other physical and chemical factors not yet recognized as being of importance.

There has been, in recent years, however, an increased ability on the part of atmospheric scientists to handle the interacting relationships of this multitude of factors, the consequence of the development of computer models capable of including all known pertinent factors, but as yet the degree to which each factor influences climatic patterns is still largely indeterminate, the result of nonlinear and probably inherently variable characteristics of the major factors. It is encouraging that models can be improved as knowledge expands, thus, it is the hope that eventually sufficient information will become available to make accurate forecasting possible, at least in terms of climatic trends over relatively long periods of time, providing always, of course, that the random element is not so overwhelming in importance that accurate predictions will always be impossible.

Recent studies suggest that appreciable warming, of as much as close to 1 °C, has already occurred as a consequence of anthropogenic influences, with the three warmest years on record taking place during the 1980s. The evidence suggests that the cause is the increasing concentrations of the greenhouse gaseous compounds in the atmosphere, carbon dioxide, methane, chlorofluorocarbons, nitrogen oxides, and water vapor. The effect of climatic warming on water vapor in the atmosphere is that of positive feedback; as warming occurs, more water can be held in the atmosphere, which in turn brings about more warming.

A global warming of a magnitude between 2–5 °C would be the greatest since the close of the Ice Age 10,000 years ago; it would result in movement of the great grain-producing belts, changing lake and ocean levels, modification or destruction of the habitat of plant and animal species, and disruption of the habitat of small surface organisms in oceanic waters that constitute the base of the food chain that supports the world's fisheries. The warmer surface of continental and oceanic regions would result in more vigorous hydrological cycles, alteration of the general circulation of atmosphere and oceans, and the predicted changes during the next few decades would far exceed natural climatic changes in historic time (Bolin, et al., 1987; Blake and Rowland, 1988; Geller, 1988; Heath, 1988; Jones et al., 1986; Kerr, 1988; Kiehl et al., 1988; Lindley, 1988; Lowe et al., 1988; Mitchell, 1988; Pearman and Fraser, 1988; Volz and Kley, 1988; Ramanathan, 1988; JOnes et al., 1988).

Under the circumstances it would be advisable to build a basic inventory of natural global baseline data, as can be had in biogeographical indices such as plant and animal range limit and community data, as well as events such as glacial movements and the chemistry of atmospheric composition, as an adjunct to global climatological records. In this manner, as much knowledge as possible will be obtained concerning the causes, the timing, and the consequences of climatic change. We know from the geological and paleoclimatological record that climatic change is inevitable, only when and how it will occur are unknown.

The effect of climatic change on vegetation can, of course, be profound. This is borne out by studies of the effect of freezing resistance in native tree species (Sakai and Weiser, 1973) and the correlation of tree species range with climate in various regions (see, e.g., Denton and Barnes, 1987), as well as the water relations of native and introduced evergreens in interior Alaska (Cowling and Kedrowski, 1980; Jarvis, 1986; Rosenzweig and Dickinson, 1986), and the direct effects of increased carbon dioxide on plants and ecosystems (Strain, 1987; Oechel and Riechers, 1986).

On average, a 1°C drop in growing season temperature means a shortening of the growing season by 10 days. A listing of some pertinent growing seasons, on the basis of existing data from nearby meteorological stations, is as follows:

Area	Growing Season*	Region
District of Mackenzie**	80	Allows black spruce reproduction
District of Mackenzie**	56	No black spruce reproduction
Ennadai Lake, SW Keewatin	50	Little or no reproduction
Churchill, Manitoba	50	Just south of tree-line
Schefferville, Quebec	65	South of tree-line
Baker Lake, NWT	0	Far north of tree-line, Keewatin
Tuktoyaktuk, NWT	20	North of tree-line, Mackenzie Delta

*Number of days with mean temperature above 10°C.
**Black and Bliss, 1980.

The study of black spruce reproduction in northwestern Mackenzie by Black and Bliss (1980) demonstrates that a critical variation in temperature for spruce reproduction would be about 1.5°C in June and 1.0°C for each of July and August. A drop in temperature of this amount would be sufficient to account for the absence of reproduction in the northern study area, a site only about 40 km from the southern stands where reproduction was observed. Thus, a shift toward the warmer would result in a movement of this magnitude in the forest border, as long as there were trees in the vicinity to provide a seed source. Or, conversely, a shift toward the cooler would prevent reproduction in the existing trees, which would then remain as relicts until old age or fire ended their existence (unless they were able to persist by layering, rather commonly noted in the spruce of forest border regions).

The alacrity with which plant species and communities respond to climatic changes has been discussed by a number of workers, who note that species differ in their ability to track environmental change, and many do so too slowly to remain in equilibrium with the prevailing environment (Davis, 1986; Davis and Botkin, 1985). Lags in response can result from competition from resident plants that continue to occupy the area although they may have lost reproductive ability, from low rates of seed dispersal, or from such traits as shade intolerance; the result can be a delay of a century or more before hardwood forest begins to replace a coniferous forest when climate warms.

On the other hand, disturbance such as fire or logging may greatly speed the transition, shortening the time from a century to a decade, and seed dispersal of most trees is sufficiently rapid to permit species limits to adjust relatively quickly to change. Human transport corridors make this even more rapid. Research is needed, however, to determine the rate at which the individual northern species can expand ranges under favorable climate conditions as a result of fire or disturbance (*Canadian Weekly Bulletin*, 1972; Heinselman, 1973; Wright and Heinselman, 1973; Rowe and Scotter, 1973; Ball, 1986; Singh and Powell, 1986; Webb, 1986).

10. Summary and Conclusions

Much of the discussion in preceding chapters has dealt with the essentially climatic relationships between the vegetation of the forest-tundra ecotone and the climatic conditions that prevail in this interesting zone, and there is, hence, little need for further discussion of the existing literature here. There is, however, some need for a codification of the data and the various observations and thoughts of those individuals who have seen the ecotone during the course of biological explorations or botanical and ecological research and who have commented in some significant way upon the apparent physiological relationships between the plants and climate in the region. Rather than discuss these observations and research results in a complete and exhaustive—perhaps exhausting—commentary, I leave to those interested in pursuing the topic further those many titles in the bibliography. Here I will summarize, more or less in chronological order, those observations that have appeared, to me, the most significant for furthering understanding of the ecological relationships existing in the ecotonal region, without, however, making any assumption that others might make the identical selection.

The following selection, therefore, is subjective, but it summarizes, for me at least, the pertinent ecological factors as they are known at the present time.

1. Certainly one of the most significant early observations on the northern tree-line in Canada was that of C. H. D. Clark (1940), whose survey of the biology of the Thelon Game Sanctuary led him to write as follows: "There is actually, however, a limit of trees beyond the timber-line, namely, the limit of occurrence of black spruce, white spruce, and tamarack as species. . . . There are, nonetheless, so many tiny clumps of spruce

that were any climatic change to occur making it possible for trees to occupy exposed situations in that region the occupation would be rapid. . . . Timber-line east of Great Slave Lake seems to be relatively stable. It is also a fairly sharp line, and capable of being located within a mile or two."

2. The physical action of permafrost upon root systems in the Far North obviously is an important ecological factor limiting development. Certainly among the first to observe this factor in some detail was Benninghoff (1952), who wrote as follows: "In the high latitudes roots are not only encased in frozen material for a great part of the year, but by repeated freezing and thawing, especially during the autumn freeze-up, they are heaved, torn, split by forces of great strength." Moreover, as a vegetational cover develops and increases in density, the depth of the active layer in summer is decreased, and thus the vegetation itself tends to create conditions that limit its own development: ". . . plants commonly generate conditions of extreme lack of drainage and greatly intensified soil frost; in short, the plants frequently destroy the very environmental conditions that favor their growth."

3. It is discerned that a distinct belt of maximum frontal frequency exists along a zone from north-central Siberia across northern Alaska and Mackenzie District to James Bay in Canada (Reed, 1960), thus confirming an earlier more general statement by Stupart (1928) that the path of the summer cyclonic storms lies to the southward of this region and seems to correlate with the northern limit of trees. The position was also confirmed by streamline fields of surface winds over Canada (Bryson, 1966) and by the position of the arctic frontal zone (Barry, 1967), although the latter does not appear to possess as definite a modal position over Quebec-Labrador as it does west of Hudson Bay, probably accounting for the greater latitudinal width of the lichen woodland (forest-tundra ecotone) in eastern Canada. Much the same coincidence between the arctic front and the forest tundra ecotone exists in Eurasia as in central and western Canada (Krebs and Barry, 1970).

4. Observations made at Great Whale River, on the east coast of Hudson Bay, in 1949 indicated that a complex of factors are involved in the limitations apparent on the advance of spruce northward in the area (Savile, 1963). The interacting factors are discerned as these: relatively low mean summer temperatures, short growing season, low soil fertility, limitations upon growth of plants on sandy areas during dry seasons, snow abrasion as well as sand abrasion where it occurs, limitations upon seed dispersal against the prevailing northwest winds, difficulty of establishing seedlings in the heavy ground cover of *Cladonia* (*Cladina*) lichens, and several rust fungi seen to be abundant enough to be serious threats to spruce development at Great Whale River.

5. Buried charcoal and associated buried podzol paleosols demonstrated that during the Holocene forests occupied positions up to 280 km north of the present day forest border in Keewatin (Bryson et al., 1965; Larsen, 1965; Nichols, 1967, 1970; Sorenson et al., 1971; Ritchie, 1984, 1988), strongly suggesting that changes in atmospheric circulation patterns have induced displacement of the mean frontal zone at times in the past and in subsequent latitudinal shifting of vegetational boundaries in the region.

6. Literature on the relationships between plant growth and temperature is, of course, exceedingly voluminous, but of the published early studies on the physiology of plant growth in the Arctic and the limitations imposed by the environment are those of Warren Wilson (1957, 1966). In the former publication it is stated: "In arctic regions, temperature is the most important factor ultimately limiting the growth and development of the vegetation. This is indicated by the general correspondence between the spatial distribution of vegetation types and temperature climates within this zone, and also by the correlation between annual temperatures and the yearly increments of growth. . . ." Summarizing, there are three temperature-dependent physiological processes that must be considered as most importantly limited by temperature: translocation, water absorption, and photosynthesis, all of which are markedly checked at low temperatures.

7. Spruce habitat studies at Ennadai Lake, southwestern Keewatin, within the northern edge of the forest-tundra ecotone, reveal that rather special site conditions account for the presence of small clumps of spruce at the north end of the lake, some 70 km (45 miles) from the closed-crown forest at the south end. The special habitat required by spruce at the north end of the lake is evidently best provided by those sites with a coarse rock substratum through which the first spring meltwater will drain, thus providing a supply of water beneath the covering of snow and even beneath a surface lens of frozen peat which may persist throughout the short summer. These are sites capable of providing adequate moisture to replace losses incurred as soon as transpiration begins in early spring. As daytime air temperatures in late May and early June climb well above freezing, and leaf temperatures without doubt rise very high due to intense insolation on clear days, the habitat required by spruce is one capable of furnishing a water supply to roots when most of the land surface is still solidly frozen (Larsen, 1965).

8. A comprehensive review of research on the northern tree-line by Hustich (1966) summarizes knowledge available to that time on the ecological relationships of the forest-tundra ecotone. In a discussion of climatic influences, Hustich described mathematical analyses that define the correlation between growth of trees in the north and climate at the nearest meteorological station, citing a paper (Hustich and Elfving, 1944) describing a study in which simple July mean temperature emerged as possessing the highest correlation with radial growth rate. Also revealed was an important lag effect, in which a very favorable or a very bad summer influenced growth not only in the current but also in subsequent years. "We have to understand the physiology of the forest-tundra tree-species to understand in a deeper sense the ecology of the tree-line."

9. Among the first, and certainly one of the most detailed of the early chemical analyses of northern forest soils, showing that nutrient deficiency is an important factor in limiting growth of far northern trees, is that of Heilman (1968). Reporting on analyses for P, K, Ca, Mg, Mn, and Zn in black spruce foliage on forests on north slopes in interior Alaska, Heilman showed that levels of N, P, and K in foliage are deficient in trees growing on peat soils derived from sphagnum moss, and the declining availability of nutrients is a primary factor in reducing tree growth in these northern areas. The increasing coldness of the soil as the moss peat accumulates appears to be sufficient

to reduce the availability of nutrients in the far northern forest soils. The complex of ecological factors limiting growth of forest trees in Alaska is the topic of Van Cleve et al. (1983) in a most thorough discussion of their long-term research program.

10. In an exhaustive review of research on the physiology of arctic and alpine plant species, much of which is a summary of their own extensive research, Billings and Mooney point out that alpine timberlines and, by inference, the northern latitudinal forest boundary, is an ecotone between the severity of the tundra environment and the relative protection of forest, the zone with the sharpest gradients and clines in both environmental parameters and genetic characteristics (Billings and Mooney, 1968). They state that such gradients and clines are usually steepest at tree-line, and thus the forest-tundra ecotone provides an approximate but reasonably acceptable lower limit to arctic and alpine conditions and adaptations to these conditions as revealed in the physiological ecology of the plant species as well as life-cycle characteristics, morphology, dormancy and germination traits, seedling establishment, photosynthesis, and primary productivity. A similarly exhaustive review of the same general subject but with emphasis on arctic and alpine plant life cycles is that of Bliss (1971).

11. Correlations between air mass types in northern Canada and the frequency with which plant species occur in vegetational communities of the region are reported by Larsen (1971), who shows that species showing highest relationships to arctic air mass frequencies are those attaining dominance in black spruce communities in the more northern areas. Similar relationships with arctic or Pacific air are shown by the species of the rock field and white spruce vegetational communities; in tussock muskeg communities the relationship is less apparent. The study shows that although variability in composition and structure of plant communities is usually relatively great, a frequent observation by plant ecologists everywhere, it is significant that northward there are communities in which many species can be clearly correlated in their frequency with the macro-climatic parameters. A most interesting group of species are seven that increase in frequency northward through the forest and then decrease northward beyond the forest. All are wide-ranging, but they attain highest frequencies in the communities of the forest-tundra ecotone, evidently uniquely adapted or simply widely tolerant of the ecotonal conditions.

12. In the first comprehensive review of the radiation balance of the boreal forest and tundra regions of central Canada, Hare and Ritchie (1972) describe the net radiation regimes of forest, forest-tundra ecotone, and tundra, relating these to productivity characteristics of the various vegetational communities present in each biotic zone, and concurring in the assumption that thermal rather than moisture conditions control plant growth in the boreal zone. They qualify their observation, however, with the caveat that snowfall and snow-cover regimes have not as yet been studied in the detail required to dismiss them as significant influences upon vegetation. Also, they add that while their studies show a matching of radiational distribution with geographical distribution of vegetation, these can be taken only as indications since there can never be firm proof by simple curve-matching. "To make the findings firm quite different methods will be needed, methods that require detailed field studies of the physiological, biochemical, and micrometeorological processes."

13. The autecology of black spruce is described in detail by Black and Bliss (1978, 1980) on the basis of studies conducted in the lower Mackenzie Valley. One of the more important findings is that minimum average monthly temperatures for June, July, and August are 11°, 14°, 11°C for adequate reproduction. Elliott (1979) confirmed earlier observations by Larsen (1965) that black spruce at the northern forest border in southwestern Keewatin was unable to produce viable seeds in at least most years, the result evidently of cold temperatures and too short a growing season. In this area, black spruce reproduces vegetatively by layering, and seedlings are rare. Payette and Gagnon (1979) and Gagnon (1982) show that black spruce persists in decumbent or krummholz form in areas beyond the forest border in Quebec and that here it reproduces vegetatively beyond the climatic limit for seed production. It has also been observed by Payette (1983) that although the forest border appears relatively stable in northern Quebec, there are indications that recent climatic trends have resulted in increased density of trees within the forests of this region. In Alaska there is also evidence that the northern forest border is advancing northward, the consequence apparently of above-average summer temperatures in the period since 1932 (Blasing and Fritts, 1973). In some areas of the interior of the Alaska Range, however, there is no evidence of any recent change in tree-line (Denton and Karlen, 1977).

14. A detailed exposition of research on the physiological ecology of plant species in alpine timberline regions of Europe by Tranquillini (1979) summarizes much of the European work on this topic, a great deal of it conducted by the author and most of it directly applicable to northern forest border ecology in North America. In summary: timberline is ultimately dependent everywhere on an increasingly unfavorable heat balance and a short growing season, upon available energy that is insufficient to allow completion of the growth cycle. The influence of increasing elevation on physiological and growth characteristics of trees at timberline was studied in the northern Rocky Mountains by Hansen-Briston (1986), who summarizes previous studies and provides the result of original research in the region.

15. There is a striking correspondence between the northern forest border ecotonal region and one or more parameters of the environmental complex that exceed the tolerance limits of the trees. Here also, in the forest-tundra ecotone, large numbers of shrub and herbaceous species show similar changes in abundance along environmental gradients, with indicator possibilities that hitherto seldom, if ever, have been taken into account. Moreover, a number of species showing high frequency correlations with air mass frequencies are among the dominant species in some of the plant communities, and the utility of this application of the continuum concept of vegetational community structure becomes apparent. The plant species are responding to a northern and southern climatic component, a combination of air mass frequencies and other factors such as the direction of the winds, wave structure of frontal zones, gradual air mass modification during movement from one region to another, solar radiation, microclimate, and so on. The dominance relationships of the species with high correlations with climatic parameters are of particular interest in the black spruce and white spruce communities and in the rock field communities; here species with highest correlations with climate are consistently among the dominant species in the sampled stands. In the white and black spruce communities, these species include *Vaccinium vitis-idaea*,

Empetrum nigrum, *Cornus canadensis*, *Linnaea borealis*, *Rubus pubescens*, *Betula glandulosa*, *Mitella nuda* (all in white spruce communities), *Vaccinium vitis idaea*, *Empetrum nigrum*, *Betula glandulosa*, *Ledum decumbens*, *Vaccinium uliginosum*, *Rubus chamaemorus*, and *Salix glauca* (of importance in the black spruce communities). Species of importance in the rock field communities are of special interest since the species highly correlated with arctic air do not attain dominance: *Salix glauca*, *S. reticulata*, *S. arctica*, *Luzula confusa*, *Arctostaphylos alpina*, *Cassiope tetragona*, *Dryas integrifolia*, *Polygonum viviparum*, *Carex atrofusca*, *Arctagrostis latifolia*, and *Eriophorum angustifolium*. Instead, these are species with strong arctic affinities and attain dominance in the rock field community only much farther north.

Thus there is a group of species with frequency values that correlate with climatic parameters, some of importance in only one community type, others present in two or more, or as many as five in the case of *Vaccinium vitis-idaea*, and all found in both forest and tundra over a wide geographic area. It appears that these species, and these communities, become of great significance in attempting correlations between vegetation and climate in northern regions. As field ecologists well know, variability in vegetation is often so great as to virtually preclude high correlations between vegetational composition and evironmental parameters. It is of some ecological significance that northward there are communities in which at least some species may be found correlated with macroclimatic parameters.

Perhaps most interesting are a group of species that correlate with opposing air mass types in different communities. Each is correlated with arctic air when it occurs in either the black or white spruce community and with air of Pacific origin (with one exception) when it is a component of the rock field or the tussock muskeg communities. These species are wide-ranging through forest and tundra. They increase in frequency northward through the forest and then decrease northward beyond the forest border in the tundra communities; in the former case, they correlate positively with arctic air, and the latter, positively with Pacific air. The species of this group include *Empetrum nigrum*, *Ledum decumbens*, *Rubus chamaemorus*, *Vaccinium uliginosum*, *Betula glandulosa*, and *Salix glauca*, for example, and they attain highest frequencies within the communities of the forest-tundra ecotone (Larsen 1965, 1971a,b, 1972a,b, 1973, 1974, 1980).

Appendix A. Survival Strategies of the Ecotonal Species

With greater numbers of species found in relatively high abundance in communities of both the boreal zone and the arctic regions, with fewer in the forest-tundra ecotonal communities, the question arises as to the strategies involved that permit adaptations to disparate environments yet prohibit all but a few from surviving in the ecotone, at least in numbers sufficient so they are recorded in quadrats of the size employed in transects of this study. There is a depauperate zone in the forest-tundra ecotone marked by a paucity of species, a zone in which only a relatively few wide-ranging species are persistently dominant (see Tables 5.1–5.9). Plant communities in both the forests to the south and the tundra to the north possess more species of sufficient abundance to show up in the transects. In the depauperate zone, there is no lack of vegetational cover over the landscape, but fewer species in total dominate the communities. There are fewer species demonstrating intermediate abundance, fewer rare species. The obvious question is, "Why should this be so?" As yet, there is perhaps no fully satisfactory answer, but there have been studies that at least seem to point the way for an approach to the question.

Studies of the survival and reproductive strategies of tundra plants, arctic and alpine, have a long history, and these are thoroughly discussed in classic papers and reviews by Britton (1957), Jeffree (1960), Warren Wilson (1957a,b, 1966a,b, 1967), Johnson et al. (1966), Johnson and Packer (1967), Billings and Mooney (1968), Bliss (1956, 1958, 1962, 1971), Billings (1974), and Wielgolaski et al. (1975). Concepts derived from more recent research programs are discussed by many workers, including Bliss (1977), Solbrig (1980), Brown et al. (1980), Bell and Bliss (1977, 1978, 1980a,b),

Addison and Bliss (1977, 1980, 1984), Sohlberg and Bliss (1984), Jackson and Bliss (1984), Tieszen (1972, 1973), Tieszen and Wieland (1975), and others. Early discussions of the causes of timberline and the physiology of timberline trees are to be found in Daubenmire (1954), Lindsay (1971), Pisek and Winkler (1958), Larcher (1957), Tranquillini (1963, 1964, 1969, 1979), Vincent (1965), Viereck (1973), and Mosquin (1966). There also has been much recent work on the physiology and growth habit of northern evergreen trees of the genera *Abies*, *Pinus*, and *Picea* that are dominant species in the alpine and northern tree-lines of North America (Hom and Oechel, 1983; Strang and Johnson, 1981; Lawrence and Oechel, 1983a,b; Strong and La Roi, 1983; Van Cleve et al., 1981; Tryon and Chapin, 1983; Viereck et al., 1983; Vowinckel et al., 1975; Yarie, 1983; Johnson, 1981; Kershaw, 1977, 1978; Lambert and Maycock, 1968; Arno, 1984; Ritchie, 1984; Rowe, 1966, 1984).

Greater variability in environmental conditions in the ecotonal zone are likely to be a significant factor in the survival of these species. For example, there is only a few degrees difference in average summer temperature between the northern and southern edges of the boreal forest, with also a correspondingly great uniformity in the airmass characteristics of low latitude arctic tundra. Yet, at the ecotone these two climatic regimes are separated by a steep gradient clearly apparent in climatological data. This gradient is, however, deceptive in the sense that there is, in fact, not so much a persistent gradient as a relatively rapid fluctuation from one regime to the other, not apparent in the averaged data: boreal one day, arctic the next, frost a possibility at any time during summer, yet short periods of calm with high temperatures and clear skies, all of which follow one another with rapidity alien to other regions, one day hot, dry, and calm, the next windy, cold, wet, intermixed daily with a few frosty hours around midnight. Plant species adapted to the boreal regime do poorly in an arctic environment, arctic plants do poorly under boreal conditions, and only a few can survive persistent rapid oscillations between the two.

For obvious reasons, most studies of ecological strategy have been conducted with temperate zone species, particularly with annuals and with some perennials possessing small well-defined bulbs. One early study was conducted on *Equisetum sylvaticum*, however, and revealed that this species, common in the northern spruce forest and a long-lived perennial, possesses an extensive rhizome which may outweigh the aerial shoots by a ratio of 100 to 1. Moreover, marked variations in growth from year to year indicate that the species is capable of a markedly plastic response to variable conditions. The conclusion drawn from the study is that phenotypic plasticity enables the species to survive marked environmental variability. This also provides assurance against major climatic changes and vegetational migrations that occur every few thousand years. Other species with similar ecological strategy are *Vaccinium* spp., *Ledum groenlandicum*, *Kalmia* spp., *Rubus* spp., and others, according to the authors (Beasleigh and Yarranton, 1974), who add that longevity in these species permits survival where microclimatic and edaphic conditions fluctuate wildly and the vegetation cover varies from zero to closed forests.

Plasticity, thus, permits survival during changing conditions. Species that lack such plasticity, although they may inhabit colder climatic regimes northward or warmer and wetter conditions southward in the boreal forest, are eliminated as competitors in forest-tundra ecotonal regions. Moreover, this is seen to be the case especially in the zone where environmental fluctuations are most pronounced, in the forest-tundra ecotonal fringe at the northern edge of the forest-tundra transition zone.

Only additional studies of the individual strategies of the species will reveal whether this hypothesis is a valid one, particularly with such species as *Ledum groenlandicum*, *L. decumbens*, *Vaccinium uliginosum*, *Rubus chamaemorus*, *Salix* spp., *Arctostaphylos alpina*, *Andromeda polifolia*, *Carex* spp., *Eriophorum* spp., and perhaps a few other wide-ranging species that tend to dominate the relatively species-poor plant communities at the northern edge of the forest and the southern fringe of tundra, the extreme northern edge of the forest-tundra ecotone (Larsen, 1965, 1971a,b, 1974, 1980).

Somewhat like repeated fire, these environmental fluctuations tend to stimulate vegetative reproduction of the prevailing dominants, to the extent that they ultimately eliminate species of lower densities and dominance (Solberg, 1980; Strang and Johnson, 1981).

It has also been noted that the long period of germination of *Ledum groenlandicum* and *L. decumbens* seed serves to minimize the influence of brief periods of unseasonably warm temperatures that may occur in fall, presumably as well as brief periods of very severe cold in late spring. The seeds, moreover, have relatively high temperature requirements for germination. This tends to delay shoot emergence until at least early summer conditions are firmly established in both soil and atmosphere. Seed of *L. decumbens* germinates more rapidly than that of *L. groenlandicum*, however, which would account for the more northern distribution of the former, extending far into tundra while the latter is more or less restricted to the northern forest and forest-tundra ecotone (Karlin and Bliss, 1983). It is undoubtedly significant that these are wide-ranging species, among the group listed above, indicating by the breadth of their geographical range that they do, indeed, possess a degree of phenotypic plasticity and perhaps genetic diversity absent in species of more restricted range and adaptation.

It has been observed that studies of the structure and dynamics of plant communities have at present no theoretical base (Austin, 1985), at least broadly so, although concepts of continuum and niche are certainly valid steps along the way (Johnson, 1977a,b). There are, however, no predictive equations describing causal relationships that establish which plant species will be found under specific habitat conditions, in which combinations, in what numbers, and whether or not inter- or intraspecific competition represent important factors establishing the structure observed. It is almost as though much of what is observed is the consequence of random events in the environment occurring concurrently with virtually random opportunities for species to gain a foothold at a given time and place, within the necessary limits imposed by the physiological tolerance of the species, existence of a sufficiently abundant supply of seed, and favorable habitat conditions.

On the other hand, it is possible to be rather sanguine over the fact that it is now possible to find descriptive, perhaps even quantitative, information on the composition and structure of hundreds of communities that have been sampled by ecologists, but the absence of a theoretical basis for this effort results in a corresponding absence of a uniform methodology for statistical description of community composition and structure, making comparison difficult, the pathway of successional change, and so on, largely conjectural. In the instance of forest-tundra ecotonal communities described above (and in Tables 5.1–5.9), in which a few wide-ranging species attain dominance over a broad longitudinal range, the observation that this is a community structure found most commonly in harsh environments is one of great interest and perhaps broad significance (Giller, 1984, pp. 121–123). Additional detailed analyses of the structure of these communities was given in previous publications (see tables, figures, and references in Larsen, 1974).

While much remains to be done, there are, nevertheless, recent studies of northern plant communities that encourage the belief that a certain degree of underlying predictability exists in the behavior of the northern species and the plant communities of northern regions, forest, ecotone, and tundra. This is true, of course, providing that it is possible to take into account the entire complex of habitat factors that affect establishment, survival, and reproduction of the native plants—and some of the principles that govern behavior are beginning to emerge. Among these are some more generalized observations on the effects of environmental variability on the ecological and evolutionary response of vegetational communities (Weins, 1977; White, 1978; Conner and Simberloff, 1986), as well as general observations on the organization of communities in terms of composition, structure, competition, pattern, process, and response to habitat conditions (Sousa, 1980; Strong, 1983; Salt, 1983; Loveless and Hamrick, 1984; Diamond and Case, 1985).

There are, moreover, growing numbers of studies of the organization and dynamics of the far northern forest communities (see refs. in Larsen, 1980; also Johnson, 1981; Jasieniuk and Johnson, 1982), as well as more general studies specifically related to niche theory (McNaughton and Wolf, 1970; Whittaker, 1970; Pianka, 1974; MacArthur, 1972; Johnson, 1977a,b, 1985; Giller, 1984). There has also accumulated a mass of information on the type and frequency of disturbance in northern forest communities such as fire in boreal forest and the boreal forest-tundra ecotone (Johnson and Rowe, 1975; Johnson, 1979, 1983; Johnson and Van Wagner, 1985; Foster, 1983), as well as in the tundra (Wein, 1976).

Also revealing are the effects of various environmental factors upon individual species, such as observed in the lichens (Maikawa and Kershaw, 1976; Kershaw, 1977; Carstairs and Oechel, 1978), in mosses (Hicklenton and Oechel, 1976, 1977), tree species (Black and Bliss, 1978), and individual shrubs and herbs (Lems, 1956; Beasleigh and Yarranton, 1974). There are also increasingly numerous studies of the environmental factors that must be taken into consideration for successful sampling and modeling of vegetational communities (Monteith, 1981; Pielou, 1984; Ritchie, 1984; Austin, 1985; Beals, 1985).

Putative explanations for plant community structure and composition have occupied plant ecologists as long as the science has been in existence, and rousing debate still occurs over which factors are most importantly involved in what has been termed theoretical plant ecology (Austin, 1986; Strong et al., 1984; Pianka, 1981; Gubbe, 1986; Tilman, 1986). While it is encouraging that descriptive plant ecology is now evolving into a theoretical stage in development, there are apparently some key bricks missing in the foundation upon which a conceptual structure might be built.

The physical chemistry of enzymatic proteins is now emerging as exceedingly significant in the environmental response of species, at least as important, perhaps more so, than the morphological, phenological, spatial, and energy relationships, all more or less traditional avenues by which explanations for the ability of species to persist in given environments under given conditions are derived. It has been a persistent source of frustration for those studying plant communities that none of these approaches, or any combination, have conferred an ability to accurately predict observed patterns of vegetation variation along environmental gradients or successional series.

The biochemical characteristics of the cellular processes, conferred by the enzymes, now offer another factor to take into consideration. There is the real probability that enzyme function is ultimately the most critical factor of all. The kinetic properties of enzymes have long been studied, but only recently have variations in these been ascribable directly to chemical composition of the enzymes. Now it is apparent that there are fine-scale adaptations of proteins to temperature and pressure, with only modest changes in amino acid sequences required to achieve adaptation of enzyme systems to new environmental regimes. There are indications that enzymatic adaptation to temperature, pressure, and pH are strongly selective, with very sharp boundaries between viable and nonviable conditions for enzymatic activity, hence survival for individual plants. This is, of course, most readily apparent in such places as the limit for tree growth on mountains, the climatic correlations of which have been revealed in some detail recently by means of a transect of meteorological recording stations located along an altitudinal gradient across timberline (Hansen-Bristow, 1986). The literature on the protein kinetics involved in enzymatic processes and their relationship to biogeography is as yet limited, but a review of research to date is revealing of the potential significance (Somero, 1986; Hochachka and Somero, 1983). Enzymatic studies of the dominant ecotonal species mentioned above, and comparison with species of more restricted distribution, promise to be most revealing of their respective survival strategies.

Appendix B. A Note on Methods

Multivariate analysis of community data has been shown to possess real promise for general descriptions of communities, delineating groups of species with similar characteristics. But there is still a place for the expression of data in relatively straightforward presentation, in which the performance of the species (as recorded in measurements of frequency, cover, or other parameters selected as appropriate) is given no analytical treatment, leaving analysis and interpretation to the reader, if so desired, using any method selected. This, of course, can be carried on only when the data are not so voluminous as to preclude such presentation or so cursory as to make it likely that the true nature of the population sampled is not represented. It seems reasonable that the data sets presented here fall in neither of the above categories, and so they are presented as obtained in field sampling without recourse to any of the more elaborate multivariate techniques now available for reduction of ecological data to one or a few major summarized descriptions. These latter can be derived by whatever manner an interested reader may prefer (Gauch, 1982; Harper, 1982; Pielou, 1984; Hall and DeAngelis, 1985).

As Gauch points out: "The goals of community ecology are partially indefinite and frequently progress by successive refinement." The data as presented here can be employed in whatever way seems appropriate for achieving more definite goals as they may arise, and by refinement as other workers add to the body of data from sampled communities in the future. This is a reasonable expectation even taking into account the remote nature of the regions in which these data have been obtained; difficult of access excepting by aircraft or, in many instances, canoe.

It is conceivable that the data herein presented can also be employed as baseline data for future comparisons, in efforts to determine the influence of various environmental situations such as disturbance or climatic change. Studies of successional relationships of vegetational communities may also be possible, although the latter would necessarily be conducted on a long-term area-wide basis since the precise locations of the sampled stands are not given here and there is no firm and unequivocal record other than the recollections of the author and rough jottings in field notebooks. The aerial photographs, however, provide records of the general locations of the transects and thus delineate the character of the vegetation at the time the photographs were taken. This did not differ appreciably from the character of the vegetation at the time the sampling transects were run and the data obtained.

Appendix C. Introgression in Black and White Spruce

It has long been an accepted observation that wherever two similar species of the same genus are found both at the edge of their range, hybridization may occur more frequently here than elsewhere. This may be the case with black and white spruce (*Picea mariana* (Mill.) B.S.P. and *P. glauca* (Moench) Voss), at least in some northern tree-line areas. In studies at Ennadai Lake, it was at times difficult to make a distinction between black and white spruce, particularly on the esker sites (Larsen, 1965), as apparent intermediate forms were present in some abundance. Twigs bearing varying degrees of pubescence were found on what might otherwise be considered black or white spruce, and trees bearing cones distinctly those of black spruce lacked pubescence on either terminal or lateral twigs. Sterigmata on the twigs were at least atypical for black spruce and more nearly resemble the projections characteristic of white spruce. The cones of many trees, on the other hand, bear an average of approximately 50 scales, and are thus intermediate between the average of 30 scales per cone for black spruce and 60–90 for white spruce as described by Scoggan (1957). These cones, moreover, are not as globose as those of typical black spruce. Reproduction by layering is common in these intermediate forms, a characteristic evidently observed in white spruce more rarely than in black spruce. Further work by taxonomic specialists might reveal more interesting details of this apparent instance of introgression. Hybridization has been observed in white and black spruce elsewhere (Heinselman, 1957; Vincent, 1965; Arno, 1984), and certain other aspects of the genetics of northern and alpine tree species might be pertinent (La Roi and Dugle, 1968; Singh and Owens, 1981; Strang and Johnson, 1981; Parker et al., 1981; Park and Fowler, 1984): see Figs C1–C3.

Figure C.1. Black spruce (*Picea mariana*) tree in the Ennadai area with normal cones.

Figure C.2. White spruce (*Picea glauca*) clump, on the south shore of Dubawnt Lake, with normally elongated cones.

It is also of interest that while spruce reproduction at the northern edge of the forest-tundra ecotone at Ennadai Lake is principally by layering (Larsen, 1965; Elliott, 1979a,b), there is an occasional seedling to be found in particularly favorable habitats. One such is that shown in Fig. C.4. Here a spruce seedling was found at the head of a grave, surrounded by wooden fencing, thus protected from winds and with a substrate open for colonization. The site was atop a low ridge at the edge of the lake, perhaps 50 yards from the lakeshore. The seedling was, however, sufficiently unique to merit permanent record in the form of a photograph.

This should not be construed as a claim that hybridization and subsequent introgression exist in the Ennadai Lake area; it is, however, believed to be a possibility that would bear a more detailed analysis than it was possible to undertake at the time. Similar studies might well be taken also elsewhere. White spruce occurs in a wide range of habitats in Alaska and elsewhere; populations at the tree limit in Alaska (Cooper, 1986) are genetically quite diverse, capable of adapting to a wide range of conditions (Alden and Loopstra, 1987). Genetic variation is also apparent in black spruce, as shown in the Maritimes by Park and Fowler (1988), Gordon (1976), and Manley (1972). Because of the natural phenological barrier (black spruce sheds pollen from 7–10 days later than white spruce), Dugle and Bols (1971) explain the rarity of hybrids, but also state that

Figure C.3. Clump of apparently black spruce (*Picea mariana*) trees, with abnormally long cones.

when even a few hybrids survive they will provide a genetic source for backcrosses and introgression for many years. It may be, too, that flowering times in northern areas overlap to a greater degree than in the south.

It should be noted that Parker and McLachlan (1978) cast doubt on the reports of *P. glauca* × *P. mariana* hybridization, due to the inability to bring it about artificially, as well as the unconfirmed nature of the reports of natural hybridization. They employed morphological characteristics in an attempt to detect natural hybrids or introgressed individuals in northwestern Ontario, without success. The same criteria were employed to cast doubt on the validity of claims of hybrids from other areas, and they state that reports of chemical analyses indicating hybridization were insufficient in terms of sample size to provide a realistic breadth of variation in the pure white and black spruce populations. In their study, they were able to select a set of primary diagnostic characters for which no overlap was observed between standard specimens of black and white spruce: cone width, bud-scale width, bud-scale length, bud length, and the ratios between them were among the measurements employed. In view of their own study, they point out, also, that the subject warrants further examination.

A number of techniques should be employed in the study if and when it is undertaken: the measurements employed by Parker and McLachlan above; those used by Roche

Figure C.4. An unusual site which evidently has been favorable for the growth of a black spruce (*Picea mariana*) seedling: the soil has been freed of surface vegetation and the fence buffers the wind and probably acts to create a rather deep snow drift in winter, which nevertheless disappears quickly in the spring. The seedling can be seen at the head of the grave. The site is on a remote hilltop in a bay toward the north end of Ennadai Lake. The grave is unmarked; it is evidently that of a child. The area is north of the forest, with only a scattering of trees in favorable sites.

(1969) in studies of individuals exhibiting characteristics of both black and white spruce on the Alaskan Highway north of latitude 57°; those used by Hills and Ogilvie (1970) in delineating the identity of a spruce found in fossil deposits on Banks Island (to the former of whom I am greatly indebted for kind comments in personal correspondence); as well as *Picea* needle-extract chromatography used by La Roi and Dugle (1967) and Dugle and Bols (1971) and terpene distribution patterns employed by Ogilvie and von Rudloff (1968).

Since viable seeds are usually not available at Ennadai Lake and other tree-line areas, other morphological characters necessarily would be used as well as chemical composition. If, in a favorable year, viable seeds were to become available, then characteristics of the seeds and seedlings would be possible. At other times, morphological traits, as well as times of flushing and dormancy assessment, could be made.

References for Appendix C

Alden, John, and Carol Loopstra. 1987. Genetic diversity and population structure of *Picea glauca* on an altitudinal gradient in interior Alaska. *Canadian Journal of Forest Research* **17**: 1519–1526.

Anderson, E. 1953. Introgressive hybridization. *Biological Reviews* **28**: 280–307.

Arno, Stephen F. 1984. *Timberline: Mountain and Arctic Forest Frontiers.* The Mountaineers Press, Seattle, Washington.

Cooper, David J. 1986. White spruce above and beyond treeline in the Arrigetch Peaks region, Brooks Range, Alaska. *Arctic* **39**(3): 247–252.

Dugle, Janet R., and Niels Bols. 1971. Variation in *Picea glauca* and *P. mariana* in Manitoba and adjacent areas. Atomic Energy of Canada Ltd., Whiteshell Nuclear Research Establishment, Report AECL-3681, Pinawa, Manitoba.

Elliott, D.L. 1979a. The stability of the northern Canadian tree limit; current regenerative capacity. Ph.D. thesis, University of Colorado, Boulder, Colorado.

Elliott, D.L. 1979b. The current regenerative capacity of the northern Canadian trees, Keewatin, N.W.T., Canada; some preliminary observations. *Arctic and Alpine Research* **11**: 243–251.

Gordon, A.G. 1952. Spruce identification by twig characteristics. *Forestry Chronicle* **28**: 43.

Gordon, Alan G. 1976. The taxonomy and genetics of *Picea rubens* and its relationship to *Picea mariana*. *Canadian Journal of Botany* **54**: 781–813.

Grant, Verne. 1971. *Plant Speciation.* Columbia University Press, New York, NY.

Heinselman, M.L. 1957. *Silvical Characteristics of Black Spruce.* Forest Service, U.S. Department of Agriculture, Lake States Forest Experiment Station, Paper No. 45, St. Paul, Minn.

Hills, L.V. 1975. Late tertiary floras of arctic Canada. *Proceedings of the Circumpolar Conference on Northern Ecology*, National Research Council of Canada, Onnata, pp. I 65–71.

Hills, L.V., and R.T. Ogilvie. 1970. *Picea banksii* n. sp. Beaufort Formation (Tertiary), northwestern Banks Island, Arctic Canada. *Canadian Journal of Botany* **48**: 457–464.

La Roi, G.H., and J.R. Dugle. 1968. A systematic and genecological study of Picea glauca and P. engelmannii, using paper chromatograms of needle extracts. *Canadian Journal of Botany* **46**: 649–687.

Larsen, James A. 1965. The vegetation of the Ennadai Lake area, N.W.T.: studies in arctic and subarctic bioclimatology. *Ecological Monographs* **35(1)**: 37–59.

Little, E.L., and S.S. Pauley. 1958. A natural hybrid between black and white spruce in Minnesota. *American Midland Naturalist* **60**: 202–211.

Manley, S.A.M. 1972. The occurrence of hybrid swarms of red and black spruce in central New Brunswick. *Canadian Journal of Forest Research* **2**: 381–391.

Ogilvie, R.T., and E. von Rudloff. 1968. Chemosystematic studies in the genus *Picea* (Pinaceae). IV. The introgression of white and Engelmann spruce as found along the Bow River. *Canadian Journal of Botany* **48**: 901–908.

Park, Y.S., and D.P. Fowler. 1984. Inbreeding in black spruce (*Picea mariana* (Mill.) B.S.P.): self-fertility, genetic load, and performance. *Canadian Journal of Forest Research* **14**: 17–21.
Park, Y.S., and D.P. Fowler. 1988. Geographic variation of black spruce tested in the Maritimes. *Canadian Journal of Forest Research* **18**: 106–114.
Parker, William H., Jack Maze, and Gary E. Bradfield. 1981. Implications of morphological and anatomical variation in Abies balsamea and A. lasiocarpa (Pinaceae) from Western Canada. *Canadian Journal of Botany* **68(6)**: 843–854.
Parker, William H., and David G. McLachlan. 1978. Morphological variation in white and black spruce: investigation of natural hybridization between *Picea glauca* and *P. mariana*. *Canadian Journal of Botany* **56**: 2514–2520.
Roche, L. 1969. A genecological study of the genus *Picea* in British Columbia. *New Phytologist* **68**: 505–554.
Scoggan, H.J. 1957. *Flora of Manitoba.* National Museum of Canada Bulletin No. 149, Ottawa.
Strang, R.M., and A.H. Johnson. 1981. Fire and climax spruce forest in central Yukon. *Arctic* **34(1)**: 60–61.
Vincent, A.B. 1965. *Black Spruce: a Review of Its Silvics, Ecology, and Silviculture.* Department of Forestry, Publication No. 1100, Ottawa, Canada.
Wright, Jonathan W. 1955. Species crossibility in spruce in relation to distribution and taxonomy. *Forest Science* **1(4)**: 319–349.

Appendix D. Nomenclature

The discerning reader will note that no single taxonomic author's terminology has been employed; the species discussed are all sufficiently common so that no difficulty should be encountered in making accurate identifications from the specific names used. Those who wish to pursue the matter, however, will find a number of publications of value (Porsild, 1945, 1951, 1955, 1957, 1964; Scoggan, 1957, 1978; Hultén, 1968; Polunin, 1948, 1959; Cody and Porsild, 1980).

Appendix E. Aerial Photographic Interpretation

It is recommended that the aerial photographs for the areas covered by the photographs included and the keys be obtained for direct inspection. The numbers of the photographs are included so they can be obtained easily from the Canadian National Air Photo Library, Department of Energy, Mines, and Resources, Ottawa, Ontario, Canada K1A 0E9. For stereographic coverage, three prints should be obtained, one print of the photograph indicated and one print each of photographs numbered one digit above and one digit below the number given, i.e., Dubawnt Lake area photograph number A 15059-20 is the number of the photograph for which the key is provided. For stereographic coverage, photographs numbered A 15059-19 and A 15059-21 should also be obtained.

In the data below, the scale refers to the conventional size of 9 × 9 inches, and adjustment should be made for the size of the photographs as they appear in this volume so that they will fit on a single page. A single portion of an adjacent photograph is included, so that some idea of the value of stereographic coverage can be obtained. Many individuals have the ability, after a little practice, to adjust the eyes so that stereographic effect is obtained without a stereographic viewer, but the latter may be required.

Numbers in the keys (see Fig. 4.2 for definitions) refer to the broad vegetational and physiographic features indicated. Thus, in the Dubawnt Lake photograph, for example, the numbers 9 and 10 in sequence are intended to mean that both rock field and tussock muskeg vegetational communities (and topographic features on which these most often occur) are most prominent in the area, but that intermediate forms are also present; other communities may also be present, but in much more restricted area. It is left to the viewer to interpret the characteristics of any small area of particular interest. This, of course, is made necessary by the intricate and small-scale patterning often present on the photographs. It is impossible to reproduce these in every case by the conventional lines and numbers employed in the key charts. When no number is present to designate the characteristics of an area, this is simply the consequence of the presence of many similar areas which have been numbered and from which accurate identification can be made.

For each of the study areas, vegetational data obtained during field work delineates structure and composition of the vegetational communities wherever they occur. Sampling methods have been described briefly in the chapter on community structure: each stand was sampled by means of a transect of 20 quadrats, each a square meter in size, and the result was a frequency count expressed as a percentage. Thus, a figure of 50 to designate frequency indicates that the species was present in half (10) of the quadrats in the transect. Quadrats were usually spaced some 30 meters apart (one stride being considered a meter). The data for stands in the boreal forest proper are presented in earlier publications, as well as the data for some tundra areas (see Larsen references 1965 to 1982 in the bibliography).

For each study area described, publications and maps of the Canadian Geological Survey have been consulted. These may not be the most recent publication(s) dealing with the area. The Canadian Geological Survey is an active agency and the reports published may subsequently take precedence over those consulted or listed in the bibliography. Those persons interested in more recent accounts, as well as more recent aerial photographs, are advised to consult recent listings of Geological Survey and Canadian National Air Photo Library publications and photographs available.

The aerial photographs in this volume, listed below with the date at which the flight occurred, are copyright as follows: © (copyright year as shown in the attached listing) Her Majesty the Queen in Right of Canada, reproduced from the collection of the National Air Photo Library with permission of Energy, Mines and Resources Canada.

Aerial Photograph Data

Number and Area (see Fig. 1.1)	Photo Number	Date Flown (D/M/Y)	Altitude (ASL*)	Scale (=1″)	Camera Focal Length
1. Canoe Lake	A 14363-27	9/8/54	35,000′	5600′	6.043″
2. Trout Lake	A 13383-162	6/8/52	35,000	5600	6.000
3. Florence Lake	A 12599-287	18/6/50	20,000	2640	6.008
4. Carcajou Lake	A 12148-467	28/7/49	20,000	2640	5.998
5. Coppermine	A 13608-172	17/6/53	35,000	5600	5.970
6. Pelly Lake	A 15048-28	15/9/55	30,000	4900	6.000
7. Snow Bunting Lake	A 15324-57	13/7/56	30,000	5000	153.44 mm
8. Curtis Lake	A 15411-150	6/8/56	30,000	5000	152.65 mm
10. Aylmer Lake	A 15704-168	21/7/57	30,000	5000	5.990″
11. Clinton Colden Lake	A 14370-103	12/8/54	30,000	5600	6.079
12. No. Artillery Lake	A 14371-36	12/8/54	35,000	5600	6.079
13. Dubawnt Lake	A 15059-20	23/9/55	30,000	5000	6.011
15. Kazan River	A 14852-39	7/8/55	30,000	5000	6.011
16. No. Ennadai Lake	A 17222-113	20/9/60	30,000	5000	5.992
19. "East Keller" Lake	A 12700-73	10/6/50	20,000	3333	6.000
20. Winter Lake	A 14062-70	29/7/54	35,000	5833	6.000

Aerial Photograph Data (*continued*)

Number and Area (see Fig. 1.1)	Photo Number	Date Flown (D/M/Y)	Altitude (ASL*)	Scale (=1″)	Camera Focal Length
21. So. Artillery Lake	A 14398-126	15/8/54	35,000	5600	6.079
22. Fort Reliance	A 14357-45	4/8/54	35,000	5600	6.079
23. So. Ennadai Lake	A 17183-201	4/9/60	30,000	4800	5.990
24. Kasba Lake	A 17777-46	6/8/62	30,000	5000	6.000

*Above sea level.

It is of interest to note that satellite imagery is now being used to locate vegetational zonation in many parts of the world, and one of the first such uses was that made in an effort to identify the Canadian forest-tundra ecotone signature on weather satellite imagery (Aldrich et al., 1971).

Appendix F

Coordinates of Study Sites Numbered in Fig. 1.1 (Except Those Numbered in Fig. 4.12; see below).

1. Trout Lake	68°50′N	138°44′W
2. Canoe Lake	68°14′N	135°56′W
3. Florence Lake	65°08′N	128°03′W
4. Carcajou Lake	64°39′N	127°52′W
5. Coppermine	67°47′N	115°10′W
6. Pelly Lake	(see listing below and Fig. 4.12)	
7. Snow Bunting Lake	(see listing below and Fig. 4.12)	
8. Curtis Lake	(see listing below and Fig. 4.12)	
9. Reindeer Depot	68°40′N	134°07′W
10. Aylmer Lake	64°26′N	108°25′W
11. Clinton Colden Lake	63°53′N	106°57′W
12. No. Artillery Lake	63°27′N	107°40′W
13. Dubawnt Lake	(see listing below and Fig. 4.12)	
14. Yathkyed Lake	62°30′N	98°38′W
15. Kazan River	(see listing below and Fig. 4.12)	
16. No. Ennadai Lake	(see listing below and Fig. 4.12)	
17. Inuvik	68°17′N	133°35′W
18. Colville Lake	67°15′N	125°40′W
19. "East Keller" Lake	63°55′N	120°28′W
20. Winter Lake	64°28′N	113°06′W
21. So. Artillery Lake	63°00′N	108°15′W
22. Fort Reliance	62°40′N	109°07′W
23. So. Ennadai Lake	(see listing below and Fig. 4.12)	
24. Kasba Lake	60°13′N	101°52′W

Coordinates of Study Sites Numbered in Fig. 4.12

1. So. Ennadai Lake	60°42′N	101°45′W
2. No. Ennadai Lake	61°15′N	100°58′W
3. Kazan River	61°22′N	100°38′W
4. Kazan River	61°29′N	100°35′W
5. Kazan River	61°42′N	100°38′W
6. Kazan River	61°48′N	100°40′W
7. Kazan River	61°57′N	100°48′W
8. Kazan River	62°00′N	100°46′W
9. Kazan River*	62°21′N	100°33′W
10. Dubawnt Lake	62°43′N	101°27′W
11. Pelly Lake	66°02′N	101°07′W
12. Snow Bunting Lake	66°10′N	94°25′W
13. Curtis Lake	66°50′N	88°55′W

*Unnamed tributary to the Kazan River.

Appendix G

Bird and Animal Species Observed in the Ennadai Lake Area

Common Name	Species	Eskimo	Remarks
Arctic Hare	*Lepus arcticus*	Ookahlick	Common in 1961; rare in 1963.
Ground Squirrel	*Citellus parryi*		Rare in 1961; none seen in 1963.
Red Squirrel	*Tamiasciurus hudsonicus*		Frequent in spruce forest at south end of Ennadai.
Bog Lemming	*Synaptomys borealis*		Skull picked up on trundra at north end of Ennadai Lake.
Collared Lemming	*Dicrostonyx groenlandicus*	Ah wing suk (Lemming)	
Red-Backed Vole	*Clethrionomys (?rutilus)*		
Meadow Vole	*Microtus pennsylvanicus*		
Wolf	*Canis lupis*	Ah mowh	Animals seen on rare occasions but tracks frequently.

Bird and Animal Species Observed in the Ennadai Lake Area *(continued)*

Common Name	Species	Eskimo	Remarks
Arctic Fox	*Alopex lagopus*	White: tsery goonie ack caw-cooktook Blue: tsery goonie ack care-necktauk	
Red Fox	*Vulpes fulva*	tsery goonie ack ow palooktook	None seen.
Shorttail Weasel	*Mustela erminea*	Tsair ee ack	Rocky areas of ridges and shorelines.
Caribou	*Rangifer arcticus*	Tukto Pangneek you wack (Caribou bull, large rack) Pangneek (Caribou bull, small rack) No wal ik (Caribou cow) No wack (Caribou calf)	
		Birds	
Common Loon	*Gavia immer*		
Red Throated Loon	*Gavia stellata*		
Horned Grebe	*Podiceps auritus*		South Ennadai.
Canada Goose	*Branta canadensis*	Tingmee ahlay oh	Spring migration.
American Pintail	*Anas acuta*		Breeding pairs.
Lesser Scaup	*Aytha affinis*		
Old Squaw	*Clangula hyemalis*	Ah an uck	Breeding pairs.
American Scoter	*Melanitta nigra*		Seen infrequently in spring; in groups during summer.
Common Merganser	*Mergus merganser*		
Red Breasted Merganser	*Mergus serrator*		
Rough Legged Hawk	*Buteo lagopus*		
Golden Eagle	*Aquila chrysaetos*		One: May 28, 1961.
Marsh Hawk	*Circus cyaneus*		Gray phase and dark phase; both observed.
Willow ptarmigan	*Lagopus lagopus*		Common.
Semipalmated Plover	*Charadrius semipalmatus*		
Ruddy Turnstone	*Arenaria interpres*		
Hudsonian Curlew	*Numenius phaeopus*		
Lesser Yellowlegs	*Tringa flavipes*		
Least Sandpiper	*Calidris minutilla*		
Stilt Sandpiper	*Micropalama himantopus*		

Bird and Animal Species Observed in the Ennadai Lake Area *(continued)*

Common Name	Species	Eskimo	Remarks
Semipalmated Sandpiper	*Calidris pusilla*		Most abundant of the sandpipers; breeding pairs common.
Dowitcher	*Limnodromus griseus*		
Sanderling	*Calidris alba*		
Northern Phalarope	*Phalaropus (Lobipes) lobatus*		
Parasitic Jaeger	*Stercorarius parasiticus*		Uncommon.
Long-Tailed Jaeger	*Stercorarius longicaudus*		Common: breeding pairs.
Herring Gull	*Larus argentatus*		Common.
Arctic Tern	*Sterna paradisaea*	Oomet koo tai luck	Common.
Snowy Owl	*Nyctea scandiaca*	Ootpeet u wack	Rare.
Short-Eared Owl	*Asio flammeus*		
Flycatcher	*Empidonax* sp.		
Northern Horned Lark	*Eremophila alpestris*		
Canada Jay	*Perisoreus canadensis*		South end of Ennadai Lake in spruce forest.
Raven	*Corvus corax*		
Boreal Chickadee	*Parus hudsonicus*		Common at the south end of Ennadai Lake.
Gray-Cheeked Thrush	*Catharus minimus*		Uncommon: breeding pairs.
American Pipit	*Anthus spinoletta*		
Northern Shrike	*Lanius excubitor*		
Blackpoll Warbler	*Dendroica striata*		Single pair seen: in small spruce grove at north end of Ennadai Lake.
Northern Water Thrush	*Seiurus noveboracensis*		
Hoary Redpoll	*Carduelis hornemanni*		
Common Redpoll	*Carduelis flammea*		
Savannah Sparrow	*Passerculus sandwichensis*		
Tree Sparrow	*Spizella arborea*		Frequent in small spruce clumps beyond tree-line; breeding pairs.
Harris Sparrow	*Zonotrichia querula*		Abundant.
White Crowned Sparrow	*Zonotrichia leucophrys*		
Lapland Longspur	*Calcarius lapponicus*	Coo pan u wack	Most abundant birds in the Ennadai Lake area.
Snow Bunting	*Plectrophenax nivalis*	Ow mow lee gah	

Appendix H. Spruce Outliers Beyond Northern Limit of Forest-Tundra

Sites north of "approximate limit of trees" as shown on Canadian Department of Mines and Technical Surveys, National Topographic Series, Artillery Lake N.W. 62/112 (New Edition 1951, of base map previous edition 1943).

			Authority
Sandy Lake	62°50′N	107°28′W	Gus D'Aoust*
Lynx Creek	62°30′N	106°47′W	Gus D'Aoust*
Lynx Lake	62°20′N	106°20′W	Gus D'Aoust*
Thelon River	62°25′N	104°51′W	Warden's Grove
Beaverhill Lake	62°50′N	104°20′W	Gus D'Aoust
Crystal Island	63°05′N	107°58′W	Personal observation
Eyeberry Lake	63°10′N	104°45′W	Gus D'Aoust
Thelon River	63°35′N	104°15′W	Gus D'Aoust
Thelon River	63°47′N	104°12′W	Gus D'Aoust
Artillery Lake	63°27′N	107°40′W	Personal observation
Ptarmigan Lake	63°35′N	107°23′W	Personal observation
Hanbury P.	63°40′N	107°05′W	Personal observation
Sifton Lake	63°45′N	106°30′W	Personal observation
Clinton Colden Lake	63°57′N	106°58′W	Personal observation
Aylmer Lake	64°02′N	109°10′W	Personal observation
Mackay Lake	63°50′N	111°00′W	In litt.

*Personal communication by Gus D'Aoust, old-time trapper who traveled regularly between Fort Reliance and Beaverhill Lake where he had a cabin.

Sites north of limit of trees, Dubawnt Lake N.W. 62/104 map:

			Authority
Kazan River	62°24′N	98°18′W	Personal observation
Kazan River	62°26′N	98°43′W	Personal observation
Kazan River	62°35′N	98°25′W	Personal observation
Kazan River	62°15′N	100°15′W	Personal observation
Kazan River	62°03′N	100°38′W	Personal observation
Kazan River	62°06′N	100°38′W	Personal observation
Nowleye Lake	62°13′N	101°19′W	Personal observation
Dubawnt Lake	62°40′N	101°42′W	Personal observation
Dubawnt Lake	62°37′N	101°15′W	Personal observation*

*Other areas at the extreme south end of Dubawnt Lake not shown in individual detail.

References

Addison, P. A. 1977. Studies on evapotranspiration and energy budgets on the Truelove Lowland. In *Truelove Lowland, Devon Island, Canada: A High Arctic Ecosystem* (L.C. Bliss, ed.), University of Alberta Press, Edmonton, Alberta, Canada, pp. 281–300.

Addison, P. A., and L. C. Bliss. 1980. Summer climate, microclimate and energy budget of polar semi-desert on King Christian Island, N.W.T. *Arctic and Alpine Research* **12**:161–170.

Addison, P. A., and L. C. Bliss. 1984. Adaptations of *Luzula confusa* to the polar semi-desert environment. *Arctic* **37(2)**:121–132.

Agassiz, L. 1845. Essai sur la géographie des animaux. *Rev. Suisse et Chron. Litt.* **8**:441–452, 538–555.

Ahti, G., G. W. Scotter, and H. Vanska. 1973. Lichens of the reindeer preserve, Northwest Territories, Canada. *Bryologist* **76**:48–76.

Alden, J., and C. Loopstra. 1987. Genetic diversity and population structure of *Picea glauca* on an altitudinal gradient in interior Alaska. *Canadian Journal of Forest Research* **17**:1519–1526.

Aldrich, S. A., F. T. Aldrich, and R. Rudd. 1971. An effort to identify the Canadian forest-tundra ecotone signature on weather satellite imagery. *Remote Sensing of the Environment* **2**:9–20.

Allen, T. F. H., and T. B. Starr. 1982. *Hierarchy: Perspectives for Ecological Complexity.* University of Chicago Press, Chicago, Ill.

Anderson, E. 1953. Introgressive hybridization. *Biological Reviews* **28**:280–307.

Anderson, J. 1856. Letters of chief factor James Anderson to Sir George Simpson, Governor-in-Chief of Rupert's Land. *Journal of the Royal Geographical Society* **26**:18–25.

Andrews, J. T. 1985. Reconstruction of environmental conditions in the eastern Canadian Arctic during the last 11,000 years. In *Critical Periods in the Quaternary Climatic History of Northern North America*. Syllogeus Series No. 55, C. R. Harington, ed., National Museums of Canada, Ottawa: pp. 423–451.

Andrews, J. T., and D. M. Barnett. 1979. Holocene (Neoglacial) moraine and proglacial lake chronology, Barnes Ice Cap, Canada. *Boreas* **8**:341–358.

Andrews, J. T., R. G. Barry, R. S. Bradley, G. H. Miller, and L. O. Williams. 1972. Past and present glaciological responses to climate in East Baffin Island. *Quaternary Research* **2**:303–314.

Andrews, J. T., and H. Nichols. 1981. Modern pollen deposition and Holocene paleotemperature reconstructions, central Northern Canada. *Arctic and Alpine Research* **13(4)**: 387–408.

Argus, G. W. 1964. Plant collections from Carswell Lake and Beartooth Island, Northwestern Saskatchewan, Canada. *The Canadian Field-Naturalist* **78(3)**:139–149.

Argus, G. W. 1965. The taxonomy of the *Salix glauca* complex in North America. *Contributions from the Gray Herbarium of Harvard University* **196**:1–142.

Argus, G. W. 1966. Botanical investigations in northeastern Saskatchewan: the subarctic Patterson-Hasbala Lakes region. *Canadian Field-Naturalist* **80(3)**:119–143.

Argus, G. W. 1973. *The Genus Salix in Alaska and the Yukon.* National Museums of Canada Publications in Botany No. 2, Ottawa.

Arno, S. F. 1984. *Timberline: Mountain and Arctic Forest Frontiers.* The Mountaineers Press, Seattle, Washington.

Arnold, C. A. 1959. Some paleobotanical aspects of tundra development. *Ecology* **40(1)**:146–148.

Atlas of Canada. 1957. Department of Mines and Technical Surveys, Geographical Branch, Ottawa, Ontario.

Auclair, A. N. D., and A. N. Rencz. 1982. Concentration, mass and distribution of nutrients in a subarctic *Picea mariana–Cladina alpestris* ecosystem. *Canadian Journal of Forest Research* **12**:947–968.

Auerbach, M., and A. Shmida. 1987. Spatial scale and the determinants of plant species richness. *Trends in Ecology and Evolution* **2(8)**:238–242.

Austin, M. P. 1985. Continuum concept, ordination methods, and niche theory. *Annual Review of Ecology and Systematics* **16**:39–61.

Austin, M. P. 1986. The theoretical basis of vegetation science. *Trends in Ecology and Evolution* **1(6)**:161–164.

Back, G. 1836. *Narrative of the Arctic Land Expedition to the Mouth of the Great Fish River and Along the Shores of the Arctic Ocean in the Years 1835–6*. London.

Baldwin, W. K. W. 1953. Botanical investigations in the Reindeer-Nueltin Lakes area, Manitoba. *National Museum of Canada Bulletin No. 128*, Ottawa, Ontario: pp. 1–33.

Ball, T. 1984. A dramatic change in the general circulation of the West Coast of Hudson Bay in 1760 A.D.: synoptic evidence based on historical records. In *Critical Periods in the Quaternary History of Northern North America*, C. R. Harington, ed., Syllogeus Series No. 55, National Museums of Canada, Ottawa: pp. 219–228.

Ball, T. V. 1986. Historical evidence and climatic implications of a shift in the boreal forest tundra transition in Central Canada. *Climatic Change* **8(2)**:121–134.

Bamberg, S. A., and J. Major. 1968. Vegetation and soils associated with calcareous parent materials. *Ecological Monographs* **38(2)**:127–167.

Barry, R. G. 1967. Seasonal location of the arctic front over North America. *Geographical Bulletin* **9(2)**:79–95.

Barry, R. G. 1981. The nature and origin of climatic fluctuations in northeastern North America. *Geographie Physique et Quaternaire* **35**:41–47.

Barry, R. G., and F. K. Hare. 1974. Arctic Climate. In *Arctic and Alpine Environments*, J. D. Ives and R. G. Barry, eds., Methuen, London, pp. 17–54.

Barry, R. G., G. M. Courtin, and C. Labine. 1981. Tundra climates. In *Tundra Ecosystems: A Comparative Analysis*, L. C. Bliss, O. W. Heal, and J. J. Moore, eds., Cambridge University Press, Cambridge, pp. 81–114.

Barry, R. G., D. L. Elliott, and R. G. Crane. 1981. The palaeoclimatic interpretation of exotic pollen peaks in Holocene records from the eastern Canadian Arctic: a discussion. *Review of Palaeobotany and Palynology* **33**:153–167.

Barry, R. G., W. H. Arundale, J. T. Andrews, R. S. Bradley, and H. Nichols. 1977. Environmental change and cultural change in the eastern Canadian Arctic during the last 5000 years. *Arctic and Alpine Research* **9(2)**:193–210.

Beadle, C. L., P. G. Jarvis, and R. E. Neilson. 1979. Leaf conductance as related to xylem water potential and carbon dioxide concentrations in Sitka spruce. *Physiologia Plantarum* **45**: 158–166.

Beals, E. 1960. Forest bird communities in the Apostle Islands of Wisconsin, *Wilson Bulletin* **72**:156–181.

Beals, E. W. 1984. Bray-Curtis ordination: An effective strategy for analysis of multivariate ecological data. In *Advances in Ecological Research*, Volume 14, A. MacFadyen and E. D. Ford, eds., pp. 1–55.

Beasleigh, W. J., and G. A. Yarranton. 1974. Ecological strategy and tactics of *Equisetum sylvaticum* during a postfire succession. *Canadian Journal of Botany* **52**:2299–2318.

Begin, C., and L. Filion. 1985. Analyse dendrochronologique d'un glissement de terrain de la region du Lac à l'eau Claire (Québec nordique). *Canadian Journal of Earth Sciences* **22**:175–182.

Belisle, J., and T. R. Moore. 1981. Characteristics of podzolised soils formed in carbonate-bearing tills in the Schefferville area. *McGill Subarctic Research Paper No. 32*, McGill University, Montreal: 85–95.

Bell, R. 1879. Report on exploration of the east coast of Hudson Bay in 1877. *Report on Progress for 1877–1978*, Geological Survey of Canada: 1c-37c.

Bell, R. 1881a. The northern limits of the principal forest trees of Canada, east of the Rocky Mountains. *7th Report of the Montréal Horticultural Society*, Witness Printing House, Montreal: 18–41.

Bell, R. 1881b. Report on Hudson's Bay and some of the lakes and rivers lying to the west of it. *Report of Progress for 1879–80*, Geological and Natural History Survey of Canada, Montreal: 1cc-113cc.

Bell, R. 1884. Observations on the geology, mineralogy, zoology, and botany of the Labrador coast, Hudson's Strait and Bay. *Report of the Geological and Natural History Survey and Museum of Canada for 1882–3–4*: 1DD-62DD.

Bell, R. 1885. Report on part of the basin of the Athabaska River, Northwest Territory. *Geological Survey of Canada Report on Progress for 1882–83–84: Part CC.*

Bell, R. 1886. Observations made on the geology, zoology, and botany of Hudson's Strait and Bay made in 1885. Geological and Natural History Survey of Canada, Annual Report, New Series I, Montreal: 1DD-27DD.

Bell, R. 1900. Preliminary report of explorations about Great Slave Lake in 1899. *Geological Survey of Canada Summary Report for 1899, Part A*: 103–110.

Bell, K. L., and L. C. Bliss. 1977. Overwintering phenology in a polar semi-desert. *Arctic* **30**:118–121.

Bell, K. L., and L. C. Bliss. 1978. Root growth in a polar semi-desert environment. *Canadian Journal of Botany* **56**:2470–2490.

Bell, K. L., and L. C. Bliss. 1980a. Plant reproduction in a high arctic environment. *Arctic and Alpine Research* **12**:1–10.

Bell, K. L., and L. C. Bliss. 1980b. Autecology of *Kobresia bellardi*: Why winter snow accumulation limits local distribution. *Ecological Monographs* **49**:377–402.

Benninghoff, W. S. 1952. Interactions of vegetation and frost phenomena. *Arctic* **5(1)**:34–44.

Billings, W. D. 1974. Arctic and alpine vegetation: Plant adaptation to cold summer climates. In *Arctic and Alpine Environments*, Jack Ives and R. G. Barry, eds., Methuen, London, pp. 403–443.

Billings, W. D., and H. A. Mooney. 1968. The ecology of arctic and alpine plants. *Biological Reviews* **43(4)**:481–529.

Billington, M. M., and V. Alexander. 1983. Site-to-site variations in nitrogenase activity in a subarctic black spruce forest. *Canadian Journal of Forest Research* **13(5)**:782– 787.

Bird, C. D. 1974a. Botanical studies in the Yukon and Northwestern Territories. Canada Department of Energy, Mines and Resources, Geological Survey of Canada, Open File 227.

Bird, C. D. 1974b. Botanical studies near the Mackenzie River, N. W. T. Canada Department of Energy, Mines and Resources, Geological Survey of Canada, Open File 225.

Bird, C. D., J. W. Thomson, A. H. Marsh, G. W. Scotter, and P. Y. Wong. 1980. Lichens from the area drained by the Peel and Mackenzie Rivers, Yukon and Northwest Territories, Canada. I. Macrolichens. *Canadian Journal of Botany* **58**:1847–1985.

Bird, J. B. 1974. Geomorphic processes in the Arctic. In *Arctic and Alpine Environments*, J. Ives and R. G. Barry, eds., Methuen, London, pp. 703–720.

Birket-Smith, K. 1941. Five hundred Eskimo words. *Report of the Fifth Thule Expedition 1921–24* **III(3)**:1–64.

Black, R. A., and L. C. Bliss. 1978. Recovery sequence of *Picea mariana-Vaccinium uliginosum* forests after burning near Inuvik, Northwest Territories, Canada. *Canadian Journal of Botany* **56**:2020–2030.

Black, R. A., and L. C. Bliss. 1980. Reproductive ecology of *Picea mariana* (Mill.) BSP., at tree line near Inuvik, Northwest Territories, Canada. *Ecological Monographs* **50(3)**:331–354.

Blake, D., and S. Rowland. 1988. Continuing worldwide increase in tropospheric methane. *Science* **239**:1129.

Blanchet, G. H. 1927. Crossing a Great Divide. *Bulletin of the Geographic Society of Philadelphia* **25**:141–153.

Blanchet, G. H. 1930. *Keewatin and Northeastern Mackenzie*. Publications of the Canadian Department of the Interior, Ottawa.

Blasing, T. J., and H. C. Fritz. 1973. Past climate of Alaska and northwestern Canada as reconstructed from tree rings. In *Climate of the Arctic*, G. Weller and S. A. Bowling, eds., 24th Alaska Science Conference, Fairbanks, pp. 48–58.

Bliss, L. C. 1956. A comparison of plant development in microenvironments of arctic and alpine tundras. *Ecological Monographs* **26**:303–337.

Bliss, L. C. 1958. Seed germination in arctic and alpine species. *Arctic* **11(3)**:180–188.

Bliss, L. C. 1962. Adaptations of arctic and alpine plants to environmental conditions. *Arctic* **15(2)**:117–144.

Bliss, L. C. 1971. Arctic and alpine life cycles. *Annual Review of Ecology and Systematics* **2**:405–438.

Bliss, L. C. 1977a. General summary: Truelove lowland ecosystem. *Truelove Lowland, Devon Island, Canada: A High Arctic Ecosystem*, L. C. Bliss, ed., University of Alberta Press, Edmonton, pp. 657–675.

Bliss, L. C., ed. 1977b. *Truelove Lowland, Devon Island, Canada: A High Arctic Ecosystem*. University of Alberta Press, Edmonton, Canada.

Bliss, L. C. 1981. North American and Scandinavian tundras and polar deserts. In *Tundra Ecosystems: A Comparative Analysis*, L. C. Bliss, O. W. Heal, and J. J. Moore, eds., Cambridge University Press, Cambridge, pp. 8–24.

Bliss, L. C. 1985. Alpine. In *Physiological Ecology of North American Plant Communities*, B. F. Chabot and H. A. Mooney, eds., Chapman and Hall, New York and London, pp. 41–65.

Bolin, B., B. R. Döös, J. Jager, and R. A. Warrick, eds., 1986. *The Greenhouse Effect, Climatic Change and Ecosystems*. John Wiley & Sons, New York.

Bond, R. R. 1957. Ecological distribution of breeding birds in the upland forests of southern Wisconsin. *Ecological Monographs* **27**:351–384

Bostock, H. S. 1970. Physiographic subdivisions of Canada. In *Geology and Economic Minerals of Canada*, R. J. W. Douglas, ed., Geological Survey of Canada Economic Geology Report No. 1, 5th Edition, Department of Energy, Mines, and Resources, Ottawa.

Boucher, P. 1664. *Histoire véritable et naturalle des moeurs et productions du pays de la Nouvelle-France, vulgairement dite le Canada*. Paris. (see Rousseau, 1966).

Boyd, W. L. 1958. Microbiological studies of arctic soils. *Ecology* **39**:332–336.

Boyd, W. L., and J. W. Boyd. 1971. Studies of soil microorganisms, Inuvik, Northwest Territories. *Arctic* **24(3)**:162–176.

Bradley, S. W., J. S. Rowe, and C. Tarnocai. 1982. *An Ecological Land Survey of the Lockhart River Map Area, Northwest Territories*. Ecological Land Classification Series No. 16, Lands Directorate, Environment Canada, Ottawa.

Bray, J. R., and J. T. Curtis. 1957. An ordination of the upland forest communities of southern Wisconsin. *Ecological Monographs* **27**:325–349.
Brinkmann, W. A. R. 1976. Surface temperature trend for the Northern Hemisphere—updated. *Quaternary Research* **6**:355–358.
Britton, M. E. 1966. Vegetation of the arctic tundra. In *Arctic Biology*, Henry P. Hansen, ed., Oregon State University Press, Corvallis, Ore.
Brown, J. 1969. Buried soils associated with permafrost. In *Pedology and Quaternary Research*, S. Pawluk, ed., University of Alberta Press, Edmonton, pp. 115–127.
Brown, J., R. Rickard, and D. Vietor. 1969. *The effect of disturbance on permafrost terrain.* Cold Regions Research and Engineering Laboratory, Special Report No. 138, U.S. Army, CRREL, Hanover, N.H.
Brown, J., P. C. Miller, L. L. Tieszen, and F. L. Bunnell. 1980. *An Arctic Ecosystem.* Dowden, Hutchinson, and Ross, Inc., Stroudsburg, Penn.
Brown, J., K. R. Everett, P. J. Webber, S. F. MacLean, Jr., and D. F. Murray. 1980. The coastal tundra at Barrow. In *An Arctic Ecosystem*, J. Brown, P. C. Miller, L. L. Tieszen, and F. L. Bunnell, eds., Dowden, Hutchinson and Ross, Inc., Stroudsburg, Penn., pp. 1–29.
Brown, R. J. E. 1943. The distribution of permafrost and its relation to air temperature in Canada and the U.S.S.R. *Arctic* **13**:163–177.
Brown, R. J. E. 1967. *Permafrost in Canada.* Map 1246A. Geological Survey of Canada. Energy, Mines, and Resources, Ottawa.
Brown, R. J. E. 1969. Factors influencing discontinuous permafrost in Canada. In *The Periglacial Environment*, T. L. Péwé, ed., McGill-Queen's University Press, Montreal, pp. 11–53.
Brown, R. J. E. 1970. *Permafrost in Canada.* University of Toronto Press, Toronto.
Brown, R. T., and P. Mickola. 1974. The influence of fruticose soil lichens upon the mycorrhizae and seedling growth of forest trees. *Acta Forest Fennica* **141**:1–22.
Bryson, R. A. 1966. Air masses, streamlines and the boreal forest. *Geographical Bulletin* **8**:228–269.
Bryson, R. A., W. M. Irving, and J. A. Larsen. 1965. Radiocarbon and soil evidence of former forests in the southern Canadian tundra. *Science* **147**:46–48.
Bryson, R. A., W. M. Wendland, J. D. Ives, and J. T. Andrews. 1969. Radiocarbon isochrones of the disintegration of the Laurentide ice sheet. *Arctic and Alpine Research* **1**:1–14.
Bunce, R. G. H., S. K. Morell, and H. E. Stel. 1975. The application of multivariate analysis to regional survey. *Journal of Environmental Management* **3**:151–166.
Bunnell, F. L., O. K. Miller, P. W. Flanagan, and R. E. Benoit. 1980. The microflora: composition, biomass, and environmental relations. In *An Arctic Ecosystem*, J. Brown, P. C. Miller, L. L. Tieszen, and F. L. Bunnell, eds., Dowden, Hutchinson, and Ross, Inc., Stroudsburg, Penn., pp. 255–290.
Busby, J. R., L. C. Bliss, and C. D. Hamilton. 1978. Microclimate control of growth rates and habitats of the boreal forest mosses, *Tomenthypnum nitens* and *Hylocomium splendens*. *Ecological Monographs* **48**:95–110.
Canadian Soil Survey Committee. 1978. *The Canadian System of Soil Classification.* Agriculture Canada Publication No. 1646, Ottawa, 164 pp.
Canadian Weekly Bulletin. 1972. Report on Northwest Territories forest fires. Canadian Department of External Affairs 27:6. Ottawa.
Carleton, T. J., and P. F. Maycock. 1978. Dynamics of the boreal forest south of James Bay. *Canadian Journal of Botany* **56**:1157–1173.
Carleton, T. J., and P. F. Maycock. 1981. Understory-canopy affinities in boreal forest vegetation. *Canadian Journal of Botany* **59**:1709–1716.
Carstairs, A., and W. C. Oechel. 1978. Effects of several microclimatic factors and nutrients on the net carbon dioxide exchange in *Cladonia alpestris* in the subarctic. *Arctic and Alpine Research* **10**:81–94.
Carter, R. N., and S. D. Prince. 1981. Epidemic models used to explain biogeographical distribution limits. *Nature* **293**:644–645.
Chapin, F. S. 1980. The mineral nutrition of wild plants. *Annual Review of Ecology and Systematics* **11**:233–260.

Chapin, F. S., III. 1983. Nitrogen and phosphorus nutrition and nutrient cycling by evergreen and deciduous understory shrubs in an Alaskan black spruce forest. *Canadian Journal of Forest Research* **13(5)**:773–781.

Chapin, F. S., III, and P. R. Tryon. 1983. Habitat and leaf habit as determinants of growth, nutrient absorption, and nutrient use by Alaskan taiga forest species. *Canadian Journal of Forest Research* **13(5)**:818–826.

Chapin, F. S. III, and Gaius R. Shaver. 1985. Arctic. In *Physiological Ecology of North American Plant Communities*, B. F. Chabot and H. A. Mooney, eds., Chapman and Hall, New York and London, pp. 16–40.

Churchill, E. S., and H. C. Hansen. 1958. The concept of climax in arctic and alpine vegetation. *Botanical Review* **24**:127–191.

Claridge, F. B., and A. M. Mirza. 1981. Erosion control along transportation routes in northern climates. *Arctic* **34(2)**:147–157.

Clarke, C. H. D. 1940. *A Biological Investigation of the Thelon Game Sanctuary.* National Museum of Canada Bulletin No. 96, Biological Series No. 25, Ottawa, 135 pp.

Clausen, J.. Tree lines and germ plasm—a study in evolutionary limitations. *Proceedings of the National Academy of Science* **50**:860–868.

Cody, W. J. 1954a. Plant records from Coppermine, Mackenzie District, N. W. T. *Canadian Field-Naturalist* **68(3)**:110–117.

Cody, W. J. 1954b. New plant records from Bathurst Inlet, N. W. T. *Canadian Field-Naturalist* **68(1)**:40.

Cody, W. J. 1965. New plant records from northwestern Mackenzie District. *Canadian Field-Naturalist* **79(2)**:79–158.

Cody, W. J., and J. G. Chillcott. 1955. Plant collections from Matthews and Muskox Lakes, Mackenzie District, N. W. T. *Canadian Field-Naturalist* **69(4)**:153–162.

Cody, W. J., and A. E. Porsild. 1980. *Vascular Plants of Continental Northwest Territories.* National Museums of Canada, Ottawa.

Cogbill, C. V. 1985. Dynamics of the boreal forests of the Laurentian Highlands, Canada. *Canadian Journal of Forest Research* **15**:252–261.

Cole, L. C. 1954. Some features of random population cycles. *Journal of Wildlife Management* **18(1)**:2–24.

Coleman, A. P. 1921. *Northeast Part of Labrador and New Quebec.* Geological Survey of Canada Memoir 124, Montreal.

Connor, E. F., and D. Simberloff. 1986. Competition, scientific method, and null models in ecology. *American Scientist* **74(2)**:155–162.

Cook, F. A. 1959a. Some types of patterned ground in Canada. *Geographical Bulletin* **13**:73–79.

Cook, F. A. 1959b. A review of the study of periglacial phenomena in Canada. *Geographical Bulletin* **13**:22–38.

Cooke, A., and C. Holland. 1978. *The Exploration of Northern Canada, 500 to 1920, a Chronology.* The Arctic History Press, Toronto, 575 pp.

Cooke, F., R. K. Ross, R. K. Schmidt, and A. J. Pakulak. 1975. Birds of the tundra biome at Cape Churchill and La Pérouse Bay. *Canadian Field-Naturalist* **89(4)**:413–422.

Cooke, H. C. 1929. Studies of the physiography of the Canadian Shield, I; mature valleys of the Labrador Peninsula. *Transactions of the Royal Society of Canada, Series 3,* **23(4)**:91–120.

Cooley, J. H., and F. B. Golley (eds.). 1984. *Trends in Ecological Research in the 1980s.* Plenum Press, New York.

Cooper, D. J. 1986. White spruce above and beyond treeline in the Arrigetch Peaks region, Brooks Range, Alaska. *Arctic* **39(3)**:247–252.

Cooper, J. G. 1859. On the distribution of the forests and trees of North America, with notes on its physical geography. *Annual Report of the Board of Regents of the Smithsonian* Inst., pp. 246–280.

Corns, I. G. W. 1974. Arctic plant communities east of the Mackenzie Delta. *Canadian Journal of Botany* **52**:1731–1745.

Corns, I. G. W. 1983. Forest community types of west-central Alberta in relation to selected environmental factors. *Canadian Journal of Forest Research* **13**:995–1010.

Corns, I. G. W., and D. J. Pluth. 1984. Vegetational indicators as independent variables in forest growth productions in west-central Alberta. *Forest Ecology and Management* **9**:13–25.
Cottam, G., and J. T. Curtis. 1956. The use of distance measures in phytosociological sampling. *Ecology* **37**:451–460.
Cottam, G., F. G. Goff, and R. H. Whittaker. 1973. Wisconsin comparative ordination. In *Ordination and Classification of Communities*, R. H. Whittaker, ed., Dr. W. Junk, Publishers, The Hague, Netherlands, p. 193–222.
Couillard, L., and S. Payette. 1985. Évolution holocène d'une tourbière à perigélisol (Québec nordique). *Canadian Journal of Botany* **63(6)**:1104–1121.
Courtin, G. M., and C. L. Labine. 1977. Microclimatological studies of the Truelove Lowland. In *Truelove Lowland, Devon Island, Canada: A High Arctic Ecosystem*, L. C. Bliss, ed., University of Alberta Press, Edmonton, Alberta, pp. 73–106.
Cowles, S. 1982. Preliminary results investigating the effect of lichen ground cover on the growth of black spruce. *Naturaliste Canadien (Review of Écology and Systematics)* **109**: 573–581.
Cowling, J., and R. A. Kedrowski. 1980. Winter water relations of native and introduced evergreens in interior Alaska. *Canadian Journal of Botany* **58**:94–99.
Craig, B. G. 1960. *Surficial Geology of North-Central District of Mackenzie.* Geological Survey of Canada Paper 60–18, Ottawa.
Craig, B. G. 1961. *Surficial Geology of Northern District of Keewatin*, Northwest Territories. Geological Survey of Canada Paper 61–5.
Craig, B. G. 1964a. *Surficial Geology of Boothia Peninsula and Somerset, King William, and Prince of Wales Islands, District of Franklin.* Geological Survey of Canada Bulletin. **63**–44:1–10.
Craig, B. G. 1964b. *Surficial Geology of East-Central District of Mackenzie.* Geological Survey of Canada Bulletin **99**:1–41.
Craig, B. G., and J. G. Fyles. 1960. *Pleistocene Geology of Arctic Canada.* Geological Survey of Canada Paper 61–10, Ottawa.
Crittenden, P. D. 1983. The role of lichens in the nitrogen economy of subarctic woodlands: Nitrogen loss from the nitrogen-fixing lichen *Stereocaulon paschale* during rainfall. In *Nitrogen as an Ecological Factor*, J. A. Less, S. McNeill, and I. H. Morison eds., Blackwell Scientific Publications, Oxford, England, pp. 43–68.
Crittenden, P. D., and K. A. Kershaw. 1978. Discovering the role of lichens in the nitrogen cycle in boreal-arctic ecosystems. *Bryologist* **18(2)**:258–267.
Crittenden, P. D., and K. A. Kershaw. 1979. Studies on lichen-dominated systems. XXII. The environmental control of nitrogenase activity in *Stereocaulon paschale* in spruce-lichen woodland. *Canadian Journal of Botany* **57**:236–254.
Crowley, T. J., D. A. Short, J. G. Mengel, and G. R. North. 1986. Role of seasonality in the evolution of climate during the last 100 million years. *Science* **231**:579–584.
Cunningham, C. M., and W. W. Shilts. 1977. *Surficial Geology of the Baker Lake Area, District of Keewatin.* Geological Survey of Canada Paper 77–1B: 311–314.
Curtis, J. T. 1959. *The Vegetation of Wisconsin.* University of Wisconsin Press, Madison, Wis.
Cwynar, L. C., and J. C. Ritchie. 1980. Arctic steppe-tundra: a Yukon perspective. *Science* **208**:1375–1376.
Dahl, E. 1951. On the relation between summer temperature and the distribution of alpine vascular plants in the lowlands of Fennoscandia. *Oikos* **3(1)**:22–52.
Daly, G. T. 1966. Nitrogen fixation by nodulated *Alnus rugosa. Canadian Journal of Botany* **44**:1607–1621.
Damman, A. W. H. 1964. *Some Forest Types of Central Newfoundland and their Relation to Environmental Factors.* Canada Department of Forests, Forest Science Monographs 8.
Damman, A. W. H. 1965. The distribution patterns of northern and southern elements in the flora of Newfoundland. *Rhodora* **67**:363–392.
Damman, A. W. H. 1971. Effect of vegetation changes on the fertility of a Newfoundland forest site. *Ecological Monographs* **41(3)**:260–269.
Daubenmire, R. F. 1954. Alpine timberlines in the Americas and their interpretation. *Butler University Botanical Studies* **11:119**–136.

Davies, O. K. 1984. Multiple thermal maxima during the Holocene. *Science* **225**:617–619.
Davis, M. B. 1986. Climatic instability, time lags, and community disequilibrium. In *Community Ecology*, J. Diamond and T. J. Case, eds., Harper and Row, New York, pp. 269–284.
Davis, M. B. 1986. Lags in the response of forest vegetation to climatic change. In *Climate-Vegetation Interactions*, C. Resenzweig and R. Dickinson, eds., Office for Interdisciplinary Earth Studies, University Corporation for Atmospheric Research, Boulder, Colo., pp. 70–71.
Davis, M. B., and D. B. Botkin. 1985. Sensitivity of cool-temperate forests and their fossil pollen record to rapid temperature change. *Quaternary Research* **23**:327–340.
Day, J. H. 1968. *Soils of the upper Mackenzie River area, Northwest Territories*. Research Branch, Canada Department of Agriculture, Ottawa.
Day, J. H., and H. M. Rice. 1964. The characteristics of some permafrost soils in the Mackenzie Valley, N. W. T. *Arctic* **17**:223–236.
Dean, W. E., J. P. Bradbury, R. Y. Anderson, and C. W. Barnosky. 1984. The variability of Holocene climatic change: Evidence from varved lake sediments. *Science* **226**:1191–1194.
Denton, G. H., and W. Karlen. 1977. Holocene glacial and treeline variation in the White River valley and Skolai Pass, Alaska. *Quaternary Research* **7**:63–111.
Denton, S. R., and B. R. Barnes. 1987. Tree species distributions related to climatic patterns in Michigan. *Canadian Journal of Forest Research* **17**:613–629.
Diamond, J., and T. J. Case, eds. 1985. *Community Ecology*. Harper and Row, New York.
Dingman, S. L., R. G. Barry, G. Weller, C. Benson, E. F. LeDrew, and C. W. Goodwin. 1980. Climate, snow cover, microclimate, and hydrology. In *An Arctic Ecosystem*, J. Brown, et al., eds., Dowden, Hutchinson, and Ross, Inc., Stroudsburg, Penn, pp. 30–65.
Dix, W. L. 1950. Lichens and hepatics of the Nueltin Lake expedition. *Bryologist* **53**:283–289.
Douglas, G. M. 1914. *Lands Forlorn*. G. P. Putnam's Sons, New York.
Douglas, L. A., and J. C. F. Tedrow. 1959. Organic matter decomposition rates in arctic soils. *Soil Science* **88(6)**:305–312.
Douglas, L. A., and J. C. F. Tedrow. 1960. Tundra soils of arctic Alaska. *9th International Congress of Soil Science*, Madison, Wis., Vol. **41**:291–304.
Dowling, D. B. 1893. Narratives of a journey in 1890, from Great Slave Lake to Beechy Lake, on the Great Fish River. *Ottawa Naturalist* **7**:85–92, 101–114.
Downs, J. A. 1964. Arctic insects and their environment. *Canadian Entomologist* **96**:279–307.
Drew, J. V., and R. E. Shanks. 1965. Landscape relationships of soils and vegetation in the forest-tundra ecotone, Upper Firth River Valley, Alaska-Canada. *Ecological Monographs* **35(3)**:285–306.
Drew, J. V., and J. C. F. Tedrow. 1957. Pedology of an arctic brown profile near Point Barrow, Alaska. *Soil Science Society of America Proceedings* **21**:336–339.
Drew, J. V., and J. C. F. Tedrow. 1962. Arctic soil classification and patterned ground. *Arctic* **15(2)**:106–116.
Drury, W. H. Jr. 1962. Patterned ground and vegetation on southern Bylot Island, Northwest Territories, Canada. *Contributions from the Gray Herbarium of Harvard University* **190**:1–111.
Dubreuil, M. A., and T. R. Moore. 1982. A laboratory study of post-fire nutrient redistribution in spruce-lichen woodlands. *Canadian Journal of Botany* **60**:2511–2517.
Dugle, J. R., and N. Bols. 1971. Variation in *Picea glauca* and *P. mariana* in Manitoba and adjacent areas. Atomic Energy of Canada Ltd., Whiteshell Nuclear Research Establishment, Report AECL-3681, Pinawa, Manitoba.
Dunbar, M. J. 1968. *Ecological Development in Polar Regions: A Study in Evolution*. Prentice-Hall, Englewood Cliffs, N.J.
Dyrness, C. T., and D. F. Grigal. 1979. Vegetation-soil relationships along a spruce forest transect in interior Alaska. *Canadian Journal of Botany* **57**:2644–2656.
Dyrness, C. T., and R. A. Norum. 1983. The effects of experimental fires on black spruce forest in interior Alaska. *Canadian Journal of Forest Research* **13(5)**:879–893.
Edlund, S. A. 1977. Vegetation types in north-central District of Keewatin. *Report of Activities, Part A, Geological Survey of Canada Paper* **77–1A**: 385–392.

Eldredge, N. 1985. *Time Frames*. Simon and Schuster, New York.
Ellenberg, H. 1952. Cited in: Extrapolation from controlled environments to the field, by L. T. Evans. In *Environmental Control of Plant Growth*, L. T. Evans, ed., 1963, Academic Press, New York.
Elliott, D. L. 1979a. *The stability of the northern Canadian tree limit: current regenerative capacity*. Ph.D. thesis, University of Colorado, Boulder, Colo.
Elliott, D. L. 1979b. The current regenerative capacity of the northern Canadian trees, Keewatin, N. W. T., Canada, some preliminary observations. *Arctic and Alpine Research* **11**:243–251.
Elliott, D. L., and S. K. Short. 1979. The northern limit of trees in Labrador: a discussion. *Arctic* **32**:201–206.
Elliott-Fisk, D. L. 1983. The stability of the northern Canadian tree limit. *Annals of the Association of American Geographers* **73**:560–576.
Ellis, H. 1748. *A Voyage to Hudson's Bay by the Dobbs Galley and California in the Years 1746 and 1747, for Discovering a North West Passage*. H. Whitridge, London.
Elton, C. 1942. *Voles, Mice, and Lemmings: Problems in Population Dynamics*. Oxford University Press, London.
Erskine, A.J. 1977. *Birds in Boreal Canada: Communities, Densities, and Adaptations*. Canadian Wildlife Services, Report Series No. 41, Ministery of Fisheries and the Environment, Ottawa.
Filion, L. 1984a. A relationship between dunes, fire and climate recorded in the Holocene deposits of Quebec. *Nature* **309(5968)**:543–546.
Filion, L. 1984b. Analyse macrofossile et pollinique de paléosols de dunes en Hudsonie, Québec nordique. *Gêographie Physique et Quaternaire* **38(2)**:113–122.
Filion, L., S. Payette, and L. Gauthier. 1985. Analyse dendroclimatique d'un krummholz à la limite des arbres, Lac Bush, Québec nordique. *Gêographie Physique et Quaternaire* **39(2)**:221–226.
Flanagan, P. W., and F. L. Bunnell. 1980. Microflora activities and decomposition. In *An Arctic Ecosystem*, J. Brown, P. C. Miller, L. L. Tieszen, and F L. Bunnell, eds., Dowden, Hutchinson, and Ross, Inc., Stroudsburg, Penn., pp. 291–334.
Flanagan, P. W., and L. Van Cleve. 1983. Nutrient cycling in relation to decomposition and organic-matter quality in taiga ecosystems. *Canadian Journal of Forest Research* **13(5)**: 795–817.
Flinn, M. A., and R. W. Wein. 1988. Regrowth of forest understory species following seasonal burning. *Canadian Journal of Botany* **66**:150–155.
Ford, M. J. 1982. *The Changing Climate: Responses of the Natural Fauna and Flora*. George Allen and Unwin, London.
Foster, D. R. 1983. The history and pattern of fire in the boreal forest of southeastern Labrador. *Canadian Journal of Botany* **61**:2459–2471.
Foster, D. R. 1984a. Phytosociological description of the forest vegetation of southeastern Labrador. *Canadian Journal of Botany* **62**:899–906.
Foster, D. R. 1984b. The dynamics of *Sphagnum* in forest and peatland communities in southeastern Labrador, Canada. *Arctic* **37(2)**:133–140.
Foster, D. R., and G. A. King. 1984. Landscape features, vegetation, and the developmental history of a patterned fen in southeastern Labrador, Canada. *Journal of Ecology* **72**: 115–143.
Franklin, J. 1823. *Narrative of a Journey to the Shores of the Polar Sea in the Years 1819, 20, 21, and 22, with an Appendix on Various Subjects Relating to Science and Natural History.* London.
Franklin, J. 1828. *Narrative of the Second Expedition to the Shores of the Polar Sea in the Years 1825, 1826, and 1827, Including an Account of the Progress of a Detachment to the Eastward by John Richardson.* London.
Fraser, E. M. 1956. *The Lichen Woodlands of the Knob Lake Area of Quebec-Labrador. McGill Subarctic Research Paper No. 1*, McGill University, Montreal.
Fraser, J. W. 1976. Viability of black spruce seed in or on a boreal forest seedbed. *Forestry Chronicle* **52**:229–231.

Fritts, H. C. 1976. *Tree Rings and Climate*. Academic Press, New York.
Gagnon, R. 1982. Fluctuations holocènes de la limit des forêts, Rivièrs aux Feuilles, Québec nordique: une analyse macrofossele. Thèse de doctorate, Université Laval, Québec.
Gagnon, R., and S. Payette. 1980. Tree-line Holocène fluctuations in northern Québec. *Fourth Symposium on the Quaternary of Québec* (Abstracts), Université Laval, Québec.
Gagnon, R., and S. Payette. 1981. Fluctuations holocènes de la limite des forêts de mélèze, rivière aux Feuilles, Noveau-QuebecL une analyse macrofossile en milieux tourbeux. *Géographie Physique et Quaternaire* **35**:57–72.
Gagnon, R., and S. Payette. 1985. Régression holocène du couvert coniférien à la limite des forêts (Québec nordique). *Canadian Journal of Botany* **63(7)**:1213–1225.
Gajewski, K. 1987. Climatic impacts on the vegetation of eastern North America during the past 2000 years *Vegetatio* **68**:179–190.
Gauch, H. G. 1982. *Multivariate Analysis in Community Ecology.* Cambridge University Press, Cambridge-New York.
Geller, M. A. 1988. Solar cycles and the atmosphere. *Nature* **332**:584–685.
Gersper, P. L., V. Alexander, S. A. Barkley, R. J. Barsdate, and P. S. Flint. 1980. The soils and their nutrients. In *An Arctic Ecosystem*, J. Brown et al., eds., Dowden, Hutchinson, and Ross, Inc., Stroudsburg, Penn., pp. 219–254.
Gilbert, H., and S. Payette. 1982. Écologie des populations d'aulne vert (Alnus crispa (Ait.) Pursh) a la limite des forêts, Québec nordique. *Géographie Physique et Quaternaire* **36(1-2)**:109–124.
Giller, P. S. 1984. *Community Structure and the Niche.* Chapman and Hall, London-New York.
Gleason, H. A. 1939. The individualistic concept of the plant association. *American Midland Naturalist* **21**:92–100.
Godmaire, A., and S. Payette. 1981. Dynamique spatio-temporelle d'une bande forestière près de la limite des forêts (riviere aux Feuilles, noveau-Québec). *Géographie Physique et Quaternaire* 35(1):73–85.
Gordon, A. G. 1952. Spruce identification by twig characteristics. *Forestry Chronicle* **28**:43.
Gordon, A. G. 1976. The taxonomy and genetics of *Picea rubens* and its relationship to *Picea mariana*, *Canadian Journal of Botany* **54**:781–813.
Grant, V. 1971. *Plant Speciation*. Columbia University Press, New York, N.Y.
Greig-Smith, P. 1964. *Quantitative Plant Ecology.* Butterworths, Washington, D.C.
Grime, J. P. 1979. *Plant Strategies and Vegetation Processes.* John Wiley & Sons, Toronto.
Grubb, P. J. 1977. The maintenance of species-richness in plant communities: the importance of the regeneration niche. *Biological Reviews* **52**:107–145.
Grubb, P. J. 1984. Some growth points in investigative plant ecology. In *Trends in Ecological Research in the 1980s*, J. H. Cooley and F. B. Golley, eds., Plenum Press, New York, pp. 51–74.
Grubb, P. J. 1986. Problems posed by sparse and patchily distributed species in species-rich plant communities. In *Community Ecology*, J. Diamond and T. J. Case, eds., Harper and Row, New York, pp. 207–225.
Gubbe, D. M. 1976. Vegetation. In *Landscape Survey, District* of Keewatin, N. W. T. D. M. Gubbe, ed., 1975 report for the Polar Gas Project, R. M. Hardy and Assoc., Ltd., Toronto, 90–244 (Dist. by Infopall, Calgary, Alberta, Canada).
Haag, R. W. 1974. Nutrient limitations to plant production in two tundra communities. *Canadian Journal of Botany* **52**:103–116.
Haag, R. W., and L. C. Bliss. 1974. Functional effects of vegetation on the radiant energy budget of boreal forest. *Canadian Geotechnical Journal* **11(3)**:374–379.
Halfman, J. D., and T. C. Johnson. 1984. Enhanced atmospheric circulation over North America during the early Holocene: Evidence from Lake Superior. *Science* **224**:61–53.
Hall, C. A. S., and D. L. DeAngelis. 1985. Models in ecology: paradigms found or paradigms lost? *Bulletin of the Ecological Society of America* **66(3)**:339–345.
Halliday, W. E. D. *A Forest Classification for Canada.* 1937. Canadian Forestry Service, Canada Department of Mines and Resources, Forestry Service Bulletin No. 89, Ottawa.
Hamburg, S. P., and C. V. Cogbill. 1988. Historical decline of red spruce populations and climatic warming. *Nature* **331**:428–431.

Hämet-Ahti, L. 1963. Zonation of the mountain birch forests in northernmost Fennoscandia. *Annales Botanici Societatis Zoologicae Botanicae 'Vanamo'* **34(4)**:1–127.
Hämet-Ahti, L. 1978. Timberline meadows in Wells Gray Park, British Columbia, and their comparative geobotanical interpretation. *Syesis* **11**:187–211.
Hämet-Ahti, L. 1979. The dangers of using the timberline as the "zero line" in comparative studies on altitudinal vegetation zones. *Phytocoenologia* **6**:49–52.
Hämet-Ahti, L. 1981. The boreal zone and its biotic subdivisions. *Fennia* **159(1)**:69–75.
Hanbury, D. T. 1900. A journey from Chesterfield Inlet to Great Slave Lake, 1898–9. *Geographical Journal* **16**:63–77.
Hanbury, D. T. 1903. Through the barren ground of northeastern Canada to the arctic coast. *Geographical Journal* **21**:178–191.
Hansell, R. I. C., P. A. Scott, R. Staniforth, and J. Svoboda. 1983. Permafrost development in the intertidal zone at Churchill, Manitoba: a possible mechanism for accelerated beach uplift. *Arctic* **36(2)**:198–203.
Hansen-Bristow, K. 1986. Influence of increasing elevation on growth characteristics at timberline. *Canadian Journal of Botany* **64**:2517–2523.
Hare, F. K. 1950. Climate and zonal divisions of the boreal forest formation in eastern Canada. *Geographical Review* **40**:615–635.
Hare, F. K. 1954. The boreal conifer zone. *Geographical Studies* **1**:4–18.
Hare, F. K. 1959. *A Photo-reconnaissance Survey Labrador-Ungava.* Canadian Department of Mines and Technical Surveys, Geographical Branch, Memoir **6**:1–83.
Hare, F. K. 1968. The Arctic. *Quarterly Journal of the Royal Meteorological Society* **94(402)**:439–459.
Hare, F. K. 1973. On the climatology of post-Wisconsin events in Canada. *Arctic and Alpine Research* **5**:169–170.
Hare, F. K. 1976. Late Pleistocene and Holocene climates: Some persistent problems. *Quaternary Research* **6**:507–517.
Hare, F. K., and J. E. Hay. 1971. Anomalies in the large-scale annual water balance over northern North America. *The Canadian Geographer* **15(2)**:79–94.
Hare, F. K., and J. C. Ritchie. 1972. The boreal bioclimates. *Geographical Review* **62**:333–365.
Hare, F. K., and R. G. Taylor. 1956. The position of certain forest boundaries in southern Labrador-Ungava. *Geographical Bulletin* **8**:51–73.
Harington, C. R., ed. 1985. *Critical Periods in the Quaternary Climatic History of North America. Vol. 5 in Climatic Change in Canada.* Syllogeus 55, National Museums of Canada, Ottawa, 482 pp.
Harms, V. L. 1974. Botanical studies in the boreal forest along the Green Lake-La Roche Road, northwestern Saskatchewan. *The Musk-Ox,* **14**:37–54.
Harper, F. 1953. Birds of the Nueltin Lake expedition, Keewatin, 1947. *American Midland Naturalist* **49**:1–116.
Harper, F. 1955. *The Barren Ground Caribou of Keewatin.* Univ. Kansas Museum of Natural History Miscellaneous Publication No. 6, pp. 1–164.
Harper, F. 1956. *The Mammals of Keewatin.* Univ. Kansas Museum of Natural History Miscellaneous Publication No. 12, pp. 1–94.
Harper, J. C. 1982. After description. In *The Plant Community as a Working Mechanism*, E. I. Newman, ed., British Ecological Society Special Publication 1, Blackwell Publishers, Oxford, pp. 11–25.
Harris, S. A. 1981. Climatic relationships of permafrost zones in areas of low winter snow cover. *Arctic* **34(1)**:64–70.
Harris, S. A. 1987. Effects of climatic change on northern permafrost. *Northern Perspectives* **15(5)**:7–9; Canadian Arctic Resources Committee, Ottawa.
Hatcher, R. J. 1963. *A Study of Black Spruce Forests in Northern Quebec.* Canada Department of Forestry, Forest Research Branch Publication No. 1018, Ottawa.
Heal, O. W., and P. M. Vitousek. 1986. Introduction (to section on ecological interactions), in *Forest Ecosystems in the Alaskan Taiga*, K. Van Cleve et al., eds., Springer-Verlag, New York, pp. 155–159.

Hearne, S. 1775. *A Journey from Prince of Wales Fort in Hudson's Bay, to the Northern Ocean.* London. Also: (1) Tyrrell, J. B., editor, edition published by Champlain Society, Toronto; (2) see reference to P. Turnor and J. B. Tyrrell below.

Heath, D. F. 1988. Non-seasonal changes in total column ozone from satellite observations. *Nature* **332**:219–227.

Heilman, P. E. 1968. Relationship of availability of phosphorus and cations to forest succession and bog formation in interior Alaska. *Ecology* **49(2)**:331–336.

Heinselman, M. L. 1957. *Silvical Characteristics of Black Spruce.* Forest Service, U.S. Department of Agriculture, Lake States Forest Experiment Station, Paper No. 45, St. Paul, Minn.

Heinselman, M. L. 1973. Fire in the virgin forests of the Boundary Waters Canoe Area, Minnesota. *Quaternary Research* **3**:329–382.

Hendrickson, O., and J. B. Robinson. 1982. *The Microbiology of Forest Soils: A Literature Review.* Canadian Forestry Service, Information Report PI-X-19, Petawawa National Forestry Institute, Petawawa, Ontario, Canada.

Hickey, J. J. 1954. Mean intervals in indices of populations. *Journal of Wildlife Management* **18(1)**:90–106.

Hicklenton, P., and W. C. Oechel. 1976. Physiological aspects of the ecology of *Dicranum fuscescens* in the subarctic. I. *Canadian Journal of Botany* **54**:1104–1109; II. *Ibid.*, **55**:2168–2177.

Hicklenton, P., and W. C. Oechel. 1977. The influence of light intensity and temperature on the field CO_2 exchange of *Dicranum fuscescens. Arctic and Alpine Research* **9**:407–419.

Hicks, S. D. 1955. Natural history survey of Coppermine. *Canadian Field-Naturalist* **69(4)**:162–166.

Hill, D. E., and J. C. F. Tedrow. 1961. Weathering and soil formation in the arctic environment. *American Journal of Science* **259**:84–101.

Hills, G. A. 1960. Regional sites research. *Forestry Chronicle* **36**:401–423.

Hills, L. V. 1975. Late Tertiary floras of Arctic Canada. *Proceedings of the Circumpolar Conference on Northern Ecology,* National Research Council of Canada, Ottawa, pp. I 65–71.

Hills, L. V., and R. T. Ogilvie. 1970. *Picea banksii* n. sp. Beaufort Formation (Tertiary), northwestern Banks Island, Arctic Canada. *Canadian Journal of Botany* **48**:457–464.

Hochachka, P. W. and G. N. Somero. 1983. *Biochemical Adaptation.* Princeton University Press, Princeton, N.J.

Holmen, K., and G. W. Scotter. 1967. *Sphagnum* species of the Thelon River and Kaminuriak Lake regions. *Bryologist* **70**:432–437.

Holmen, K., and G. W. Scotter. 1971. Mosses of the Reindeer Preserve, Northwest Territories, Canada. *Lindbergia* **1–2**:34–56.

Hom, J. L., and W. C. Oechel. 1983. The photosynthetic capacity, nutrient content, and nutrient use efficiency of different needle-age classes of black spruce (*Picea mariana*) found in interior Alaska. *Canadian Journal of Forest Research* **13(5)**:834–839.

Hopkins, D. M. 1959. Some characteristics of the climate in forest and tundra regions in Alaska. *Arctic* **12**:215–220.

Hopkins, D. M., ed. 1967. *The Bering Land Bridge.* Stanford University Press, Stanford, Cal.

Hopkins, D. M., and R. S. Sigafoos. 1951. *Frost Action and Vegetation Patterns on Seward Peninsula, Alaska.* U.S. Geological Survey Bulletin 974-C, Washington, D.C.

Hopkins, D. M., and R. S. Sigafoos. 1954. Role of frost thrusting in the formation of tussocks. *American Journal of Science* **252**:55–59.

Horton, D. H., D. H. Vitt, and N. G. Slack. 1979. Habitats of circumboreal-subarctic *Sphagna*: 1. A quantitative analysis and review of species in the Caribou Mountains, northern Alberta. *Canadian Journal of Botany* **57**:2283–2317.

Hultén, E. 1968. *Flora of Alaska and Neighboring Territories.* Stanford University Press, Stanford, California.

Hustich, I. 1939. Notes on the coniferous forest and tree limit on the east coast of Newfoundland-Labrador. *Acta Geographica* **7(1)**:1–77.

Hustich, I. 1949a. Phytogeographical regions of Labrador. *Arctic* **2**:36–42.

Hustich, I. 1949b. On the forest geography of the Labrador Peninsula. *Acta Geographica* **10(2)**:1–63.

Hustich, I. 1950. Notes on the forests on the east coast of Hudson Bay and James Bay. *Acta Geographica* **11**:3–83.
Hustich, I. 1951a. *Forest-botanical notes from Knob Lake area in the interior of Labrador-Peninsula.* National Museum of Canada Bulletin 123, Ottawa, pp. 166–217.
Hustich, I. 1951b. The lichen woodlands in Labrador and their importance as winter pastures for domesticated reindeer. *Acta Geographica* **12**:1–48.
Hustich, I. 1952. The boreal limits of conifers. *Arctic* **6**:149–162.
Hustich, I. 1957. On the phytogeography of the subarctic Hudson Bay lowland. *Acta Geographica* **16(1)**:1–48.
Hustich, I. 1962. A comparison of the floras on subarctic mountains in Labrador and in Finnish Lapland. *Acta Geographica* **17(2)**:1–24.
Hustich, I. 1965. A black spruce feather moss forest in the interior of the Labrador peninsula. *Acta Geographica* **10**:1–63.
Hustich, I. 1966. On the forest-tundra and the northern tree-lines. *Annales of the University of Turku, Kevo Subarctic Research Station* **3**:7–47.
Hustich, I. 1979. Ecological concepts and biogeographical zonation in the North: The need for a generally accepted terminology. *Holarctic Ecology* **2**:208–217.
Ives, J. D. 1974. Biological refugia and the nunatak hypothesis. In *Arctic and Alpine Environments*, Jack D. Ives and R. G. Barry, eds., Methuen, pp. 605–636.
Jackson, L. E., and L. C. Bliss. 1984. Phenology and water relations of three plant life forms in a dry tree-line meadow. *Ecology* **65(4)**:1302–1314.
Jacoby, G. C., Jr. 1983. A dendroclimatic study in the forest-tundra ecotone on the east shore of Hudson Bay. In *Tree-Line Ecology. Proceedings of the Northern Quebec Tree Line Conference*, P. Morisset and Serge Payette, eds., Nordicana **47**:95–99.
Jacoby, G. C., and E. R. Cook. 1981. Past temperature variations inferred from a 400-year tree-line chronology from Yukon Territory, Canada. *Arctic and Alpine Research* **13(4)**:409–418.
Jarvis, P. G. 1986. Stomatal control of transpiration. In *Climate-Vegetation Interactions*, C. Rosenzweig and R. Dickinson, eds., Office for Interdisciplinary Earth Studies, University Corporation for Atmospheric Research, Boulder, Colo., pp. 20–23.
Jarvis, P. G., G. B. James, and J. J. Landsberg. 1976. Coniferous forest. In *Vegetation and Atmosphere, Vol. II*, J. L. Monteith, ed., Academic Press, New York, pp. 171–238.
Jasieniuk, M. A., and E. A. Johnson. 1979. A vascular flora of the Caribou Range, Northwest Territories, Canada. *Rhodora* **81**:249–274.
Jasieniuk, M. A., and E. A. Johnson. 1982. Peatland organization and dynamics in the western subarctic, Northwest Territories, Canada. *Canadian Journal of Botany* **60(12)**:2581–2593.
Jeffree, E. P. 1960. A climatic pattern between latitudes 40° and 70°N and to probable influence on biological distributions. *Proceedings of the Linnean Society of London* **171**:89–121.
Jeglum, J. K. 1974. Relative influence of moisture, aeration, and nutrients on vegetation and black spruce growth in northern Ontario. *Canadian Journal of Forest Research* **4**:114–126.
Jehl, J. R., and D. J. T. Hussell. 1966. Effects of weather on reproductive success of birds at Churchill, Manitoba. *Arctic* **19**:185–191.
Jehl, J. R., Jr., and B. A. Smith. 1970. *Birds of the Churchill Region, Manitoba*. Manitoba Museum of Man and Nature, Special Publication 1, Winnipeg.
Jenny, H. 1941. *Factors of Soil Formation: A System of Quantitative Pedology.* McGraw-Hill Book Co., New York.
Johansen, F. 1919. The forest's losing fight in arctic Canada. *Canadian Forestry Journal* **15**:303–305.
Johnson, A. W., L. A. Viereck, R. E. Johnson, and M. Melchoir. 1966. Vegetation and flora. In *Environment of the Cape Thompson Region, Alaska*. N. J. Wilimovsky and J. N. Wolfe, eds., U.S. Atomic Energy Commission, Oak Ridge, Tenn., pp. 277–354.
Johnson, E. A. 1975. Buried seed populations in the subarctic forest east of Great Slave Lake, Northwest Territories. *Canadian Journal of Botany* **53**:2933–2941.
Johnson, E. A. 1977a. A multivariate analysis of the niches of plant populations in raised bogs. I. Niche dimensions. *Canadian Journal of Botany* **55(9)**:1201–1210.
Johnson, E. A. 1977b. A multivariate analysis of the niches of plant populations in raised bogs. II. Niche width and overlap. *Canadian Journal of Botany* **55(9)**:1211–1220.

Johnson, E. A. 1979. Fire recurrence in the subarctic and its implications for vegetational composition. *Canadian Journal of Botany* **57(12)**:1374–1379.
Johnson, E. A. 1981. Vegetation organization and dynamics of lichen woodland communities in the Northwest Territories, Canada. *Ecology* **62(1)**:200–215.
Johnson, E. A. 1983. The role of history in determining vegetational composition—an example in the western subarctic. In *Tree-Line Ecology*, P. Morisset and S. Payette, eds., Collection Nordicana, No. 47, Univ. Laval, Quebec, pp. 133–140.
Johnson, E. A. 1985. Disturbance: the process and the response. An epilogue. *Canadian Journal of Forest Research* **15**:292–293.
Johnson, E. A., and C. E. Van Wagner. 1985. The theory and use of two fire history models. *Canadian Journal of Forest Research* **15**:214–220.
Johnson, P. L., and W. D. Billings. 1962. The alpine vegetation of the Beartooth Plateau in relation to cryopedogenic processes and patterns. *Ecological Monographs* **32**:105–135.
Johnson, P. L., and T. C. Vogel. 1966. *Vegetation of the Yukon Flats region, Alaska*. U.S. Army Material Command, Cold Regions Research and Engineering Report No. 209, Hanover, N.H.
Johnston, G. H., and R. J. E. Brown. 1965. Stratigraphy of the Mackenzie River delta, Northwest Territories, Canada. *Geological Society of America Bulletin* **76(1)**:103–112.
Jones, P. D., T. M. L. Wigley, and P. B. Wright. 1986. Global temperature variations between 1861 and 1984. *Nature* **322**:430–434.
Jones, P. D., T. M. L. Wigley, C. K. Folland, D. E. Parker, J. K. Angell, S. Lebedeff, and J. E. Hansen. 1988. Evidence for global warming in the past decade. *Nature* 332:790.
Jordan, R. 1975. Pollen diagrams from Hamilton Inlet, central Labrador, and their implications for the northern Maritime Archaic. *Arctic Anthropology* **12**:106–112.
Jozsa, L. A., M. L. Parker, P. A. Bramhall, and S. G. Johnson, 1984. *How climate affects tree growth in the boreal forest*. Information Report NOR-X-255, Northern Forest Research Center, Canadian Forestry Service, Edmonton, Alberta, Canada.
Kalnicky, R. A. 1974. Climatic change since 1950. *Annals of the Association of American Geographers* **64(1)**:100–112.
Karlin, E. F., and L. C. Bliss. 1983. Germination ecology of *Ledum groenlandicum* and *Ledum palustre* spp. *decumbens*. *Arctic and Alpine Research* **15(3)**:397–404.
Karlin, E. F., and L. C. Bliss. 1984. Variation in substrate chemistry along microtopographical and water-chemistry gradients in peatlands. *Canadian Journal of Botany* **62**:142–153.
Kauppi, P., and M. Posch. 1985. Sensitivity of boreal forests to possible climatic warming. *Climatic Change* **7**:45–54.
Kay, P. A. and J. T. Andrews. 1983. Re-evaluation of pollen-climate transfer functions in Keewatin, Northern Canada. *Annals of the Association of American Geographers* **73**:550–559.
Kelly, P. M., P. D. Jones, C. B. Sear, B. S. G. Terry, and R. K. Tavakol. 1982. Variations in surface air temperatures: Part II. Arctic regions, 1881–1980. *Monthly Weather Review* **110**:71–83.
Kerr, R. A. 1988. Stratospheric ozone is decreasing. *Science* **239**:1489–1491.
Kershaw, G. P. 1984. Tundra plant communities of the Mackenzie mountains, Northwest Territories: floristic characteristics of long-term surface disturbance. In *Northern Ecology and Resource Management*, R. Olson, F. Geddes, and R. Hastings, eds., University of Alberta Press, Edmonton, pp. 239–309.
Kershaw, K. A. 1977. Studies on lichen-dominated systems. XX. An examination of some aspects of the northern boreal lichen woodlands in Canada. *Canadian Journal of Botany* **55**:393–410.
Kershaw, K. A. 1978. The role of lichens in boreal tundra transition areas. *Bryologist* **81**: 294–306.
Kershaw, K. A. 1985. *Physiological Ecology of Lichens*. University of Cambridge Press, Cambridge, England.
Kershaw, K. A., and W. R. Rouse. 1971. Studies on lichen-dominated systems. I. The water relations of *Cladonia alpestris* in spruce-lichen woodland in northern Ontario. *Canadian Journal of Botany* **49**:1389–1399.
Kessell, S. R. 1979. Phytosociological inference and resource management. *Environmental Management* **3(1)**:29–40.

Kiehl, J. T., B. A. Boville, and B. P. Briegleb. 1988. Response of a general circulation model to a prescribed Antarctic ozone hole. *Nature* **332**:501–504.
Kindle, E. M. 1924. *Geography and Geology of Lake Melville District, Labrador Peninsula.* Geological Survey of Canada Memoir 141, Ottawa.
King, R. 1836. *Narrative of a Journey to the Shores of the Arctic Ocean.* Two volumes. London.
Krause, H. H., S. Rieger, and S. A. Wilde. 1959. Soils and forest growth on different aspects in the Tanana watershed of interior Alaska. *Ecology* **40**:492–495.
Krebs, C. J. 1964a. *The Lemming Cycle at Baker Lake, Northwest Territories, During 1959–62.* Arctic Institute of North America Technical Paper No. 15, Montreal, Canada.
Krebs, C. J. 1964b. Spring and summer phenology at Baker Lake Keewatin, N. W. T., during 1959–1962. *Canadian Field-Naturalist* **78**:25–27.
Krebs, J. S., and R. G. Barry. 1970. The arctic front and the tundra-taiga boundary in Eurasia. *Geographical Review* **60(4)**:548–554.
Kukla, G. L., J. K. Angell, J. Korshover, H. Dronia, M. Hoshiai, J. Namias, M. Rodewald, R. Yamamoto, and T. Iwashima. 1977. New data on climatic trends. *Nature* **270**:573–580.
Kuyt, E. 1980. Distribution and breeding biology of raptors in the Thelon River area, Northwest Territories, 1957–1969. *Canadian Field-Naturalist* **94(2)**:121–130.
Lachenbruch, A. H., and B. V. Marshall. 1986. Changing climate: Geothermal evidence from permafrost in the Alaskan Arctic. *Science* **234**:689–696.
LaMarch, V. C., Jr. 1973. Holocene climatic variations inferred from treeline fluctuations in the White Mountains, California. *Quaternary Research* **3**:632–660.
Lamb, H. F. 1985. Palynological evidence for postglacial change in the position of tree limit in Labrador. *Ecological Monographs* **55(2)**:241–258.
Lamb, H. H. 1966. *The Changing Climate: Selected Papers of H. A. Lamb*, reprinted 1972, Methuen & Co., London.
Lamb, H. H. 1977. *Climate, Present and Future, Vol. II. Climatic History and the Future.* Methuen & Co., London.
Lambert, J. D. H. 1968. *The Ecology and Successional Trends of Tundra Plant Communities in the Low Arctic Subalpine Zone of the Richardson and British Mountains of the Canadian Western Arctic.* Ph.D. Thesis, University of British Columbia, Vancouver.
Lambert, J. D. H. 1972. Plant succession on tundra mudflows: preliminary observations. *Arctic* **25(2)**:99–106.
Lambert, J. D. H., and P. F. Maycock. 1968. The ecology of terriculous lichens of the northern conifer-hardwood forests of central eastern Canada. *Canadian Journal of Botany* **46**:1043–1078.
LaRoi, G. H. 1967. Ecological studies in the boreal spruce-fir forests of the North America taiga. I. Analysis of the vascular flora. *Ecological Monographs* **37(3)**:229–253.
LaRoi, G. H., and J. R. Dugle. 1968. A systematic and genecological study of *Picea glauca* and *P. engelmannii*, using paper chromatograms of needle extracts. *Canadian Journal of Botany* **46**:649–687.
Larsen, J. A. 1965. The vegetation of the Ennadai Lake area, N. W. T.: Studies in Arctic and Subarctic bioclimatology. *Ecological Monographs* **35(1)**:37–59.
Larsen, J. A. 1971a. Vegetation of the Fort Reliance and Artillery Lake area, N. W. T. *Canadian Field-Naturalist* **85(2)**:147–167.
Larsen, J. A. 1971b. Vegetational relationships with air mass frequencies: boreal forest and tundra. *Arctic* **24(3)**:177–194.
Larsen, J. A. 1972a. Observations of well-developed podzols on tundra and of patterned ground within forested boreal regions. *Arctic* **25**:153–154.
Larsen, J. A. 1972b. The vegetation of northern Keewatin. *Canadian Field-Naturalist* **86(1)**: 45–72.
Larsen, J. A. 1972c. Growth of spruce at Dubawnt Lake, Northwest Territories. *Arctic* **25(1)**:59.
Larsen, J. A. 1973. Plant communities north of the forest border, Keewatin, Northwest Territories. *Canadian Field-Naturalist* **87**:241–248.
Larsen, J. A. 1974. Ecology of the northern continental forest border. In *Arctic and Alpine Environments*, J. D. Ives and R. G. Barry, eds., Methuen & Co., London, pp. 341–370.

Larsen, J. A. 1980. *The Boreal Ecosystem.* Academic Press, New York.
Larsen, J. A. 1982. *Ecology of the Northern Lowland Bogs and Conifer Forests.* Academic Press, New York.
Larsen, J. A., and R. G. Barry. 1974. Palaeoclimatology. In *Arctic and Alpine Environments*, J. D. Ives and R. G. Barry, eds., Methuen & Co., London, pp. 253–276.
Larsen, J. A., and J. F. Lahey. 1958. Influence of weather upon a ruffed grouse population. *Journal of Wildlife Management* **22(1)**:63–70.
Lawrence, D. B., and L. Hulbert. 1950. Growth stimulation of adjacent plants by lupine and alder on recent glacier deposits in southeastern Alaska. *Bulletin of the Ecological Society of America* **31**:58.
Lawrence, D. B., R. E. Schoenike, A. Quispel, and G. Bond. 1967. The role of *Dryas drummondi* in vegetation development following ice recession at Glacier Bay, Alaska, with special reference to its nitrogen fixation by root nodules. *Journal of Ecology* **55**:793–813.
Lawrence, W. T., and W. C. Oechel. 1983a. Effects of soil temperature on the carbon exchange of taiga seedlings. I. Root respiration. *Canadian Journal of Forest Research* **13(5)**: 840–849.
Lawrence, W. T. 1983b. Effects of soil temperature on the carbon exchange of taiga seedlings. II. Photosynthesis, respiration, and conductance. *Canadian Journal of Forest Research* **13(5)**:850–859.
Lee, H. A. 1959. *Surficial Geology of Southern District of Keewatin and the Keewatin Ice Divide*, Northwest Territories. Geological Survey of Canada Bulletin 51, pp. 1–42.
Lee, H. A., B. G. Craig, and J. G. Fyles. Keewatin ice divide. *Bulletin of the Geological Society of America 65, part 2*, pp. 1760–1761.
Légère, A., and S. Payette. 1981. Ecology of a black spruce (*Picea mariana*) clonal population in the hemiarctic zone, northern Quebec: population dynamics and spatial development. *Arctic and Alpine Research* **13(3)**:261–276.
Lems, K. 1956. Ecological study of peat bogs of eastern North America. III. Notes on the behavior of *Chamaedaphne calyculata. Canadian Journal of Botany* **34**:197–207.
Lescarbot, M. 1612. *Relation Desniere de ce qui s'est Passe au Voyage de Sieur de Poutrincourst en La Nouvelle-France Despuis 20 Mois Enca.* Paris. (See Rousseau, 1966.)
Lewis, H., 1966. Speciation in flowering plants. *Science* **152**:167–172.
Lindley, D. 1988. CFCs cause part of global ozone decline. *Nature* **332**:293.
Lindsay, J. H. 1971. Annual cycle of leaf water potential in *Picea engelmannii* and *Abies lasiocarpa* at timberline in Wyoming. *Arctic and Alpine Research* **3**:131–138.
Lindsey, A. A. 1952. Vegetation of the ancient beaches above Great Bear and Great Slave Lakes. *Ecology* **33**:535–549.
Little, E. L., and S. S. Pauley. 1958. A natural hybrid between black and white spruce in Minnesota. *American Midland Naturalist* **60**:202–211.
Lord, C. S. 1953. *Geological Notes on Southern District of Keewatin, Northwest Territories.* Geological Survey of Canada Paper 53–22.
Löve, A., and D. Löve. 1974. Origin and evolution of the arctic and alpine floras. In *Arctic and Alpine Environments*, J. D. Ives and R. G. Barry, eds., Methuen & Co., London, pp. 571–604.
Loveless, M. D., and J. L. Hamrick. 1984. Ecological determinants of genetic structure in plant populations. *Annual Review of Ecology and Systematics* **15**:65–95.
Low, A. P. 1888. Report on explorations in James Bay and country east of Hudson Bay, drained by the Big, Great Whale, and Clearwater Rivers. *Annual Report of the Geological Survey of Canada, for 1887–88*: 1J-94J.
Low, A. P. 1896. Report on explorations in the Labrador Peninsula, along the East Main, Koksoak, Hamilton, Manicuagan and portions of other rivers, in 1892–93–94–95. *Annual Report of the Geological Survey of Canada, for 1895 (New Series) 8:* 1L-387L.
Low, A. P. 1898. Report on a traverse of the northern part of the Labrador Peninsula from Richmond Gulf to Ungava Bay. *Annual Report of the Geological Survey of Canada, for 1896, 9:* 1L-43L.
Low, A. P. 1899. Report on an exploration of part of the south shore of Hudson Strait and Ungava Bay. *Annual Report of the Geological Survey of Canada, for 1898*, 11:1L-47L.

Low, A. P. 1902. Report on an exploration of the east coast of Hudson Bay from Cape Wolstenholme to the south end of James Bay. *Report of the Geological Survey of Canada*, No. 778:5D-84D.

Low, A. P. 1903. Report on the geology and physical character of the Nastapoka Islands, Hudson Bay. *Report of the Geological Survey of Canada* No. 819:4DD-28DD.

Lowe, D. C., C. A. M. Brenninkmeijer, M. R. Manning, R. Sparks, and G. Wallace. 1988. Radiocarbon determination of atmospheric methane at Baring Head, New Zealand. *Nature* **332**:522–525.

Lucas, R. E., and J. F. Davis. 1961. Relationships between pH values of organic soils and availabilities of 12 plant nutrients. *Soil Science* **92**:177–182.

MacArthur, R. H. 1972. *Geographical Ecology.* Harper and Row, New York.

Mackay, J. R. 1963. *The Mackenzie Delta Area, N. W. T.* Canadian Department of Mines and Technical Surveys, Geography Branch Memoirs, No. 8, Ottawa.

Mackay, J. R. 1967. Permafrost depths, lower Mackenzie Valley, Northwest Territories. *Arctic* **20**:21–26.

Mackenzie, A. 1801 (1971). *Voyages from Montreal on the River St. Lawrence . . . to the Frozen and Pacific Oceans . . . in the years 1789 and 1793.* London. Reprinted by M. G. Hurtig, Ltd., Edmonton, in 1971.

Macoun, J. M. 1895. List of plants known to occur on the east coast and in the interior of the Labrador Peninsula. *Annual Report of the Geological Survey of Canada, VIII, N. S. L., Appendix VI*, Ottawa.

Macoun, J. A., and T. Holm. 1921. Vascular plants. *Report of the Canadian Arctic Expedition 1913–1918, Volume 5, Botany, Part A.*, King's Printer, Ottawa: 1A-50A.

Macpherson, J. B. 1985. The postglacial development of vegetation in Newfoundland and eastern Labrador-Ungava: Synthesis and implications. In *Critical Periods in the Quaternary Climatic History of Northern North America*, C. R. Harington, ed., Syllogeus Series No. 55, National Museums of Canada, Ottawa, pp. 267–280.

Maikawa, E., and K. A. Kershaw. 1976. Studies on lichen-dominated systems. XIX. The postfire recovery sequence of black spruce-lichen woodland in the Abitau Lake Region, N. W. T. *Canadian Journal of Botany* **54**:2679–2687.

Maini, J. S. 1966. Phytoecological study of sylvo-tundra at Small Tree Lake, N. W. T. *Arctic* **19**:220–243.

Makinen, Y., and P. Kallio. 1980. Preliminary checklist of the vascular plants of the Schefferville area of the Quebec-Labrador peninsula. McGill Subarctic Research Paper No., 30, Montreal, pp. 17–37.

Manning, T. H. 1948. Notes on the country, birds, and mammals west of Hudson Bay between Reindeer and Baker Lakes. *Canadian Field-Naturalist* **62(1)**:1–28.

Marchand, Peter J. 1987. *Life in the Cold.* University Press of New England, Hanover and London.

Margalef, Ramon. 1968. *Perspectives in Ecological Theory.* University of Chicago Press, Chicago.

Marr, J. W. 1948. Ecology of the forest-tundra ecotone on the east coast of Hudson Bay. *Ecological Monographs* **18**:117–144.

May, R. M. ed. 1981. *Theoretical Ecology, Principles and Applications.* Blackwell & Co., London.

Maycock, P. F., and J. T. Curtis. 1960. The phytosociology of boreal conifer-hardwood forests of the Great Lakes region. *Ecological Monographs* **30**:1–35.

Maycock, P. F. 1963. The phytosociology of the deciduous forests of extreme southern Ontario. *Canadian Journal of Botany* **41**:379–438.

Maycock, P. F., and B. Matthews. 1966. An arctic forest in the tundra of northern Quebec. *Arctic* **19(2)**:114–144.

McConnell, R. G. 1891. Report on an exploration in the Yukon and Mackenzie basins, N. W. T. *Annual Report of the Geological Survey of Canada for 1888–89, 4 (N.S.)*, Part D.

McFadden, J. D. 1965. *The Interrelationships of Lake Ice and Climate in Central Canada.* Technical Report No. 20 (Nonr 1202(07)), Department of Meteorology, University of Wisconsin, Madison, Wis.

McIntosh, R. F. 1967. The continuum concept of vegetation. *Botanical Review* **33(2)**:130–187.

McKay, G. A., B. F. Findlay, and H. A. Thompson. 1970. A climatic perspective of tundra areas. In *Productivity and Conservation in Northern Circumpolar Lands*, W. A. Fuller and P. G. Kevan, eds., International Union for Conservation of Nature and Natural Resources, Morges, Switzerland.

McLaren, M., and P. L. McLaren. 1981. Relative abundance of birds in boreal and subarctic habitats of northwestern Ontario and northeastern Manitoba. *Canadian Field-Naturalist* **95(4)**:418–427.

McNaughton, S. J., and L. L. Wolf. 1970. Dominance and niche in ecological systems. *Science* **167(3915)**:131–139.

McTaggart-Cowan, I. 1981. Wildlife Conservation Issues in Northern Canada. *Canadian Environmental Advisory Council Report No. 11*, Department of the Environment, Ottawa.

Melville, R. 1966. Continental drift, Mesozoic continents, and the migrations of the Angiosperms. *Nature* **211**:116–120.

Meredith, T. C., and L. Müller-Wille. 1982. The caribou of Nouveau-Quebec, an important biological resource: Economic aspects of Naskapi utilization. *Naturaliste Canadien* **109**: 947–952.

Millbank, J. W. 1978. The contribution of nitrogen-fixing lichens to the nitrogen status of their environment. In *Environmental Role of Nitrogen-Fixing Blue-Green Algae and Asymbiotic Bacteria*, U. Granhall, ed., Ecological Bulletin 26, Swedish Natural Science Council, Stockholm, pp. 260–265.

Mitchell, J. F. B. 1988. Local effects of greenhouse gases. *Nature* **332**:399–400.

Monteith, J. L. 1981. Climatic variation and the growth of crops. *Quaterly Journal of the Royal Meteorological Society* **107**:749–774.

Moore, T. R. 1974. Pedogenesis in a subarctic environment: Cambrian Lake, Quebec. *Arctic and Alpine Research* **6(3)**:281–291.

Moore, T. R. 1976. Sesquioxide-cemented soil horizons in northern Quebec: their distribution, properties, and genesis. *Canadian Journal of Soil Science* **56**:333–344.

Moore, T. R. 1978a. Soil formation in northeastern Canada. *Annals of the Association of American Geographers* **68**:518–534.

Moore, T. R. 1978b. Soil development in arctic and subarctic areas of Quebec and Baffin Island. *Proceedings of the Third York Quaternary Symposium*, pp. 379–411.

Moore, T. R. 1980. The nutrient status of subarctic woodland soils. *Arctic and Alpine Research* **12**:147–160.

Moore, T. R. 1981. Controls on the decomposition of organic matter in subarctic woodland soils. *Soil Science* **131(2)**:107–113.

Moore, T. R. 1982. Nutrients in subarctic woodland soils. *Naturaliste Canadien* **109**:523–529.

Morin, A., and S. Payette. 1984. Expansion récente de mélèze a la limite des forêts (Québec nordique). *Canadian Journal of Botany* **62(7)**:1404–1408.

Morisset, P., and S. Payette, eds., 1983. *Tree-Line Ecology*, Collection Nordicana No. 47, Centre d'Etudes Nordique, University Laval, Quebec.

Morisset, P., S. Payette, and J. DeShaye. The vascular flora of the northern Quebec-Labrador peninsula: phytogeographical structure with respect to the tree-line. In *Tree-Line Ecology*, P. Morisset and S. Payette, eds., Collection Nordicana No. 47, Centre d'Etudes Nordique, University Laval, Quebec, pp. 141–152.

Moss, E. H. 1953. Forest communities in northwest Alberta. *Canadian Journal of Botany* **31**:212–252.

Mott, R. J. 1985. Late-glacial climatic change in the Maritime Provinces. In *Critical Periods in the Quaternary Climatic History of Northern North America*, C. R. Harington, ed., Syllogeus Series No. 55, National Museums of Canada, Ottawa, pp. 281–300.

Mowat, F. M., and A. H. Lawrie. 1955. Bird observations from southern Keewatin and the interior of northern Manitoba. *Canadian Field-Naturalist* **69**:93–116.

Müller, F. 1962. Analysis of some stratigraphic observations and radiocarbon dates from two pingos in the Mackenzie Delta region. *Arctic* **15**:279–288.

Neilson, R. P. 1986. High-resolution climatic analysis and Southwest biogeography. *Science* **232**:27–34.

Nichols, H. 1967a. Pollen diagrams from subarctic central Canada. *Science* **155(3770)**:1665–1668.
Nichols, H. 1967b. Central Canadian palynology and its relevance to northwestern Europe in the late Quaternary period. *Review of Palaeobotany and Palynology* **2**:231–243.
Nichols, H. 1967c. The post-glacial history of vegetation and climate at Ennadai Lake, Keewatin, and Lynn Lake, Manitoba. *Eiszeitalter and Gegenwart* **18**:176–197.
Nichols, H. 1968. Pollen analysis, paleotemperatures, and the summer position of the arctic front in the post-glacial history of Keewatin, Canada. *Bulletin of the American Meteorological Society* **49(4)**:155–167.
Nichols, H. 1969. The late Quaternary history of vegetation and climate at Porcupine Mountain and Clearwater Bog, Manitoba. *Arctic and Alpine Research* **1(3)**:155–167.
Nichols, H. 1970. Late Quaternary pollen diagrams from the Canadian Arctic Barren Grounds at Pelly Lake, Northern Keewatin, N. W. T. *Arctic and Alpine Research* **2(1)**:43–61.
Nichols, H. 1976. Historical aspects of the northern Canadian treeline. *Arctic* **29**:38–47.
Nichols, H. 1974. Arctic Northern America palaeo-ecology: The recent history of vegetation and climate deduced from pollen analysis. In *Arctic and Alpine Environments*, J. D. Ives and R. G. Barry, eds., Methuen, London, pp. 537–567.
Nicholson, F. H. 1976. Permafrost thermal amelioration tests near Schefferville, Quebec. *Canadian Journal of Earth Sciences* **13**:1694–1705.
Nicholson, F. H. 1978a. Permafrost modification by changing the natural energy balance. *Proceedings of the 3rd. International Conference on Permafrost*, Vol. 1, pp. 61–67.
Nicholson, F. H. 1978b. Permafrost distribution and characteristics near Schefferville, Quebec: Recent studies. *Ibid.*, Vol. 1, pp. 427–433.
Nicholson, H. M., and T. R. Moore. 1977. Pedogenesis in an iron-rich subarctic environment, Schefferville, Quebec. *Canadian Journal of Soil Science* **57**:35–45.
Nilsson, C., and L. Kullman. 1981. *Studies in Boreal Plant Ecology.* Wahlenbergia 7, University of Umeå, Umeå, Sweden.
Nordenskjöld, O., and L. Mecking. 1928. *The Geography of the Polar Regions.* American Geographical Society Special Publication No. 8, New York.
Norment, C. J. 1985. Observations on the annual chronology for birds in the Warden's Grove area, Thelon River, Northwest Territories. *Canadian Field-Naturalist* **99(4)**:471–483.
Oechel, W. C., and W. T. Lawrence. 1985. Taiga. In *Physiological Ecology of North American Plant Communities*, B. F. Chabot and H. A. Mooney, eds., Chapman and Hall, London and New York, pp. 66–94.
Oechel, W. C., and G. H. Riechers. 1986. Impacts of increasing CO_2 on natural vegetation, particularly the tundra. In *Climate-Vegetation Interactions*, C. Rosenzweig and R. Dickinson, eds., Office for Interdisciplinary Earth Studies, University Corporation for Atmospheric Research, Boulder, Colo., pp. 36–42.
Ogilvie, R. T., and E. von Rudloff. 1968. Chemosystematic studies in the Genus *Picea* (Pinaceae). IV. The introgression of white and Engelmann spruce as found along the Bow River. *Canadian Journal of Botany* **48**:901–908.
Olsen, P. E. 1986. A 40-million year lake record of early Mesozoic orbital climatic forcing. *Science* **234**:842–848.
Orloci, L., and W. Stanek. 1979. Vegetation survey of the Alaska Highway, Yukon Territory: Types and gradients. *Vegetatio* **41**:1–56.
Park, Y. S., and D. P. Fowler. 1984. Inbreeding in black spruce (*Picea mariana* (Mill.) B. S. P.): Self-fertility, genetic load, and performance. *Canadian Journal of Forest Research* **14**:17–21.
Parker, W. H., J. Maze, and G. E. Bradfield. 1981. Implications of morphological and anatomical variation in *Abies balsamea* and *A. lasiocarpa* (*Pinaceae*) from Western Canada. *Canadian Journal of Botany* **68(6)**:843–854.
Payette, S. 1974. Classification écologique des formes de croissance de *Picea glauca* (Moench) Voss et de *Picea mariana* (Mill.) B. S. P. en milieux subarctiques et subalpins. *Naturaliste Canadien* **101**:893–903.
Payette, S. 1975. La limite septentrionale des forêts sur las côte orientale de la baie d'Hudson. *Naturaliste Canadien* **102**:317–329.

Payette, S. 1983. The forest tundra and present tree-lines of the northern Quebec-Labrador peninsula. In *Tree-line Ecology*, P. Morisset and S. Payette, eds., Nordicana No. 47, University of Laval, pp. 3–23.
Payette, S., and F. Boudreau. 1984. Evolution postglaciaire des hauts sommets alpins et subalpins de la Gaspésie. *Canadian Journal of Earth Sciences* **21(3)**:319–335.
Payette, S., and L. Filion. 1975. Écologie de la limite septentrionale des forêts maritimes, baie d'Hudson. *Naturaliste Canadien* **102**:783–802.
Payette, S., and L. Filion. 1984. White spruce expansion at tree line and recent climatic change. *Canadian Journal of Forest Research* **15**:241–251.
Payette, S., and R. Gagnon. 1979. Tree-line dynamics in Ungava peninsula, northern Québec. *Holarctic Ecology* **2**:239–248.
Payette, S., and R. Gagnon. 1985. Late Holocene deforestation and tree regeneration in the forest-tundra of Quebéc. *Nature* **313(6003)**:570–572.
Payette, S., and R. Lajeunesse. 1980. Les combes a neige de la Rivière aux Feuilles (Noveau-Québec): Indicateurs paleoclimatiques holocenes. *Geographie Physiques et Quaternary* **34**:209–220.
Payette, S., J. DeShaye, and H. Gilbert. 1982. Tree seed populations at the treeline in Riveère aux Feuilles area, northern Québec, Canada. *Arctic and Alpine Research* **14(3)**:215–221.
Payette, S., L. Filion, L. Gauthier, and Y. Boutin. 1985. Secular climatic change in old-growth tree-line vegetation of northern Quebec. *Nature* **315(6015)**:135–138.
Pearman, G., and P. J. Fraser. 1988. Sources of increased methane. *Nature* **332**:489–490.
Persson, A. 1961. Mire and spring vegetation in an area north of Lake Tornetrask, Torne Lappmark Sweden. I. Description of the vegetation. *Opera Botanica* **6(1)**.
Persson, H. 1980. Spatial distribution of fine-root growth, mortality, and decomposition in a young Scots pine stand in Central Sweden. *Oikos* **34**:77–87.
Persson, T. 1980. *Structure and Function of Northern Coniferous forests: an Ecosystem Study.* Swedish Natural Science Research Council, Ecological Bulletin No. 32, Stockholm, Sweden.
Peterson, E. B., M. M. Peterson, and R. D. Kabzems. 1983. *Impact of Climatic Variation in the Boreal Forest Zone: Selected References.* Information Report NOR-X-254, Northern Forest Research Centre, Canadian Forestry Service, Edmonton, Alberta.
Pettapiece, W. W. 1974. A hummocky permafrost soil from the subarctic of northwestern Canada and some influences of fire. *Canadian Journal of Soil Science* **54**:343–355.
Pettapiece, W. W. 1975. Soils of the Subarctic in the lower Mackenzie basin. *Arctic* **28**:35–53.
Pettapiece, W. W. 1984. Some considerations of soil development in northwestern Canada and some ecological relationships. In *Northern Ecology and Resource Management*, R. Olson, R. Hastings, and F. Geddes, eds., University of Alberta Press, Edmonton, Alberta.
Petzold, D. E., and T. Mulhern. 1987. Vegetational distributions along lichen-dominated slopes of opposing aspect in the eastern Canadian Arctic. *Arctic* **40(3)**:221–224.
Pianka, E. R. 1974. *Evolutionary Ecology.* Harper and Row, New York.
Pickett, S. T. A., and P. S. White, eds. 1985. *The Ecology of Natural Disturbance and Patch Dynamics.* Academic Press, New York.
Pielou, E. C. 1966. Species-diversity in the study of ecological succession. *Journal of Theoretical Biology* **10**:370–383.
Pielou, E. C. 1969. *An Introduction to Mathematical Ecology.* John Wiley, New York.
Pielou, E. C. 1974. *Population and Community Ecology.* Gordon and Breach Science Publishers, New York.
Pielou, E. C. 1975. *Ecological Diversity.* Wiley Interscience, New York.
Pielou, E. C. 1984. *The Interpretation of Ecological Data.* John Wiley and Sons, New York.
Pielou, E. C. 1986. Assessing the diversity and composition of restored vegetation. *Canadian Journal of Botany* **64**:1344–1348.
Pike, W. 1892. *The Barren Ground of Northern Canada.* London.
Polunin, N. 1934–35. The vegetation of Akpatok Island. *Journal of Ecology* **22**:337–395; **23**:161–206.
Polunin, N. 1940. *Botany of the Canadian Eastern Arctic. I. Pteridophyta and Spermatophyta.* Bulletin No. 92, National Museum of Canada, Ottawa.

Polunin, N. 1948. *Botany of the Canadian Eastern Arctic. III. Vegetation and Ecology.* Bulletin No. 104, National Museum of Canada, Ottawa.
Polunin, N. 1959. *Circumpolar Arctic Flora*, Oxford University Press, Oxford, England.
Pomeroy, J. W. 1985. An identification of environmental disturbances from road developments in subarctic muskeg. *Arctic* **38(2)**:104–111.
Ponomareva, V. V. 1969. *Theory of Podzolization.* U.S. Department of Agriculture and National Science Foundation, Israel Program for Scientific Translations, Jerusalem. Trans. from Russian.
Porsild, A. E. 1943. Materials for a flora of the continental Northwest Territories of Canada. *Sargentia* **4**:72–83.
Porsild, A. E. 1945. *The Alpine Flora of the East Slope of the Mackenzie Mountains, Northwest Territories.* National Museum of Canada Bulletin No. 101, Ottawa.
Porsild, A. E. 1950a. *Vascular Plants of Nueltin Lake, Northwest Territories.* National Museum of Canada Bulletin No. 146, Ottawa.
Porsild, A. E. 1950b. Flora. In *Canada's Western Northland*, Department of Mines and Resources, Ottawa, pp. 130–141.
Porsild, A. E. 1951a. *Botany of Southeastern Yukon Adjacent to the Canol Road.* Canada National Museum Bulletin No. 121, Ottawa.
Porsild, A. E. 1951b. Plant life in the Arctic. *Canadian Geographical Journal* **42**:120–145.
Porsild, A. E. 1955. *The Vascular Plants of the Western Canadian Arctic Archipelago.* National Museum of Canada Bulletin No. 135, Ottawa.
Porsild, A. E. 1957. *Illustrated Flora of the Canadian Arctic Archipelago*, National Museum of Canada Bulletin No. 146, Ottawa.
Porsild, A. E. 1958. Geographical distribution of some elements in the flora of Canada. *Geographical Bulletin* **11**:57–77.
Porsild, A. E. 1974. *Materials for A Flora of Central Yukon Territory.* National Museum of Canada Publications in Botany No. 4, Ottawa.
Prest, V. K. 1970. Quaternary geology of Canada. In *Geology and Economic Minerals of Canada*, R. J. E. Douglas, ed., Department of Energy, Mines and Resources, Ottawa, pp. 676–764.
Price, D. W. 1972. *The Periglacial Environment, Permafrost, and Man.* American Geographical Society Research Paper No. 14, Washington.
Pruitt, W. O. 1978. *Boreal Ecology.* Edward Arnold, London.
Pruitt, W. O. 1960. Animals in the snow. *Scientific American* **202(1)**:61–68.
Pruitt, W. O. 1966. Ecology of terrestrial mammals. In *Environment of the Cape Thompson Region, Alaska*, Norman J. Wilimovsky and John N. Wolfe, eds., U.S. Atomic Energy Commission, Oak Ridge, Tenn., pp. 519–564.
Pruitt, W. O. 1970. Some ecological aspects of snow. In *Ecology of the Subarctic Regions: Proceedings of the Helsinki Symposium*, United Nations Educational, Scientific, and Cultural Organization, Paris, France, pp. 83–99.
Pullen, W. J. S. 1852. *Notes in British Arctic Blue Book*, Vol. 50. British Parlementary Session Papers, Feb. 3-July 1, 1852.
Ragotzkie, R. A. 1962. *Operation Freeze-up: An Aerial Reconnaissance of Climate and Lake Ice in central Canada.* Technical Report No. 10 (Nonr 1202(7)), Department of Meteorology, University of Wisconsin, Madison.
Ramanathan, V. 1988. The greenhouse theory of climate change: a test by an inadvertent global experiment. *Science* **240**:293–299.
Rapp, A. 1970. Some geomorphological processes in cold climates. *Ecology of the Subarctic Regions*, UNESCO, Paris, pp. 105–114.
Rasmussen, K. 1927. Across Arctic America. *Narrative of the Fifth Thule Expedition.* New York-London, pp. 1–338.
Raup, H. M. 1936. Phytogeographic studies in the Athabaska-Great Slave Lake Region. *Journal of the Arnold Arboretum* **17**:180–315.
Raup, H. M. 1941. Botanical problems in boreal America. *The Botanical Review* **7**:147–248.
Raup, H. M. 1943. Willows of the Hudson Bay region and the Labrador Peninsula. *Sargentia* **4**:81–127.

Raup, H. M. 1946. Phytogeographical studies in the Athabasca-Great Slave Lake Region. *Journal of the Arnold Arboretum* **27(1)**:1–85. Harvard University, Jamaica Plain, Mass.
Raup, H. M. 1947a. Some natural floristic areas in boreal America. *Ecological Monographs* **17(2)**:222–235.
Raup, H. M. 1947b. *The Botany of Southwestern Mackenzie. Sargentia 6.* Publications of the Gray Herbarium, Harvard University. Cambridge, Mass.
Raup, H. M. 1959. The willows of boreal western America. *Contributions of the Gray Herbarium, Harvard University,* No. CLXXXV: 3–96, Cambridge, Mass.
Raup, H. M., and C. S. Denny. 1950. *Photo Interpretation of the Terrain Along the Southern Part of the Alaska Highway.* U.S. Geological Survey Bulletin 963-D, Washington, D.C.
Reed, R. J. 1960. Principal frontal zones of the northern hemisphere in winter and summer. *Bulletin of the American Meteorological Society* **41(11)**:591–598.
Remmert, H. 1980. *Arctic Animal Ecology.* Springer-Verlag, Berlin-New York.
Rencz, A., and A. N. D. Auclair. 1978. Biomass distribution in a subarctic *Picea mariana–Cladonia alpestris* woodland. *Canadian Journal of Forest Research* **8**:168–176.
Rencz, A. and A. N. D. Auclair. 1980. Dimension analysis of various components of black spruce in subarctic lichen woodland. *Canadian Journal of Forestry Research* **10**:491–497.
Retzer, J. L. 1956. Alpine soils of the Rocky Mountains. *Journal of Soil Science* **7**:22–32.
Retzer, J. L. 1974. Alpine soils. In *Arctic and Alpine Environments*, J. D. Ives and R. Barry, Methuen, London, pp. 771–802.
Richardson, C., D. L. Tilton, J. A. Kadlec, J. P. M. Chamie, and W. A. Wentz. 1978. Nutrient dynamics of northern wetland ecosystems. In *Freshwater Wetlands*, R. E. Good, D. F. Whigham, and R. L. Simpson, eds., Academic Press, New York, pp. 217–241.
Richardson, J. 1851. *Arctic Searching Expedition . . . In Search of the Discovery Ships under Command of Sir John Franklin.* London, two volumes.
Rieger, S. 1983. *The Genesis and Classification of Cold Soils.* Academic Press, New York.
Ritchie, J. C. 1959. The Vegetation of Northern Manitoba. III. Studies in the Subarctic. Arctic Institute of North America Technical Paper 3, Montreal.
Ritchie, J. C. 1960a. The vegetation of northern Manitoba. IV. The Caribou Lake region. *Canadian Journal of Botany* **38(2)**:185–197.
Ritchie, J. C. 1960b. The vegetation of northern Manitoba. V. Establishing the major zonation. *Arctic* **13**:211–229.
Ritchie, J. C. 1960c. The vegetation of northern Manitoba. VI. The lower Hayes River region. *Canadian Journal of Botany* **38**:769–788.
Ritchie, J. C. 1962. *A Geobotanical Survey of Northern Manitoba.* Arctic Institute of North America Technical Paper No. 9, Calgary.
Ritchie, J. C. 1972. Pollen analysis of Late-Quaternary sediments from the arctic treeline of the Mackenzie River Delta Region, Northwest Territories, Canada. In Y. Vasari, H. Hyvärinen, and S. Hicks, eds., *Climatic Changes in Arctic Areas during the Last Ten Thousand Years* (University of Oulu, Oulu), pp. 253–271.
Ritchie, J. C. 1974. Modern pollen assemblages near the arctic tree line, Mackenzie Delta region, Northwest Territories. *Canadian Journal of Botany* **52**:381–96.
Ritchie, J. C. 1976. The late-Quaternary vegetational history of the Western Interior of Canada. *Canadian Journal of Botany* **54**:1793–1818.
Ritchie, J. C. 1977. The modern and late-Quaternary vegetation of the Campbell-Dolomite uplands near Inuvik, N. W. T., Canada. *Ecological Monographs* **47**:401–23.
Ritchie, J. C. 1982. The modern and Late-Quaternary vegetation of the Doll Creek area, North Yukon, Canada. *New Phytologist* **90**:563–603.
Ritchie, J. C. 1984. *Past and Present Vegetation of the Far Northwest of Canada.* University of Toronto Press, Toronto.
Ritchie, J. C. 1988. *Post-Glacial Vegetation of Canada.* Cambridge University Press, New York.
Ritchie, J. C., and L. C. Cwynar. 1976. Palaeobotany report. In *North Yukon Research Programme, University of Toronto, 1976 Annual Report.*
Ritchie, J. C. 1982. The Late-Quaternary vegetation of the North Yukon. In Hopkins **1982**: 113–26.

Ritchie, J. C., and F. K. Hare. 1971. Late-Quaternary Vegetation and climate near the arctic tree line of northwestern North America. *Quaternary Research* **1**:331–41.
Ritchie, J. C., and S. Lichti-Federovich. 1967. Pollen dispersal phenomena in arctic-subarctic Canada. *Review of Palaeobotany and Palynology* **3**:255–66.
Ritchie, J. C. 1968. Holocene pollen assemblages from the Tiger Hills, Manitoba. *Canadian Journal of Earth Sciences* **5**:873–80.
Ritchie, J. C., and G. A. Yarranton. 1978. The late-Quaternary history of the boreal forest of central Canada. *Journal of Ecology* **66**:199–212.
Ritchie, J. C., J. Cinq-Mars, and L. C. Cwynar. 1982. L'environnement tardiglaciaire du Yukon septentrional, Canada. *Géographie physique et quaternaire* **36**:241–50.
Ritchie, J. C., L. C. Cwynar, and R. W. Spear. 1983. Evidence for northwest Canada for an early Holocene Milankovitch thermal maxima. *Nature* **305**:126–128.
Ritchie, J. C., K. A. Hadden, and K. Gajewski. 1987. Modern pollen spectra from lakes in arctic western Canada. *Canadian Journal of Botany* **65(8)**:1605–1613.
Ritchie, J. C., and G. M. McDonald. 1986. Patterns of post-glacial spread of white spruce. *Journal of Biogeography* **13**:527–540.
Rosenzweig, C., and R. Dickinson. 1986. *Climate-Vegetation Interactions.* Proceedings of a workshop held at NASA Goddard Space Flight Center, Greenbelt, Md., 1946. Office for Interdisciplinary Earth Studies, University Corporation for Atmospheric Research, Box 3000, Boulder, Colo., 80307.
Rousseau, J. 1948. The vegetation and life zones of George River, Eastern Ungava, and the welfare of the natives. *Arctic* **1**:93–96.
Rousseau, J. 1952. Les zones biologiques de la peninsule Quebec-Labrador et l'hemiarctique. *Canadian Journal of Botany* **30**:436–474.
Rousseau, J. 1966a. Le flore de la reviere George, Noveau-Quebec. *Naturaliste Canadien* **93**:11–60.
Rousseau, J. 1966b. The movement of plants under the influence of man. In *The Evolution of Canada's Flora*, R. L. Taylor and R. A. Ludwig, eds., University of Toronto Press, Toronto, pp. 81–99.
Rousseau, J. 1968. The vegetation of the Quebec-Labrador Peninsula between 55° and 60° N. *Naturaliste Canadien* **95**:469–563.
Rowe, J. S. 1961. The level-of-integration concept and ecology. *Ecology* **42**:420–427.
Rowe, J. S. 1966. Phytogeographic zonation: An ecological appreciation. In *The Evolution of Canada's Flora*, R. L. Taylor and R. A. Ludwig, eds., University of Toronto Press, Toronto, pp. 12–27.
Rowe, J. S. 1971. Why classify forest land? *The Forestry Chronicle* **47(3)**:1–5.
Rowe, J. S. 1972. *Forest Regions of Canada.* Canadian Forestry Service, Publication No. 1300, Ottawa.
Rowe, J. S. 1984. Lichen woodland in northern Canada. In *Northern Ecology and Resource Management*, R. Olson, R. Hastings, F. Geddes, eds., the University of Alberta Press, Edmonton, pp. 225–237.
Rowe, J. S., and G. W. Scotter. 1973. Fire in the boreal forest. *Quaternary Research* **3**:444–464.
Rowe, J. S., and J. W. Sheard. 1981. Ecological land classification: a survey approach. *Environmental Management* **5(5)**:451–464.
Sager, P. E., and A. D. Hasler. 1969. Species diversity in lacustrine phytoplankton. I. The components of the index of diversity from Shannons formula. *American Naturalist* **103(929)**:51–59.
Sakai, A., and C. J. Weiser. 1973. Freezing resistance of trees in North America with reference to tree regions. *Ecology* **54(1)**:118–126.
Salt, G. W. 1983. Roles: Their limits and responsibilities in ecological and evolutionary research. *American Naturalist* **122**:697–705.
Savile, D. B. O. 1956. Known dispersal rates and migratory potentials as clues to the origin of the North American biota. *American Midland Naturalist* **56(2)**:434–453.
Savile, D. B. O. 1963. Factors limiting the advance of spruce at Great Whale River, Quebec. *Canadian Field-Naturalist* **77(2)**:95–97.

Savile, D. B. O. 1972. *Arctic Adaptations in Plants.* Canadian Department of Agriculture Research Branch Monograph No. 6, Plant Research Institute, Ottawa.

Schneider, S. H., and R. Londer. 1984. *The Coevolution of Climate and Life.* Sierra Club Books, San Francisco.

Scoggan, H. J. 1957. *Flora of Manitoba.* National Museum of Canada Bulletin No. 149, Ottawa.

Scoggan, H. J. 1978. *The Flora of Canada*, Vols. 1–4. National Museums of Canada, Ottawa.

Scotter, G. W. 1966. A contribution to the flora of the Eastern Arm of Great Slave Lake, Northwest Territories. *Canadian Field-Naturalist* **80(1)**:1–18.

Scotter, G. W., and J. W. Thomson. 1966. Lichens of the Thelon River and Kaminuriak Lake regions, N. W. T. *Bryologist* **69**:497–502.

See, M. G., and L. C. Bliss. 1980. Alpine lichen-dominated communities in Alberta and Yukon. *Canadian Journal of Botany* **58(20)**:2148–2170.

Shilts, W. W. 1974. Physical and chemical properties of unconsolidated sediments in permanently frozen terrain, District of Keewatin. *Geological Survey of Canada Paper 74*–1, **Part A**:229–235.

Shilts, W. W., and W. E. Dean. 1975. Permafrost features under an arctic lake, District of Keewatin, Northwest Territories. *Canadian Journal of Earth Sciences* **12**:649–662.

Short, S. K., and H. Nichols. 1977. Holocene pollen diagrams from subarctic Labrador-Ungava: vegetational history and climatic change. *Arctic and Alpine Research* **9(3)**:265–290.

Simmons, H. G. 1913. A survey of the phytogeography of the arctic American archipelago. *Kungl. Fysiogr. Sallsk. Handl.* N.F. **24(19)**:1–183, University of Arssk, Lunds.

Simonson, R. W. 1959. Outline of a generalized theory of soil genesis. *Soil Science Society of American Proceedings* **23**:152–156.

Singh. T., and J. M. Powell. 1986. Climatic variation and trends in the boreal forest region of Western Canada. *Climatic Change* **8(3)**:267–278.

Slaughter, C. W., and L. A. Viereck. 1986. Climatic characteristics of the taiga in interior Alaska. In *Forest Ecosystems in the Alaskan Taiga*, K. Van Cleve, F. S. Chapin III, P. W. Flanagan, L. A. Viereck, and C. T. Dyrness, Springer-Verlag, New York, pp. 9–21.

Small, E. 1972a. Photosynthetic rates in relation to nitrogen recycling as an adaptation to nutrient deficiency in peat bog plants. *Canadian Journal of Botany* **50**:2227–2233.

Small, E. 1972b. Ecological significance of four critical elements in plants of raised sphagnum peat bogs. *Ecology* **53**:498–503.

Sohlberg, E. H., and L. C. Bliss. 1984. Microscale pattern of vascular plant distribution in two high arctic plant communities. *Canadian Journal of Botany* **62**:2033–2042.

Soil Survey Staff. 1951. *Soil Survey Manual.* U.S. Department of Agriculture Handbook 18, Washington.

Somero, G. N. 1986. Protein adaptation and biogeography: threshold effects on molecular evolution. *Trends in Ecology and Evolution* **1(5)**:124–127.

Sorenson, C. J. 1973. Interrelationships between soils and climate and between paleosols and paleoclimates: forest-tundra ecotone, north central Canada. Ph.D. Thesis, Department of Geography, University of Wisconsin, Madison.

Sorenson, C. J. 1977. Reconstructed Holocene bioclimates. *Annals of the Association of American Geographers* **67**:214–222.

Sorenson, C. J., J. C. Knox, J. A. Larsen, and R. A. Bryson. 1971. Paleosols and the forest border in Keewatin, N. W. T. *Quaternary Research* **1**:468–473.

Soule, M. E., and B. A. Wilcox, eds. 1980. *Conservation Biology: An Evolutionary Approach.* Sinauer Association, London.

Sousa, W. P. 1980. The response of a community to disturbance: The importance of successional age and species life histories. *Oecologia* **45**:72081.

Sousa, W. P. 1984. The role of disturbance in natural communities. *Annual Reviews of Ecology and Systematics* **15**:353–391.

Spatt, P. D., and M. C. Miller. 1981. Growth conditions and vitality of *Sphagnum* in a tundra community along the Alaskan pipeline haul road. *Arctic* **34(1)**:48–54.

Spectzman, L. A. 1959. *Vegetation of the Arctic Slope of Alaska*, U.S. Geological Survey Professional Paper 302-b, U.S. Government Printing Office, Washington, D.C.

Stanek, W., K. Alexander, and C. S. Simmons. 1981. *Reconnaissance of Vegetation and Soils Along the Dempster Highway, Yukon Territory.* I. Vegetation Types. Canadian Forestry Service, Pacific Forest Research Center, Report BC-X-217, Victoria, B.C.

Stanek, W. 1982. *Reconnaissance of Vegetation and Soils Along the Dempster Highway, Yukon Territory: II. Soil Properties as Related to Vegetation.* Canadian Forestry Service, Pacific Forest Research Center, Report BC-X-236, Vancouver, B.C.

Strain, B. R. 1987. Direct effects of increasingly atmospheric CO_2 on plants and ecosystems. *Trends in Ecology and Evolution* **2(1)**:18–21.

Strang, R. M., and A. H. Johnson. 1981. Fire and climax spruce forest in central Yukon. *Arctic* **34(1)**:60–61.

Strong, D. R. Jr. 1983. Natural variability and the manifold mechanisms of ecological communities. *American Naturalist* **122(5)**:636–660.

Strong, D. L., D. Simberloff, L. G. Abele, and A. B. Thistle, eds., *Ecological Communities: Conceptual Issues and the Evidence.* Princeton University Press, Princeton, N.J.

Strong, W. L., and G. H. La Roi. 1983. Root-system morphology of common boreal forest trees in Alberta, Canada. *Canadian Journal of Forest Research* **13**:1164–1173.

Stupart, R. F. 1928. The influence of arctic meteorology on the climate of Canada especially. In *Problems of Polar Research*, American Geographical Society Special Publications No. 7.

Szeicz, G., D. E. Petzold, and R. G. Wilson. 1979. Wind in a subarctic forest. *Journal of Applied Meteorology* **18**:1268–1274.

Szilard, L. 1960. The control and formation of specific proteins in bacteria and animal cells. *Proceedings of the National Academy of Sciences* **46**:277–292.

Tamm, C. O., ed. 1976. *Man and the Boreal Forest.* Ecological Bulletin No. 21, Swedish Natural Science Research Council, Stockholm, Sweden.

Tarnocai, C. 1977. Soils of North-Central Keewatin. *Soil Science Society of America Proceedings* **30**:381–387.

Tarnocai, C. 1984. Characteristics of soil temperature regimes in the Inuvik area. In *Northern Ecology and Resource Management*, R. Olson, R. Hastings, and F. Geddes, eds., University of Alberta Press, Edmonton, pp. 19–38.

Taverner, P. A., and G. M. Sutton. 1934. The birds of Churchill, Manitoba. *Annals of the Carnegie Museum*, New York, pp. 1–83.

Tedrow, J. C. F. 1965. Concerning genesis of the buried organic matter in tundra soil. *Soil Science Society of America Proceedings* **29(1)**:89–90.

Tedrow, J. C. F. 1966. Polar desert soils. *Soil Science Society of America Proceedings* **30**:381–387.

Tedrow, J. C. F. 1968. Pedogenic gradients of the polar regions. *Journal of Soil Science* **19(1)**: 197–204.

Tedrow, J. C. F. 1977. *Soils of the Polar Landscapes.* Rutgers University Press, New Brunswick, New Jersey.

Tedrow, J. C. F., and D. E. Hill. 1955. Arctic brown soil. *Soil Science* **80**:265–275.

Tedrow, J. C. F., J. V. Drew, D. E. Hill, and L. A. Douglas. 1958. Major genetic soils of the Arctic Slope of Alaska. *Journal of Soil Science* **9**:33–45.

Tedrow, J. C. F., and J. E. Cantlon. 1958. Concepts of soil formation and classification in arctic regions. *Arctic* **11**:166–179.

Tedrow, J. C. F., and H. Harries. 1960. Tundra soil in relation to vegetation, permafrost, and glaciation. *Oikos* **2**:237–249.

Thomas, R. D. 1977. A Brief Description of the Surficial Materials of North-Central Keewatin, Northwest Territories. *Canadian Geological Survey Paper 77-iB*:315–317.

Thompson, D. Q. 1955. The role of food and cover in population fluctuations of the brown lemming at Point Barrow, Alaska. *Transactions of the 20th North American Wildlife Conference*, Washington, pp. 166–176.

Thompson, D. C. 1980. A classification of the vegetation of the Boothia Peninsula and the northern District of Keewatin, N. W. T. *Arctic* **33(1)**:73–99.

Thompson, D. 1787–1788. In *The Western Interior of Canada*, 1964, J. Warkenton, ed.

Manuscript journals of David Thompson, in the Department of Public Records and Archives of Ontario (Vol. 21, No. 52:131–143), McClelland and Stewart, Ltd., Toronto.
Thomson, J. W. 1967. *The Lichen Genus Cladonia in North America.* University of Toronto Press, Toronto.
Thomson, J. W. 1984. *American Arctic Lichens I, The Macrolichens.* Columbia University Press, New York.
Thomson, J. W., G. W. Scotter, and T. Ahti. 1969. Lichens of the Great Slave Lake region, N. W. T., Canada. *Bryologist* **72**:137–177.
Tieszen, L. L., and N. K. Wieland. 1975. Physiological ecology of arctic and alpine photosynthesis and respiration. In *Physiological Adaptation to the Environment*, J. Vernberg, ed. International Educational Publications, New York, pp. 157–200.
Tranquillini, U. 1979. *Physiological Ecology of the Alpine Timberline.* Springer-Verlag, Berlin.
Tryon, P. R., and F. S. Chapin, III. 1983. Temperature control over root growth and root biomass in taiga forest trees. *Canadian Journal of Forest Research* **13(5)**:827–833.
Tukhanen, S. 1984. A circumboreal system of climatic-phytogeographical regions. *Acta Botanica Fennica* **127**:1–50.
Turesson, G. 1930. The selective effect of climate upon the plant species. *Hereditas* **14**:99–152.
Turnor, P. 1934. *Journals of Samuel Hearne*, J. B. Tyrrell, ed., Champlain Society, Toronto.
Tyrrell, J. B. 1896. *Report on the Country Between Athabaska Lake and Churchill River.* Annual Report of the Geological Survey of Canada for 1894, Volume 8, Ottawa.
Tyrrell, J. B. 1897. *Report on the Doobaunt, Kazan and Ferguson Rivers and the northwest coast of Hudson Bay:* Annual Report of the Geological Survey of Canada, New Series, Vol. IX, for 1896, Part F, 1–218.
Tyrrell, J. W. 1897. Plants (exclusive of algae and fungi) collected by J. W. Tyrrell. Annual Report of the Geological *Survey of Canada*, Vol. IX, Part F, Appendix.
Ugolini, F. C., and J. C. F. Tedrow. 1963. Soils of the Brooks Range, Alaska. *Soil Science* **96**:121–127.
Van Cleve, K., and L. A. Viereck. 1981. Forest succession in relation to nutrient cycling in the boreal forest of Alaska. In *Forest Succession*, D. C. West, H. H. Shugart, and D. B. Botkin, eds., Springer-Verlag, New York, pp. 185–211.
Van Cleve, K., R. Barney, and R. Schlentner. 1981. Evidence of temperature control of production and nutrient cycling in two interior Alaska black spruce ecosystems. *Canadian Journal of Forest Resarch* **11**:258–273.
Van Cleve, K., F. S. Chapin III, P. W. Flanagan, L. A. Viereck, and C. T. Dyrness. 1986. *Forest Ecosystems in the Alaskan Taiga.* Ecological Studies Vol. 57, Springer-Verlag, New York.
Van Everdingen, R. O. 1978. Frost mounds at Bear Rock, near Fort Norman, Northwest Territories, 1975–1976. *Canadian Journal of Earth Sciences* **15**:263–276.
Van Eyk, D. W., and S. C. Zoltai. 1975. *Terrain sensitivity, Mackenzie Valley and Northern Yukon.* Environmental-Social Committee, Northern Pipelines, Task Force North, Oil Development Report No. 74–44, Calgary, Ottawa.
Viereck, L. A. 1957. The flora of Gerin Mountain, central Quebec-Labrador. M.Sc. thesis, University of Colorado, Boulder.
Viereck, L. A. 1965. Relationship of white spruce to lenses of perennially frozen ground, Mount McKinley National Park, Alaska. *Arctic* **18(4)**:262–267.
Viereck, L. A. 1973. Wildfire in the taiga of Alaska. *Quaternary Research* **3**:465–495.
Viereck, L. A. 1975. Forest ecology of the Alaskan taiga. In *Circumpolar Conference on Northern Ecology*, National Research council of Canada, Ottawa, pp. 1–73.
Viereck, L. A. 1979. Characteristics of tree-line communities in Alaska. *Holarctic Ecology* **2**:228–238.
Viereck, L. A., and C. T. Dyrness. 1979. *Ecological Effects of the Wickersham Dome Fire Near Fairbanks, Alaska.* U.S. Department of Agriculture, Forest Service General Technical Report PNW-90, Portland, Oregon.
Viereck, L. A., and C. T. Dyrness. 1980. A Preliminary Classification System for Vegetation of Alaska. *U.S. Forest Service, General Technical Report PNW-106*, Pacific Northwest Forest and Range Experiment Station, Anchorage.

Viereck, L. A., and L. A. Schandelmeier. 1980. *Effects of Fire in Alaska and Adjacent Canada, a Literature Review.* U.S. Department of the Interior, Bureau of Land Management, Alaska Technical Report 6, Anchorage, Alaska.

Viereck, L. A., K. Van Cleve, and C. T. Dyrness. 1986. Forest ecosystem distribution in the taiga environment. In *Forest Ecosystems in the Alaskan Taiga*, R. Van Cleve, F. S. Chapin III, P. W Flanagan, L. A. Viereck, and C. T. Dyrness, eds., Springer-Verlag, New York, pp. 22–43.

Viereck, L. A., C. T. Dyrness, K. Van Cleve, and M. J. Foote. 1983. Vegetation, soils, and forest productivity in selected forest types in interior Alaska. *Canadian Journal of Forest Research* **13(5)**:703–720.

Vincent, A. B. 1965. *Black Spruce: A Review of its Silvics, Ecology, and Silviculture.* Department of Forestry, Canada, Publication No. 1100, Ottawa.

Volz, A., and D. Kley. 1988. Evaluation of the Montsauris series of ozone measurements made in the nineteenth century. *Nature* **332**:240–242.

Vowinckel, T., W. C. Oechel, and W. G. Boll. 1975. The effect of climate on the photosynthesis of *Picea mariana* at the subarctic treeline. I. Field measurements. *Canadian Journal of Botany* **53**:604–620.

Walker, M. 1984. *Harvesting the Northern Wild.* Outcrop, Ltd., The Northern Publishers, Yellowknife.

Wardle, P. 1974. Alpine timberlines. In *Arctic and Alpine Environments*, J. D. Ives and R. Barry, eds., Methuen, London, pp. 341–370.

Warren Wilson, J. W. 1957a. Observations on the temperature of arctic plants and their environment. *Journal of Ecology* **45**:499–531.

Warren Wilson, J. W. 1957b. Arctic plant growth. *Advancement of Science* **53**:383–388.

Warren Wilson, J. W. 1966a. Effect of temperature on the assimilation rate. *Annals of Botany* (N.S.) **30(120)**:753–761.

Warren Wilson, J. W. 1966b. An analysis of plant growth and its control in arctic environments. *Annals of Botany* (N.S.) **30(119)**:383–402.

Warren Wilson, J. W. 1967. Ecological data on dry matter production by plants and plant communities. In *The Collection and Processing of Field Data*, E. F. Bradley and O. T. Denmean, eds., Wiley Interscience, New York.

Webb, T., III. 1986. Vegetational change in eastern North America from 18,000 to 500 yr B.P. In *Climate-Vegetation Interactions*, C. Rosenzweig and R. Dickinson, eds., Office for Interdisciplinary Earth Studies, University Corporation for Atmospheric Research, Boulder, Colo., pp. 63–69.

Wein, R. W. 1976. Frequency and characteristics of arctic tundra fires. *Arctic* **29**:213–222.

Wein, R. W., R. R. Riewe, and I. R. Methven, eds., 1984, *Resources and Dynamics of the Boreal Zone.* Association of Canadian University for Northern Studies, Ottawa.

Weins, J. A. On competition and variable environments. *American Scientist* **65**:590–597.

West, G. C., and B. B. DeWolfe. 1974. Populations and energetics of taiga birds near Fairbanks, Alaska. *The Auk* **91(4)**:757–775.

Wheeler, E. P. 1935. The Nain-Okak section of Labrador. *Geographical Review* **25**:240–254.

Whittaker, R. H. 1973. Direct gradient analysis. In *Ordination and Classification of Communities*, R. H. Whittaker, ed., Dr. W. Junk, Publishers, The Hague, Netherlands, pp. 1–52.

Whittaker, R. H. 1975. *Communities and Ecosystems*, Second edition. Macmillian Publishing Co., New York, N.Y.

Wiggins, I. L., and J. H. Thomas. 1962. *A Flora of the Alaskan Arctic Slope.* Arctic Institute of North America Special Publication No. 4, University of Toronto Press, Toronto.

Wiken, E. B., D. M. Welch, G. R. Ironside, and D. G. Taylor. 1981. *The Northern Yukon: An Ecological Land Survey.* Ecological Land Classification Series No. 6, Lands Directorate, Environment Canada, Vancouver, B.C., and Ottawa.

Wilde, S. A. 1946. *Forest Soils and Forest Growth.* Chronica Botanica, Waltham, Mass.

Wilde, S. A. 1958. *Forest Soils.* Ronald Press, New York.

Wilde, S. A., and H. H. Krause. 1960. Soil-forest types of the Yukon and Tanana Valleys in subarctic Alaska. *Journal of Soil Science* **6**:22–38.

Wilde, S. A., and A. L. Leaf. 1955. The relation between the degree of soil podzolization and the composition of ground cover vegetation. *Ecology* **36**:19–22.
Wilde, S. A., and G. W. Randall. 1951. Chemical characteristics of ground water in forest and marsh soils of Wisconsin. *Transactions of the Wisconsin Academy of Sciences, Arts and Letters, Madison, Wis.* **40**:251–259.
Wilde, S. A., G. K. Voigt, and R. S. Pierce. 1954. The relationship of soils and forest growth in the Algoma District of Ontario, Canada. *Journal of Soil Science* **5**:22–38.
Wilson, E. O. 1984. *Biophilia.* Harvard University Press, Cambridge, Mass.
Wilton, W. C. 1964. *The Forests of Labrador.* Canadian Department of Forestry, Forest Research Branch, Contribution No. 610, Ottawa.
Woodward, F. I. 1987. *Climate and Plant Distribution.* Cambridge University Press, Cambridge.
Wright, G. M. 1952. *Reliance: Northwest Territories.* Geological Survey of Canada Paper 51–26, Ottawa.
Wright, H. E., and M. L. Heinselman, eds. 1973. Introduction (to special section on ecological role of fire), *Quaternary Research* **3**:319–328.
Yamamoto, R. 1980. Change of global climate during recent 100 years. In *Proceedings of the Technical Conference on Climate*, World Climate Program, WMO Publication No. **578**: 360–375.
Yarie, J. 1983. Environmental and successional relationships of the forest communities of the Porcupine River drainage, Interior Alaska. *Canadian Journal of Forest Research* **13(5)**: 721–728.
Zoltai, S. C. 1965. Glacial features of the Quetico-Nipigon area, Ontario. *Canadian Journal of Earth Sciences* **2**:247–269.
Zoltai, S. C. 1971. Structure of subarctic forests on hummocky permafrost terrain in Northwestern Canada. *Canadian Journal of Forest Research* **5**:1–9.
Zoltai, S. C. 1972. *Geomorphology of the Amisk Lake Area, Saskatchewan.* Environment Canada, Northern Forest Research Center, Report NOR-X-16, Edmonton.
Zoltai, S. C., and W. W. Pettapiece. 1974. Tree distribution on perennially frozen earth hummocks. *Arctic and Alpine Research* **6**:403–411.
Zoltai, S. C., and C. Tarnocai. 1974. *Soils and Vegetation on Hummocky Terrain.* Environmental-Social Program, Task Force on Northern Development, Report No. 74–5, Ottawa.
Zoltai, S. C., and C. Tarnocai. 1975. Perennially frozen peatlands in the western Arctic and Subarctic of Canada. *Canadian Journal of Earth Sciences* **12**:28–41.
Zoltai, S. C., and J. D. Johnson. 1977. *Reference Sites for vegetation-soil Studies, Northern Keewatin.* Canada Forest Service, Northern Forest Research Center. (Unpubl. mimeo), Edmonton.
Zoltai, S. C., and J. D. Johnson. 1978. *Vegetation-Soil Relationships in the Keewatin District.* Canadian Forestry Service, ESCOM Report No. AI-25, Environment Canada, Ottawa.

Index

(Species named in text are included in the index; those listed in the tables are not included here. Where "spp." is noted after a genus name, reference is to the various species in the genus.)

Ecological Studies

Volume 56
Resources and Society
A Systems Ecology Study of the Island of Gotland, Sweden
By James J. Zucchetto and Ann-Mari Jansson
1985. X. 248 p., 70 figures. cloth
ISBN 0-387-96151-8

Volume 57
Forest Ecosystems in the Alaskan Taiga
A Synthesis of Structure and Function
Edited by K. Van Cleve, F.S. Chapin III, L.A. Viereck, C.T. Dyrness and P.W. Flanagan
1986. X, 240p., 81 figures. cloth
ISBN 0-387-96251-4

Volume 58
Ecology of Biological Invasions of North America and Hawaii
Edited by H.A. Mooney and J.A. Drake
1986. X, 320p., 25 figures. cloth
ISBN 0-387-96289-1

Volume 59
Acid Deposition and the Acidification of Soils and Waters
An Analysis
By J.O. Reuss and D.W. Johnson
1986. VIII, 120p., 37 figures. cloth
ISBN 0-387-96290-5

Volume 60
Amazonian Rain Forests
Ecosystem Disturbance and Recovery
Edited by Carl F. Jordan
1987. X, 133p., 55 figures. cloth
ISBN 0-387-96397-9

Volume 61
Potentials and Limitations of Ecosystem Analysis
Edited by E.-D. Schulze and H. Zwolfer
1987. XII, 435p., 141 figures. cloth
ISBN 0-387-17138-X

Volume 62
Frost Survival of Plants
By A. Sakai and W. Larcher
1987. XII, 321p., 200 figures, 78 tables. cloth
ISBN 0-387-17332-3

Volume 63
Long-Term Forest Dynamics of the Temperate Zone
By Paul A. Delcourt and Hazel R. Delcourt
1987. XIV, 450p., 90 figures, 333 maps. cloth
ISBN 0-387-96495-9

Volume 64
Landscape Heterogeneity and Disturbance
Edited by Monica Goigel Turner
1987. XII, 241p., 56 figures. cloth
ISBN 0-387-96497-5

Volume 65
Community Ecology of Sea Otters
Edited by G.R. van Blaricom and J.A. Estes
1987. X, 280p., 71 figures. cloth
ISBN 3-540-18090-7

Volume 66
Forest Hydrology and Ecology at Coweeta
Edited by W.T. Swank and D.A. Crossley, Jr.
1987. XIV, 512p., 151 figures. cloth
ISBN 0-387-96547-5

Volume 67
Concepts of Ecosystem Ecology
A Comparative View
Edited by L.R. Pomeroy and J.J. Alberts
1988. XII, 384p., 93 figures. cloth
ISBN 0-387-96686-2

Volume 68
Stable Isotopes in Ecological Research
Edited by P.W. Rundel, J.R. Ehleringer and K.A. Nagy
1989. XVI, 544p., 164 figures. cloth
ISBN 0-387-96712-5

Volume 69
Vertebrates in Complex Tropical Systems
Edited by M.L. Harmelin-Vivien and F. Bourliere
1989. XI, 200p., 17 figures. cloth
ISBN 0-387-96740-0

Volume 70
The Northern Forest Border in Canada and Alaska
By James A. Larsen
1989. XVI. 272p., 73 figures. cloth
ISBN 0-387-96753-2